R for Political Data Science

Chapman & Hall/CRC
The R Series

Series Editors
John M. Chambers, Department of Statistics, Stanford University, California, USA
Torsten Hothorn, Division of Biostatistics, University of Zurich, Switzerland
Duncan Temple Lang, Department of Statistics, University of California, Davis, USA
Hadley Wickham, RStudio, Boston, Massachusetts, USA

Recently Published Titles

Geocomputation with R
Robin Lovelace, Jakub Nowosad, Jannes Muenchow

Advanced R, Second Edition
Hadley Wickham

Dose Response Analysis Using R
Christian Ritz, Signe Marie Jensen, Daniel Gerhard, Jens Carl Streibig

Distributions for Modelling Location, Scale, and Shape
Using GAMLSS in R
Robert A. Rigby, Mikis D. Stasinopoulos, Gillian Z. Heller and Fernanda De Bastiani

Hands-On Machine Learning with R
Bradley Boehmke and Brandon Greenwell

Statistical Inference via Data Science
A ModernDive into R and the Tidyverse
Chester Ismay and Albert Y. Kim

Reproducible Research with R and RStudio, Third Edition
Christopher Gandrud

Interactive Web-Based Data Visualization with R, plotly, and shiny
Carson Sievert

Learn R
Pedro J. Aphalo

Using R for Modelling and Quantitative Methods in Fisheries
Malcolm Haddon

R For Political Data Science: A Practical Guide
Francisco Urdinez and Andrés Cruz

For more information about this series, please visit: https://www.crcpress.com/Chapman--HallCRC-The-R-Series/book-series/CRCTHERSER

R for Political Data Science
A Practical Guide

Edited by
Francisco Urdinez
Andrés Cruz

CRC Press
Taylor & Francis Group
Boca Raton London New York

CRC Press is an imprint of the
Taylor & Francis Group, an **informa** business
A CHAPMAN & HALL BOOK

First edition published 2021
by CRC Press
6000 Broken Sound Parkway NW, Suite 300, Boca Raton, FL 33487-2742

and by CRC Press
2 Park Square, Milton Park, Abingdon, Oxon, OX14 4RN

First issued in paperback 2022

© 2021 Taylor & Francis Group, LLC
CRC Press is an imprint of Taylor & Francis Group, an Informa business

No claim to original U.S. Government works

Reasonable efforts have been made to publish reliable data and information, but the author and publisher cannot assume responsibility for the validity of all materials or the consequences of their use. The authors and publishers have attempted to trace the copyright holders of all material reproduced in this publication and apologize to copyright holders if permission to publish in this form has not been obtained. If any copyright material has not been acknowledged please write and let us know so we may rectify in any future reprint.

Except as permitted under U.S. Copyright Law, no part of this book may be reprinted, reproduced, transmitted, or utilized in any form by any electronic, mechanical, or other means, now known or hereafter invented, including photocopying, microfilming, and recording, or in any information storage or retrieval system, without written permission from the publishers.

For permission to photocopy or use material electronically from this work, access www.copyright.com or contact the Copyright Clearance Center, Inc. (CCC), 222 Rosewood Drive, Danvers, MA 01923, 978-750-8400. For works that are not available on CCC please contact mpkbookspermissions@tandf.co.uk

Trademark notice: Product or corporate names may be trademarks or registered trademarks, and are used only for identification and explanation without intent to infringe.

Publisher's Note
The publisher has gone to great lengths to ensure the quality of this reprint but points out that some imperfections in the original copies may be apparent.

Visit the Taylor & Francis Web site at
http://www.taylorandfrancis.com

and the CRC Press Web site at
http://www.crcpress.com

ISBN 13: 978-0-367-81883-8 (pbk)
ISBN 13: 978-0-367-81889-0 (hbk)
ISBN 13: 978-1-003-01062-3 (ebk)

DOI: 10.1201/9781003010623

Typeset in LMRoman
by Nova Techset Private Limited, Bengaluru & Chennai, India

Contents

Preface	**ix**
Who will find this book useful?	ix
About the book	x
What to expect from the book	xii
Book structure	xii
Prerequisites	xiv
How to use the textbook in a methods course?	xiv
Contributors	**xix**
I Introduction to R	**1**
1 Basic R	**3**
1.1 Installation	3
1.2 Console	4
1.3 Script	5
1.4 Objects (and functions)	6
2 Data Management	**15**
2.1 Introduction to data management	15
2.2 Describing a dataset	17
2.3 Basic operations	19
2.4 Chain commands	28
2.5 Recode values	30
3 Data Visualization	**37**
3.1 Why visualize my data?	37
3.2 First steps	40
3.3 Applied example: Local elections and data visualization	48
3.4 To continue learning	66
4 Data Loading	**71**
4.1 Introduction	71
4.2 Different dataset formats	73
4.3 Files separated by delimiters (.csv and .tsv)	73
4.4 Large tabular datasets	84

II Models   87

5 Linear Models   89
- 5.1 OLS in R   90
- 5.2 Bivariate model: simple linear regression   96
- 5.3 Multivariate model: multiple regression   103
- 5.4 Model adjustment   108
- 5.5 Inference in multiple linear models   109
- 5.6 Testing OLS assumptions   110

6 Case Selection Based on Regressions   131
- 6.1 Which case study should I select for qualitative research?   133
- 6.2 The importance of combining methods   145

7 Panel Data   147
- 7.1 Introduction   147
- 7.2 Describing your panel dataset   152
- 7.3 Modelling group-level variation   158
- 7.4 Fixed vs. random effects   161
- 7.5 Testing for unit roots   163
- 7.6 Robust and panel-corrected standard errors   169

8 Logistic Models   173
- 8.1 Introduction   173
- 8.2 Use of logistic models   174
- 8.3 How are probabilities estimated?   176
- 8.4 Model estimation   182
- 8.5 Creating tables   186
- 8.6 Visual representation of results   190
- 8.7 Measures to evaluate the fit of the models   200

9 Survival Models   209
- 9.1 Introduction   209
- 9.2 How do we interpret hazard rates?   212
- 9.3 Cox's model of proportional hazards   213
- 9.4 Estimating Cox Models in R   215
- 9.5 Tools to interpret and present hazard ratios   226

10 Causal inference   235
- 10.1 Introduction   235
- 10.2 Causation and causal graphs   237
- 10.3 Measuring causal effects   239
- 10.4 DAGs and statistical associations   241
- 10.5 Backdoors and *do*-calculus   243
- 10.6 Drawing and analyzing DAGs   247
- 10.7 Making adjustments   256
- 10.8 Caveats   272

III Applications — 275

11 Advanced Political Data Management — 277
- 11.1 Introduction — 277
- 11.2 Merging datasets — 279
- 11.3 Fuzzy or inexact join of data — 284
- 11.4 Missing values' management — 288
- 11.5 Imputation of missing values — 295

12 Web Mining — 307
- 12.1 Introduction — 307
- 12.2 Ways to do web scraping — 309
- 12.3 Web scraping in R — 310
- 12.4 Using APIs and extracting data from Twitter — 317

13 Quantitative Analysis of Political Texts — 327
- 13.1 Analysis of political hashtags — 328
- 13.2 Wordfish — 340
- 13.3 Structural Topic Modeling — 346

14 Networks — 357
- 14.1 Introduction — 357
- 14.2 Basic concepts in a network — 358
- 14.3 Network datasets — 360
- 14.4 Graphic presentation of a network — 362
- 14.5 Measures of centrality — 366

15 Principal Component Analysis — 375
- 15.1 Introduction — 376
- 15.2 How PCA works — 377
- 15.3 Basic notions in R — 378
- 15.4 Dimensionality of the concept — 383
- 15.5 Variation of the concept — 389

16 Maps and Spatial Data — 395
- 16.1 Introduction — 395
- 16.2 Spatial Data in R — 398
- 16.3 Spatial Data Management — 401
- 16.4 Mapping in R — 407
- 16.5 Inference from Spatial Data — 413

IV Bibliography and Index — 425

Bibliography — 427

Index — 437

Preface

Who will find this book useful?

R for Political Data Science: A Practical Guide is a book that you can use as a reference textbook when you are breaking your neck using R. That's why we decided that it should be more applied than theoretical, and considered both political science and international relations issues. Hence the subtitle of the book: "A Practical Guide". A great advantage of the book is that it uses in each chapter the most up-to-date and simple option available. In addition, it occupies whenever possible the tidyverse, the group of packages that has revolutionized the use of R recently due to its simplicity.

R differs from other statistical softwares used in social sciences in that there is not one way of learning it. R is a language, thus a same task can be done in various forms, similar to a sentence in grammar. In the past, users would learn R without the tidyverse. Over the years there has been a huge improvement in performing tasks with much more simplicity, and tidyverse has played a huge role in this. In this book, we have selected sixteen topics that political scientists use frequently and turned them to R. To do so, we have chosen the most reliable and easiest packages to date.

About the authors

This book is edited by Francisco Urdinez, Assistant Professor at the Institute of Political Science[1] of the Pontifical Catholic University of Chile, and Andrés Cruz, Adjunct Instructor at the same institution. Most of the authors who contributed with chapters to this volume are political scientists affiliated to the Institute of Political Science[2] of the Pontifical Catholic University of Chile, and many are researchers and collaborators of the Millennium Data Foundation Institute[3], an institution that aims at gathering, cleaning and analyzing public data to support public policy. Andrew Heiss is affiliated with Georgia State University Andrew Young School of Policy Studies and he joined this project contributing with a chapter on causal inference. Above all, all the authors are keen users of R.

---

[1] See www.cienciapolitica.uc.cl/
[2] See www.cienciapolitica.uc.cl/
[3] See https://imfd.cl/es/

About the book

This book was born doing political data analysis. That is to say, it is the son of practice. Therefore its nature is applied, and it has its focus on being a toolbox for the reader. *R for Political Data Science: A Practical Guide* is intended to be a reference textbook for political scientists working with data from Latin America. It could be equally useful for a university student pursuing her PhD studies, a political consultant living in Mexico City, or a public official in Brasilia, all of whom need to transform their datasets into substantive and easily interpretable conclusions. If you don't specialize in Latin America, but still do political data analysis, the book will be useful as well.

The book was born from the need to have reference material to teach R. Working together in the Quantitative Data Analysis II chair at the Political Science Institute of the Catholic University of Chile, we found there was no introductory textbook written for learners. *R for Political Data Science: A Practical Guide* has as its imaginary audience the Latin American political scientist, whether she is an undergraduate or graduate student, or already in the market. Therefore, all exercises are based on Latin American datasets. However, political scientists based in other regions will find the book usefull as well.

The authors of the chapters are young political scientists (under 35) who had to deal with applied problems and had more or less steep learning curves in the use of R during their doctoral and early teaching years. Some of the authors migrated from other data analysis software such as Stata, others have started their experience directly with R. Some are skilled users, others are functional users, but all have in common that they have applied knowledge that will be useful to anyone who wants an easy-to-follow guide book for R.

Latin American universities have made great efforts to provide their political science students with literacy training in statistical and data analysis tools, something that was rare until ten years ago. Today, the top five universities in the region, according to the *Times Higher Education*[4] ranking, have courses in quantitative data analysis in their political science programs. Some departments, such as the Department of Political Science of the University of São Paulo, which co-organizes IPSA's summer school in political methodology, the Institute of Political Science of the Catholic University of Chile, which organizes its summer school in mixed political methodology, or the Division of Political Studies of the CIDE, have made efforts to expose their students to North American and European professors who have many decades of quantitative tradition in their programs. The good thing is that little by little, methodologists born and trained in Latin America are beginning to appear. We understand that, at present, no political scientist can go out into the job market without knowing how to

---

[4]See https://www.timeshighereducation.com/digital-editions/latin-america-university-rankings-2018-digital-edition

Preface

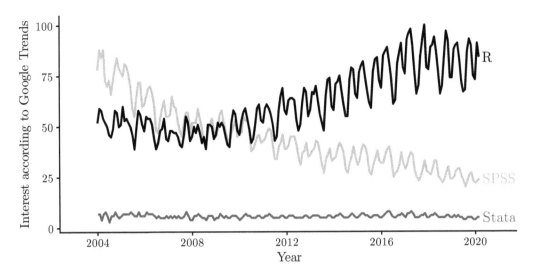

FIGURE 1: Created by the authors using the `ggplot2` package and data extracted from Google Trends. The data correspond to monthly hits in the "Science" category.

use quantitative analysis software with ease, and hence, we strive to aid those with this demand.

Right now, R is probably the best option in the market for statistical data analysis. This may come as a surprise to a reader fresh out of a time machine: ten years ago, or perhaps less, R was simply seen as the free alternative to commercial programs, which could perform "quality" quantitative analysis. However, this has changed dramatically in recent years. Figure 1 shows Google search trends for the most commonly used statistical software in science. R has moved to a market position that SPSS held 15 years ago, and niche programs —such as Stata— are less and less sought after. The trend suggests that R will become increasingly popular in Latin American political science, following a global trend.

The free software model on which R is based —with permissive copyright licenses, which make virtually all tools freely available to the public, both for use and reformulation— finally paid off. An active community of developers has anchored itself in R, adding new functionalities that have given it elegance, simplicity and flexibility. R no longer only shines in the generation of statistical models, but today it is home to a vast universe of tools that allow the user to import, order, transform, visualize, model and communicate the data that interests them, without having to change tools.

This is the technological innovation that we want to bring to the reader interested in political analysis, in the hope that it will contribute to optimize the process between the question that takes away your sleep and its solution. We know that the reader —and perhaps your university— will appreciate the kind coincidence that the best software available in terms of quality is also the best for your pocket.

Last but not least, the cover of the book was designed using the same tools in R that you will learn how to use in this book. If you want to replicate it you can access the code in the following link[5].

Francisco Urdinez and Andrés Cruz.

Santiago de Chile, November 2019.

What to expect from the book

Quantitative data analysis is one of the many tools that researchers have for addressing the questions that interest us, whether in the professional world or in academia (or "for the love of art"). This is why *R for Political Data Science: A Practical Guide* has a strong emphasis on applied political science examples. To use trite, idealized text examples about little cars or imaginary islands would be disrespectful to the reader, who is eager to use the tools of this book in the political research questions he or she finds important. Instead, we want to show the potential of these tools by putting our hands on real data, with research that colleagues have already done, and thus address the particular difficulties of translating data analysis into political questions. It will be good if you join us throughout the book with 'RStudio' open on your computer; nothing better than learning together.

The book takes advantage from the enormous progress that has been made in recent years from the creation of the language of `tidyverse`[6] and the publication of *R for Data Science*, which can be accessed for free in Spanish thanks to the collaborative work of its users at the following link[7]. All the work done by Garrett Grolemund and Hadley Wickham in facilitating the syntax of R is an excellent invitation to those who have never used statistical software or who want to move away from other alternatives such as Stata or SPSS.

Book structure

#	Chapter	Author(s)
I. Introduction to R		
1	Basic R	Andrés Cruz
2	Data Management	Andrés Cruz
3	Data Visualization	Soledad Araya
4	Data Loading	Soledad Araya and Andrés Cruz

---

[5]See https://gist.github.com/arcruz0/80ded0dce9fa28af1caa6d4d7f2da1bf
[6]See https://www.tidyverse.org/
[7]See http://es.r4ds.hadley.nz/

#	Chapter	Author(s)
II. Models		
5	Linear Models	Inés Fynn and Lihuen Nocetto
6	Cased Selection Based on Regressions	Inés Fynn and Lihuen Nocetto
7	Panel Data	Francisco Urdinez
8	Logistic Models	Francisco Urdinez
9	Survival Models	Francisco Urdinez
10	Causal Inference	Andrew Heiss
III. Applications		
11	Advanced Political Data Management	Andrés Cruz and Francisco Urdinez
12	Web Mining	Gonzalo Barría
13	Quantitative Analysis of Political Texts	Sebastián Huneeus
14	Networks	Andrés Cruz
15	Principal Component Analysis	Caterina Labrin and Francisco Urdinez
16	Maps and Spatial Data	Andrea Escobar and Gabriel Ortiz

The book is organized in three thematic sections. Within the sections, each chapter strives to solve specific problems, balancing theory and practice in R. At the beginning of the chapter we recommend literature that will help you go deeper into the subject if you are interested. We also invite you to reproduce the chapters with your own data as a practical exercise.

Part I is dedicated to data management. Ideally, the reader won't be looking at a dataset with a "I don't understand anything" face. We will introduce R from its installation and learn how to get the most out of it to obtain data, know it in depth, transform it according to the questions that interest us and represent it graphically in both functional and attractive ways.

Part II is the econometric section. We will see how to answer political questions from a statistical perspective —we can always answer them from our intuition, although this is usually less serious and scientific. In general, the section deals with statistical models that try to explain and predict the variation of certain (dependent) variables according to how other (independent) variables vary. We will explore different types of models according to the different forms of dependent variables commonly found in the political arena and the nature of our data. We will review how to interpret results and present them in a clear and appealing manner, how to choose between models, and how to simply test some of the statistical assumptions needed so that results are not biased. It should be noted that this is not an econometric book, of course, so for each model we will refer to more advanced work in theoretical terms, so that you can go deeper on your own if you feel you should use a specific model to answer his questions of interest.

Finally, in Part III we will leave the ideal world and go into the resolution of problems that make us break our heads. Whether it is because a colleague lent us his dataset that looks like a surrealist work of art and we have to clean it up, or simply because the difficulty of the problems we face leaves what we learned at the beginning of the book short, here we will present a mix of tools for the reader to integrate into his daily workflow. These have been selected from our experience and are what we believe are the most necessary in the practice of political data analysis.

Prerequisites

This book is designed for students who are motivated rather than brilliant: quantitative data analysis requires above all tenacity and curiosity. It is highly desirable for the reader to have basic notions of college mathematics, probability, and/or statistics before reading this book, even though we strive to keep it as simple as possible in those subjects. In terms of hardware, virtually any modern computer with Internet access will do, as the tools we will use are rather light. All the software we will use is free. The book was last tested under R's version 4.0.2.

In Chapter 1, "Basic R", it is explained how to install all the software that will be used in the book. If you already know a little about how R works, you should know that in subsection 1.4.3.3 you will install two packages that are already installed for the remainder of the book: `tidyverse` and `politicalds` (a package specially created to support the book).

How to use the textbook in a methods course?

If you find this textbook an interesting pedagogical resource, and you want to use it to structure your courses in quantitative methods or data analysis in political science, we offer you two alternatives. We suggest you prepare a semester-long course composed of 10 classes. The chapters in sections II and III offer readings that are strongly recommended to give robustness to the contents of this book. Remember that this is not a methodology textbook, but instead an applied R textbook, therefore, it is important that the other readings cover the theoretical part of each topic in depth.

First option: Semester-long course in political data analysis

> This is a layout for a course with emphasis on the student learning R, and quickly drawing descriptive conclusions from the data being used. Note: before starting the course, students must have read Chapter 1 "Basic R" and must have installed R, RStudio, the `tidyverse` library and the textbook's package. It is important to always bring ready-made scripts to class so that students

can quickly get to work without suffering from coding, which is something they will learn gradually at home. Another option to consider is RStudio Cloud[8], which allows students to work directly in the browser.

- *Class 1.* Use Chapter 2 to work through the presidential approval dataset to cover `mutate()`, `arrange()`, `filter()`, `rename()`, `select()` and `skim()`. It is essential that in this first class the student leaves the class feeling that she has learned something substantive from the dataset. It is advisable to offer the written script to avoid frustration with the syntax.

- *Class 2.* Continue with Chapter 2. Using `summarize()` by groups, the student should begin to interpret averages, maximum and minimum values and compare them among interest groups. You can ask students to create a dichotomous variable to compare two interest groups. Emphasize the potential of combining `mutate()` with `if_else()` and don't pause at the notion of pivoting data. It is advisable to offer the written script to avoid frustration with the syntax, the student will be able to put it into practice doing the exercises of the chapter at home.

- *Class 3.* Begin Chapter 3, introducing the layered world of `ggplot2`. Provide a script using the dataset from the previous two classes, namely the presidential approval dataset, and leave the dataset used in the chapter for the student to analyze it at home. Exemplify the idea of aesthetic mappings with `geom_boxplot()`, `geom_bar()` and `geom_line()`, and combine all three with `patchwork`. With what students learned previously, create categorical variables to group countries, which can then be used to introduce the idea of facets. End the class by demonstrating options for `ggplot2` themes.

- *Class 4.* Continue with Chapter 3 starting with `geom_histogram()`. Combine this tool with `skim()` and `filter()`. Again, provide a written script, but allow students to gradually become involved in the syntax of the commands. Incorporate supplementary options such as `labs()` and `scale_x_continuous()`. Combine `geom_point()` and `geom_smooth()` to analyze the correlation between various variables. Invite students to draw conclusions from the data and, if desired, take the opportunity to explore the concept of correlation and its calculation in more depth.

- *Class 5.* Cover Chapter 4 in one class. Loading data can be frustrating, so it is important to prepare a practical exercise to take to class and emphasize the importance of working with RStudio Projects. Create Excel and .csv versions of the presidential approval dataset and exemplify how they are imported into your project. Then repeat the process with versions of your dataset in Stata and SPSS. Finally, have students load a dataset and answer questions about the data using what they learned in the previous four classes. Now, students are ready to write their own code from scratch.

---

[8]See https://rstudio.cloud

- *Class 6.* Using the content of Chapter 12, teach students to extract data from Twitter using APIs with `rtweet`. Exemplify the step-by-step use of the `search_tweets()`, `get_followers()` and `lookup_users()` functions, and then leave a space for each student to replicate this example using their own topic of interest. With the material taught in class 5, ask students to save their datasets to use in the next class.

- *Class 7.* Using the contents in Chapter 12, analyze the dataset created in the previous class with the `tidytext` package, introducing word frequency analysis and tf-idf. Take up again what was covered in class two about combining `mutate()` with `if_else()` to group tweets by variables of interest. At this point in the course, it is important that the student tries to write code on his own, and it is good to demand the use of aesthetic options in the figures created with `ggplot2`.

- *Class 8.* Go deeper into text analysis, focusing on simple topic modeling. Demonstrate how to use the `tidytext`, `lubridate` and `stm` packages together with a previously created example, and then invite students to apply these functions to their datasets.

- *Class 9.* In this class you will go back to classes 3 and 4 in order to advance the content in Chapter 16. Start by explaining the logic of a shapefile, and exemplify how to visualize it with `geom_sf()`. For a first experience with the topic we recommend a shapefile of Latin America with its political boundaries. Using the presidential approval dataset, illustrate how to add variables to the shapefile using `left_join()`. In this class you should make sure that the joining variable matches both datasets as well as possible. We don't want to cover fuzzy join issues, for example.

- *Class 10 .* This class covers LISAs and space cluster ideas using `spdep`. We recommend also covering the utility of `geom_point()`, which can be exemplified using the `pinochet` and `naturalearthhires` packages. To conclude the course, it would be good to ask for an R Markdown that synthesizes this exercise and incorporates as many of the tools learned in previous classes as possible.

Second option: Semester-long course in introduction to quantitative methods

This is a layout for a course where the emphasis is on Ordinary Least Squares. Before starting the course students should have read chapter 1 "Basic R" and have RStudio and the textbook package installed. It is important for the first classes to bring ready-to-use scripts so that the students can quickly get to work without suffering from coding, which is something they will learn at home.

- *Class 1.* Use Chapter 2 to work through the presidential approval dataset to cover `mutate()`, `arrange()`, `filter()`, `rename()`, `select()`, and `skim()`. Using `summarize()` by groups, the student should begin to interpret averages, maximum and minimum values and compare them among interest groups. You can ask them to create a dichotomous variable to compare two interest groups. Emphasize the potential of combining `mutate()` with `if_else()` and do not stop at `pivot`. It is essential that in this first class the student leaves feeling that she has learned something substantive from the data. It is advisable to offer a written script to avoid frustration with the syntax and also to introduce the students to R Markdown.

- *Class 2.* Begin Chapter 3 by introducing the layered world of `tidyverse`. Start with `geom_histogram()`. Combine this tool with `skim` and `filter`. Again, offer a prepared script, but allow students to gradually get involved in the syntax of the commands. You can incorporate the complementary options such as `labs()` and `scale_x_continuous()`. Combine `geom_point()` and `geom_smooth()` to analyze the correlation between various variables. Introduce the concept of ordinary least squares with an applied example using `geom_smooth()`. Finish the class by showing options for `themes`.

- *Class 3.* Quickly cover Chapter 4 with a hands-on exercise and elaborate on the importance of working with RStudio Projects. Create Excel and .csv versions of the presidential approval dataset and exemplify how they are imported into your project. If the students have experience with Stata, explain `haven()`. Provide students with the .csv version of the Chapter 3 township database, show `ggcorrplot()`, and advance to Chapter 5 by quickly covering `lm()` with a univariate regression, and `texreg()`. Students will have two weeks to work at home with the Huber et al. welfare dataset and write a Markdown.

- *Class 4.* This class will take a detailed look at Chapter 5, starting with multivariate regressions. Explain `factor()`, `list()` and elaborate on the interpretation of coefficients. Expand on this topic with `prediction()` and illustrate its usefulness with `ggplot()`. From this class on, allow time for students to work alone or in pairs to write their own code.

- *Class 5.* Dedicate this class to cover model fitting and OLS assumptions, as well as the use of logarithmic transformations into dependent and independent variables and the implementation of robust standard errors. Chapter 5 covers all these topics in detail.

- *Class 6.* With the same dataset used to cover the contents of classes 4 and 5, cover the contents of Chapter 6 in an applied way, asking students to apply `broom()` to select cases from regressions. It may be helpful for students to do exercise 6C. in class and hand in a Markdown with their answers.

- *Class 7.* Focus on Chapter 10 that looks at DAGs, introducing `ggdag()`, `dagitty()`, `matchit()` and using the already known `broom()` and `texreg()`. Discuss the Neyman-Rubin causal model. To address the class, we recommend that an example similar to the one the author uses in the chapter is developed, and then that students work on the basis of the previous class to design their own DAG.

- *Class 8.* Cover the contents of Chapter 7 on panel data, featuring `lubridate()` and `plm()`. Spend a portion of the class presenting `devtools`, whose functions serve to visualize results and are well complemented by `prediction()`. Cover fixed and random effects models, `phtest()`, unit roots and time correlation tests and the incorporation of robust panel-corrected standard errors. Reinforce the use of `texreg` to export tables for use in academic papers.

- *Class 9.* The last two classes can be dedicated to offering the student tools to complement what they have learned up to this point. Use Chapter 15 to introduce students to the creation of indexes that can be incorporated into models as both dependent and independent variables.

- *Class 10.* Finish the course using Chapter 11 with a discussion on code standardization and database linking using `left_join()` and demonstrating the use of `countrycode`, `stringdist`, and `inexact`. Ask students to do an exercise in class. Then discuss missing data and imputation by presenting `naniar` and `mice` and ask students to do another class exercise using the package.

Acknowledgments

We are grateful to all the prestigious scholars who shared their data to illustrate our chapters. We would like to thank Rob Calver, CRC Press R Series editor, for his patience and good vibes. Vaishali Singh, Editorial Assistant at Taylor & Francis Group, provided us with great editorial assistance. Annie Sophia and Elissa Rudolph from NovaTechset were very helpful during the proof-editing of the book. We also want to thank Matías Piña, Vicente Quintero, Eliana Jung and Beatriz Urrutia, four great research assistants, for assisting with the corrections and edition of the book and Daniel Alcatruz, Javiera Venegas, Santiago Olivella, Laura Levick and Carsten Schulz for their feedback on various chapters.

Contributors

Soledad Araya
Political Science Institute
Pontificia Universidad Católica de Chile
Santiago, RM, Chile

Gonzalo Barría
Political Science Institute
Pontificia Universidad Católica de Chile
Santiago, RM, Chile

Andrés Cruz
Political Science Institute
Pontificia Universidad Católica de Chile
Santiago, RM, Chile

Andrea Escobar
Political Science Institute
Pontificia Universidad Católica de Chile
Santiago, RM, Chile

Inés Fynn
Political Science Institute
Pontificia Universidad Católica de Chile
Santiago, RM, Chile

Andrew Heiss
Andrew Young School of Policy Studies
Georgia State University
Atlanta, GA

Sebastián Huneeus
Political Science Institute
Pontificia Universidad Católica de Chile
Santiago, RM, Chile

Caterina Labrin
Political Science Institute
Pontificia Universidad Católica de Chile
Santiago, RM, Chile

Lihuen Nocetto
Political Science Institute
Pontificia Universidad Católica de Chile
Santiago, RM, Chile

Gabriel Ortiz
Political Science Institute
Pontificia Universidad Católica de Chile
Santiago, RM, Chile

Francisco Urdinez
Political Science Institute
Pontificia Universidad Católica de Chile
Santiago, RM, Chile

Part I

Introduction to R

1
Basic R

Andrés Cruz[1]

1.1 Installation

R

R is a programming language specially developed for statistical analysis. One of its main characteristics, as it was suggested in the preface, is that it is *open source*: besides being free, it means that the licenses that legally protect R are permissive. Under these licenses, thousands of developers around the world have contributed their two cents to the usability and attractiveness of R. In *R for Political Data Science: A Practical Guide* we will use this diversity to our advantage.

Installing R is easy, whether you use Windows, Mac or Linux. You just have to access to https://cran.r-project.org/ and follow the instructions for download and installation.

RStudio

As we stated, R is a programming language. To put in in a colloquial way, it is an orderly way of asking the computer to perform certain operations. This means that we can use R exclusively from a console or a terminal —the black screens hackers use in the movies. Although this has certain appeal (among them, to resemble a hacker), in general we want friendlier interfaces. Here is where RStudio appears in our life, the most popular program for using R. Once installed, all our analysis will occur within RStudio, which is also open source and regularly updated by a team of programmers.

To install RStudio, it is necessary to have R. Download and installation is accessible for Windows, Mac and Linux. The link is https://www.rstudio.com/products/rstudio/download/#download.

---

[1] Institute of Political Science, Pontificia Universidad Católica de Chile. E-mail: arcruz@uc.cl. Twitter: @arcruz0.

Exercise 1A. We will wait while you install R and RStudio! Go ahead, you will need it to work with us throughout the book.

If you were able to download and install R and RStudio, you just have to open the latter to start working. You will find a screen as the following Figure 1.1.

FIGURE 1.1: Four panels of the basic interface of RStudio.

RStudio is divided in four panels (console, script, objects and miscellaneous), which we will address below. The idea of this section is to familiarize the beginner reader with the basis of R.

Also, we recommend you edit some configurations of RSutdio before starting, which in our experience will improve your learning experience[2]:

- General > Uncheck "Restore .RData into workspace at startup"
- General > Save Workspace to .RData on exit > Never

1.2 Console

The bottom left panel of RSutdio is our space for direct communication with the computer, where we demand it to perform specific tasks, using the "R language". We

---

[2]The idea of these changes is that every RStudio session starts from zero so misunderstandings can be avoided. This is consistent with RStudio Projects, which we will review in a moment.

will call these demands *commands*. Let's try using or "running" a command that performs a basic arithmetic operation:

```
2 + 2
## [1] 4
```

A tip for the console is that with your up and down arrow keys you can navigate through the record of used commands.

1.3 Script

The top left panel of RStudio can be described as a "command log". Although the console can be useful for some commands, complex analyses will require that we carry a log of our code.

For writing a new script it is as easy as pressing `Ctrl + Shift + N` or going to `File > New File > R Script` (using keyboard shortcuts is a great idea, and not only for the "hacker factor"). The white screen of a new script is similar to a blank notepad, with the characteristic that every line should be thought as a command. Note that writing a command in the script and pressing `Enter` does nothing but a paragraph break. For running the command in a line, you have to press `Ctrl + Enter` (if you have Mac, `Cmd + Enter`). It is possible to select multiple lines/commands at once and running them all with `Ctrl + Enter`.

Writing only code in our scripts is not enough, since we usually want to write explanatory comments as well. This is not only relevant for group work (foreign code can be unintelligible without clear guidance), but also denotes attention for your future "you". On various occasions, we have to check code that we wrote a couple of months ago, not understanding anything, and think unlovely things about ourselves. When writing commands, R recognizes that everything following a number sign (#) is a comment. Thus, there are two ways of writing commentaries, as "sterile commands" or as an appendix of functional commands:

```
# This is a sterile command. R knows that it is only a comment!


2 + 2 # This is an appendix-command, it does not modify the code
## [1] 4
```

For saving a script, all you have to do is press `Ctrl + S` or click `File > Save`.

1.4 Objects (and functions)

This is the top right panel of RStudio. Although it has three tabs ("Environment", "History" and "Connections"), the big star is "Environment", which functions as a record of the objects we create as we work. One of the main features of R is that it allows storing objects and then running commands with them. The way to create an object is by using an arrow `<-`, so that `name_of_the_object <- content`. We will call this an *assignment*. For example:

```
object_1 <- 2 + 2
```

After running this, a new object will appear in the `Environment` panel, `object_1`. This contains the result of 2+2. It is possible to ask R what the content of an object is just by running its name as if it was a command:

```
object_1
## [1] 4
```

Crucially, objects can be inserted into other commands by referring to its contents. For example:

```
object_1 + 10
## [1] 14
```

It is also possible to reassign the objects: if we get bored of `object_1` as a 4, we can assign it any value we want. Text (character) values are also valid, they have to be written in quotation marks:

```
object_1 <- "democracy"
```

```
object_1
## [1] "democracy"
```

Erasing an object is also a simple task. Although it can sound like losing our hard work, having a clean and easy-reading "Environment" tab is often worthwhile. For that, we have to use the `rm()` function. You can also use `rm(list = ls())` to erase all the objects.

1.4.1 Vectors

Until now we have learned the most basic objects of R, which contain a single value. **vectors** are more complex objects[3]. Creating a vector is simple, we just need to insert its components between a c(), separated by commas:

```
vector_1 <- c(15, 10, 20)
vector_1
## [1] 15 10 20
```

A basic need when creating vectors is inserting number sequences. A simple way of doing this is with colons (:). For example, let's examine the following vector:

```
vector_2 <- c(9, 7:10, 2, 14)
vector_2
## [1] 9 7 8 9 10 2 14
```

We can select specific elements of a vector by using its positions:

```
vector_2[2] # it gives us the second element
## [1] 7
vector_2[4:6] # it gives us the fourth, fifth and sixth element.
## [1] 9 10 2
```

1.4.2 Functions

Look at the following example command:

```
2 + sqrt(25) - log(1) # equivalent to 2 + 5 + 0
## [1] 7
```

R interprets that sqrt(25) is the squared root of 25, while log(1) is the natural logarithm of 1. Both sqrt() and log() are **functions** of R. In simple terms, a function is a procedure that can be outlined as in Figure 1.2:

sqrt() assumes a numeric value as an input and delivers its squared root as an output. log() assumes that same input, but delivers its natural logarithm. c(), a function we previously used, assumes different unique values as inputs and delivers a vector that contains them.

It is because of vectors that functions in R begin to shine and depart from the basic qualities of a calculator (that, broadly speaking, is what we have seen until now in R,

---

[3] Technically speaking, the previous objects are vectors of longitude 1 for R.

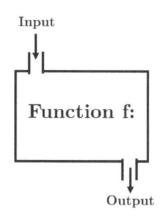

FIGURE 1.2: Function outline. Adapted from Wikimedia Commons: https://commons.wikimedia.org/wiki/File:Function_machine2.svg

nothing impressive). Let's examine other functions that extract useful information from a vector. What does each one do?

```
mean(vector_1)    # mean
## [1] 15
median(vector_1)  # median
## [1] 15
sd(vector_1)      # standard deviation
## [1] 5
sum(vector_1)     # sum
## [1] 45
min(vector_1)     # minimum value
## [1] 10
max(vector_1)     # maximum value
## [1] 20
log(vector_1)     # natural logarithm
## [1] 2.7 2.3 3.0
exp(vector_1)     # exponential
## [1] 3.3e+06 2.2e+04 4.9e+08
length(vector_1)  # length (amount of values)
## [1] 3
sort(vector_1)    # ...
## [1] 10 15 20
```

You can infer that `sort()`, the last function of the previous list, orders the vector from smallest to largest:

```
sort(vector_1)
## [1] 10 15 20
```

1.4 Objects (and functions)

What happens if we wanted to order it from largest to smallest? This allows us to introduce *arguments*, parts of the functions that modify its behavior. Now, we will add the argument `decreasing = TRUE` into the previous command, which achieves our goal:

```
sort(vector_1, decreasing = TRUE)
## [1] 20 15 10
```

A concept that has to do with both objects and functions is the one of *missing values*, a topic that we will explore in more detail in Chapter 11. The datasets we work with can have missing values for various reasons: coding errors, governments that hide information, data not collected yet, among other. Whatever the case, we need to take this into account when doing our analysis. R registers missing values as `NA` (not available). Note that this is not a character value with the letters "N" and "A", but a distinct type of value. Let's add a missing value to our previous vector:

```
vector_1_with_na <- c(vector_1, NA)
vector_1_with_na
## [1] 15 10 20 NA
```

How should functions react to this new vector? By default, most R operations performed with missing values fail (returning `NA`), alerting you that they cannot compute what you need. For example:

```
mean(vector_1_with_na)
## [1] NA
```

Perhaps in some cases you would like to let R know that it should ignore the missing values in the vector and simply carry on with the operation. In most functions, you can specify this with the argument `na.rm = TRUE`:

```
mean(vector_1_with_na, na.rm = TRUE)
## [1] 15
```

Another possibility is to do this NA omission *ex ante*, by modifying the vector. A useful function for this case is `na.omit()`, which returns the vector without any missing values:

```
na.omit(vector_1_with_na)
## [1] 15 10 20
## attr(,"na.action")
## [1] 4
## attr(,"class")
## [1] "omit"
```

Finally, a common function for dealing with NAs is is.na(), which allows you to check which values of a vector are missing, something that will be useful later on (for example, for filtering datasets):

```
is.na(vector_1_with_na)
## [1] FALSE FALSE FALSE  TRUE
```

Exercise 1B. Look at the following graph (Figure 1.3) and create a vector with the minimum wages for the countries of the region. What is its mean? And its median?

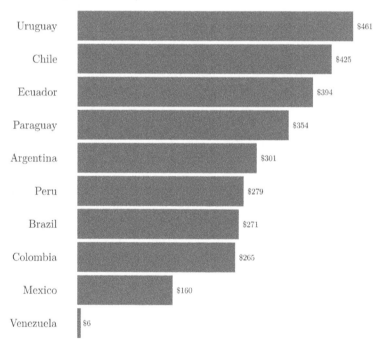

FIGURE 1.3: Minimum wage in Latin America, January of 2019. Adapted from Statista: https://es.statista.com/grafico/16576/ajuste-de-los-salarios-minimos-en-latinoamerica).

1.4 Objects (and functions)

1.4.3 Files / graphs / packages / help

The bottom right quadrant of the RStudio screen contains a mix of different uses, which we will examine below.

1.4.3.1 Files and Projects of RStudio

This tab is a window to the files that we have on our work directory. It functions as a small manager, and it allows us to move them, rename them and copy them.

Concerning files, one of the great recent innovations of R are RStudio Projects. The developers of RStudio realized that their users had scripts and others R files scattered all over their hard drives, with no order. That is why they implemented the philosophy of "one project, one folder". "One project, one folder" is as important as it sounds: the idea is that each project we work on is self-sufficient, that it includes all of what we need for working (scripts, datasets, etc.).

Projects can be managed from the top right corner in R. Did you see the three tabs, "Environment", "History" and "Connections"? Well, look a little higher and you will see the RStudio Projects logo. Here you have to be careful and note that creating or opening a new project will restart your R session, erasing all the work you did not save. Since you have not created a project yet, your session should say "Project: (None)". By clicking in "New Project", three options appear:

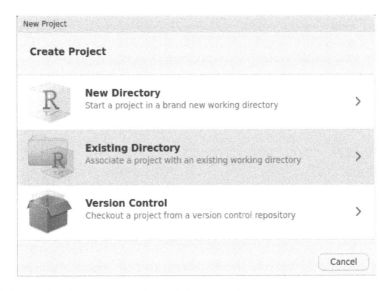

FIGURE 1.4: Creating a new Project in RStudio.

Once you create a project and you start working on it, all links to the files will be local. For example, if your project folder contains a "data" subfolder with an "example.csv" file on it, the reference to the file will simply be "data/example.csv". Remember the motto: "one project, one folder".

We recommend that you create a RStudio project for each chapter of the book that you want to follow with code!

1.4.3.2 Graphs

Here are the graphs we create with R. In Chapter 3, data visualization, you will learn how to master them.

1.4.3.3 Installing packages

One of the main features we have highlighted about R is its versatility. Its open source philosophy motivates developers to bring new features into the R community. In general, they do it through packages, which users can install as an additional appendix to R. These packages contain new functions, datasets, and features. The tabs of RStudio we reviewed allows us to access the installed packages.

Installing a package is fairly simple through the `install.packages()` function. Now, we will install the `tidyverse` package, which will be central for our next analyses. The `tidyverse` is a compilation that includes some of the best modern packages for data analysis in R.

```
install.packages("tidyverse")
```

Everytime that a user opens a new session in R, this opens as "factory new". That is, it does not only open without objects, but also with only the basic packages that allow R to work. Then, we have to load the extra packages we want to use. It is like when you buy a *smartphone* and you download the applications you will use according to your everyday needs. The most common way of doing this is through the `library()` function, as seen below. Note that this time `tidyverse` is not in quotation marks[4]

```
library(tidyverse)
```

Additionally, to get the most out of this book you must install our complementary package, `politicalds`. This will give you access to the datasets that will be used in the different chapters of this book, plus some support functions. Installation is slightly different, because it is a work in progress. To obtain it, first you must have installed the `remotes` package (Hester et al., 2020), which will allow you to use packages stored in GitHub, a software development platform.

```
install.packages("remotes")
```

---

[4]This is the most common way of using `library()` by convention. The command will work with quotation marks, although it is not very common to see it like this: `library("tidyverse")`. Following the conventions is a great idea, and that is why we recommend you to omit the quotation marks!

1.4 Objects (and functions)

Once the `remotes` package is loaded, its function `install_github()` will enable you to install the package of this book:

```
library(remotes)
install_github("arcruz0/politicalds")
```

Note that one of us has the username "arcruz0" in GitHub, where the package `politicalds` is stored. Now it is installed on your system! Every time you need it in a R session, you have to load it with `library()`:

```
library(politicalds)
```

1.4.3.4 Help

Searching for help is essential when programming in R. Look at the figure 1.5: this RStudio tab opens help files that we need and which we can search. Functions have a help file for each of them. For example we can access the help file of `sqrt()` through the command `help(sqrt)` (`?sqrt` also works). Packages as a whole also contain help files, which are more comprehensive. For example, to see the help file of `tidyverse` we just need to call upon the argument "package": `help(package=tidyverse)`. Also, help files from packages and functions from packages are only available when the corresponding packages have been loaded.

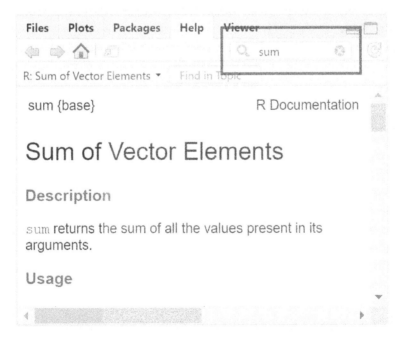

FIGURE 1.5: Help file of the `sum()` function.

Inside the help file for a package we can seek functions or doubts about specific commands, in the quadrant of the image that we have pointed out in red.

Exercise 1C. Install and look for help for the `ggparliament` package. It is recommended you play around with all the functions we saw from the R ecosystem. We will briefly mention `ggparliament` later on, in Chapter 3 of data visualization.

2
Data Management

Andrés Cruz[1]

Packages you need to install

- `tidyverse` (Wickham, 2019), `politicalds` (Urdinez and Cruz, 2020), `skimr` (Waring et al., 2020).

2.1 Introduction to data management

Tabular datasets are the quintessential form of saving information in social sciences. Its power relies on the ability to register multiple dimensions of information for every observation we are interested in. For example, for every representative in Congress we can know their gender, age, percentage of attendance at room, number of bills presented, etc.:

TABLE 2.1: Example dataset of representatives.

representative	gender	age	attendance	num_bills_presented
Estuardo, Carlos	M	48	68	8
Cerna, Marta	F	57	74	3

You are probably familiarized with these types of datasets thanks to Microsoft Excel or Google Sheets. The first row is the **header**, which indicates what data is registered in the cells of that column. The first column in this case is our **identity column**: when consulting "representative", we can know what observation each row refers to. Thus, over this small dataset we can say that the **observational unit** is the representative, of which we have information in four variables: gender, age, attendance at room and numbers of bills presented.

In the following chapter, you will learn how to modify tabular datasets like the one in the example. You will learn how to order datasets, filter its observations, create

---

[1] Institute of Political Science, Pontificia Universidad Católica de Chile. E-mail: arcruz@uc.cl. Twitter: @arcruz0.

new variables, generate summaries, change names, recode values and modify the structure of the dataset. All of these operations are an initial step for any analysis or visualization: it is estimated that 80% of the data analysis process is invested in modifying and cleaning our data for its optimal usage (Dasu & Johnson, 2003 in Wickham, 2014).

2.1.1 Our dataset

In this chapter, we will use data from Reyes-Housholder (2019), with some additions from the World Development Indicators by the World Bank, collected by Quality of Government[2]. The author arguments that, in the face of similar corruptions scandals, women presidents in Latin America suffer steeper falls in approbation in comparison to their male counterparts.

Let's start by loading the `tidyverse` package, one of the main packages in the book, which will hand us useful tools for working with our dataset.

```
library(tidyverse)
```

Now, let's load our dataset into the R session. We can do this with ease with the package of this book by using the `data()` function. The dataset is called `approval`:

```
library(politicalds)
data("approval")
```

Now, we can start working with the dataset. You can check that it was loaded correctly utilizing the `ls()` command (or by looking at the Environment tab in RStudio):

```
ls()
## [1] "approval"
```

The following are the dataset variables:

TABLE 2.2: Dataset variables

Variable	Description
country	Country
year	Year
quarter	Quarter
president	President
pres_sex	President's sex

---
[2]See https://qog.pol.gu.se/data

TABLE 2.2: Dataset variables (*Continued*)

net_approv	Net presidential approval (% approval - % disapproval)
gdp	Gross Domestic Product of the country, constant 2011 dollars and adjusted by Purchasing Power Parity (PPP)
pop	Population
exec_corr	Corruption of the Executive Power, according to V-Dem. From 0 to 100 (greater number means more corruption)
unemp	Unemployment rate
gdp_growth	Growth of the GDP

2.2 Describing a dataset

For approaching the newly loaded dataset we have several options. We can, just as before, type its name as a command for a quick summary:

```
approval
## # A tibble: 1,020 x 11
##   country  year quarter president      president_gender net_approval   gdp
##   <chr>   <dbl>   <dbl> <chr>          <chr>                   <dbl> <dbl>
## 1 Argent~  2000       1 Fernando~      Male                     40.1  14.0
## 2 Argent~  2000       2 Fernando~      Male                     16.4  14.0
## 3 Argent~  2000       3 Fernando~      Male                     24.0  14.0
## # ... with 1,017 more rows, and 4 more variables
```

We can also use the `glimpse()` function to obtain a summary from another perspective, looking into the first observations in each variable:

```
glimpse(approval)
## Rows: 1,020
## Columns: 11
## $ country          <chr> "Argentina", "Argentina", "Argentina", "Arg...
## $ year             <dbl> 2000, 2000, 2000, 2000, 2001, 2001, 2001, 2...
## $ quarter          <dbl> 1, 2, 3, 4, 1, 2, 3, 4, 1, 2, 3, 4, 1, 2, 3...
## $ president        <chr> "Fernando de la Rúa", "Fernando de la Rúa",...
## $ president_gender <chr> "Male", "Male", "Male", "Male", "Male", "Ma...
## $ net_approval     <dbl> 40.1, 16.4, 24.0, -18.3, -7.0, -20.1, -19.4...
## $ gdp              <dbl> 14, 14, 14, 14, 14, 14, 14, 25, 25, 25,...
## $ corruption       <dbl> 5.5e+11, 5.5e+11, 5.5e+11, 5.5e+11, 5.3e+11...
## $ population       <dbl> 3.7e+07, 3.7e+07, 3.7e+07, 3.7e+07, 3.7e+07...
## $ unemployment     <dbl> 15, 15, 15, 15, 18, 18, 18, 18, 18, 18,...
## $ gdp_growth       <dbl> -0.8, -0.8, -0.8, -0.8, -4.4, -4.4, -4.4, -...
```

An alternative that allows us to see the full dataset is the `View()` function, akin to clicking our object in the "Environment" tab in RStudio:

```
View(approval)
```

We can obtain a quick summary of the variables in our dataset by using the `skimr::skim()` function, as shown in Figure 2.1.

```
> skimr::skim(approval)
-- Data Summary ------------- values
Name                        approval
Number of rows              1020
Number of columns           11

Column type frequency:
  character                 3
  numeric                   8

Group variables             None

-- Variable type: character --------------------------------------------------
  skim_variable    n_missing complete_rate  min  max empty n_unique whitespace
1 country                  0             1    4   11     0       17          0
2 president                0             1   11   38     0       70          0
3 president_gender         0             1    4    6     0        2          0

-- Variable type: numeric ----------------------------------------------------
  skim_variable n_missing complete_rate    mean      sd        p0       p25      p50      p75     p100 hist
1 year                  0             1 2.01e 3 4.32e 0   2.00e 3   2.00e 3  2.01e 3  2.01e 3  2.01e 3
2 quarter               0             1 2.50e 0 1.12e 0   1.00e 0   1.75e 0  2.50e 0  3.25e 0  4.00e 0
3 net_approval          0             1 1.53e 1 2.78e 1  -6.58e 1  -6.57e 0  1.60e 1  3.76e 1  8.69e 1
4 gdp                   0             1 4.33e 1 2.69e 1   2.13e 0   1.96e 1  4.00e 1  6.66e 1  9.44e 1
5 corruption            0             1 4.11e11 6.88e11   1.74e10   4.01e10  9.20e10  4.01e11  3.14e12
6 population            0             1 3.14e 7 4.76e 7   3.03e 6   5.87e 6  1.31e 7  3.03e 7  2.04e 8
7 unemployment          0             1 7.04e 0 3.61e 0   1.30e 0   4.38e 0  6.41e 0  8.90e 0  2.05e 1
8 gdp_growth            0             1 3.77e 0 3.54e 0  -1.09e 1   2.31e 0  4.06e 0  5.63e 0  1.83e 1
```

FIGURE 2.1: Skim of our dataset.

Note: Every time you see that after a package we use :: it means that within that package we are calling for a specific function to be used. For instance, in the example above we call the `skim()` function from the `skimr` package.

Obtaining the tabulation from one of the columns in our dataset, a common function for categorical variables, is an easy task thanks to the `count()` function. For example, we can check that the countries-years-quarters with women as presidents are a minority in the region, 98 out of 1020:

```
count(approval, president_gender,
      sort = T) # order from highest to lowest by n
## # A tibble: 2 x 2
##   president_gender     n
##   <chr>            <int>
## 1 Male               922
## 2 Female              98
```

2.3 Basic operations

Next, we will look at some basic operations for our dataset, that as a whole will allow us to do major editing in the structure and contents (Wickham and Grolemund, 2016, ch. 5). This subsection utilizes the tools from the `dplyr` package, which is available when loading `tidyverse`.

2.3.1 Select columns

Sometimes we want to work with only an extract of the variables in a dataset. For this, exists the `select()` function. Let's select only the countries' column:

```
select(approval, country)
## # A tibble: 1,020 x 1
##    country
##    <chr>
## 1 Argentina
## 2 Argentina
## 3 Argentina
## # ... with 1,017 more rows
```

The first argument of the previous function (`approval`) is the dataset in which we want to execute the operation. The next argument indicates which columns to select (`country`). All the functions for basic operations that we will see in this subsection follow the same logic: the first argument is always the dataset in which we will operate, while the rest designates how we want to execute the operation.

Remember that previous code did not create any new object, it is just a command that we are executing in the console. If we wanted to create a new object, we would have to assign it, using the operator `<-`:

```
reduced_approval <- select(approval, country)
reduced_approval
## # A tibble: 1,020 x 1
##    country
##    <chr>
## 1 Argentina
## 2 Argentina
## 3 Argentina
## # ... with 1,017 more rows
```

We can select multiple columns at once, separated by commas:

```
select(approval, country, year, unemployment)
## # A tibble: 1,020 x 3
```

```
##   country     year unemployment
##   <chr>      <dbl>        <dbl>
## 1 Argentina   2000           15
## 2 Argentina   2000           15
## 3 Argentina   2000           15
## # ... with 1,017 more rows
```

Let's suppose that we wanted the first five variables of the dataset. The following are three ways of obtaining the same result, although we recommend the second one, since it is brief but explicit:

```
select(approval, country, year, quarter, president, net_approval)
select(approval, country:net_approval) # recommended way
select(approval, 1:5)
```

```
## # A tibble: 1,020 x 6
##   country   year quarter president         president_gender net_approval
##   <chr>    <dbl>   <dbl> <chr>             <chr>                   <dbl>
## 1 Argentina 2000       1 Fernando de la R~ Male                     40.1
## 2 Argentina 2000       2 Fernando de la R~ Male                     16.4
## 3 Argentina 2000       3 Fernando de la R~ Male                     24.0
## # ... with 1,017 more rows
```

The `select()` command can also help us reorder the columns. Let's suppose that we wanted the variable `president` to be first. We can rearrange the variables, obtaining the same dataset with a new order for the columns:

```
select(approval, president, country:year, net_approval:unemployment)
## # A tibble: 1,020 x 8
##   president   country  year net_approval   gdp corruption population
##   <chr>       <chr>   <dbl>        <dbl> <dbl>      <dbl>      <dbl>
## 1 Fernando~   Argent~  2000         40.1  14.0    5.52e11   37057452
## 2 Fernando~   Argent~  2000         16.4  14.0    5.52e11   37057452
## 3 Fernando~   Argent~  2000         24.0  14.0    5.52e11   37057452
## # ... with 1,017 more rows, and 1 more variable
```

This method is tedious, especially for datasets with multiple variables. There is a function that can be useful for these scenarios, called `everything()`. In this case, it will select the `president` column and *everything else*:

```
select(approval, president, everything())
## # A tibble: 1,020 x 11
##   president   country  year quarter president_gender net_approval   gdp
##   <chr>       <chr>   <dbl>   <dbl> <chr>                   <dbl> <dbl>
## 1 Fernando~   Argent~  2000       1 Male                     40.1  14.0
## 2 Fernando~   Argent~  2000       2 Male                     16.4  14.0
```

2.3 Basic operations

```
## 3 Fernando~ Argent~  2000       3 Male                           24.0  14.0
## # ... with 1,017 more rows, and 4 more variables
```

Another helpful function for `select` is `starts_with()`, which allows us to select columns according to patterns in its names. For example, the following will select all the columns that start with the "gdp" prefix.

```
select(approval, starts_with("gdp"))
## # A tibble: 1,020 x 2
##      gdp gdp_growth
##    <dbl>      <dbl>
## 1   14.0       -0.8
## 2   14.0       -0.8
## 3   14.0       -0.8
## # ... with 1,017 more rows
```

2.3.2 Rename columns

We can change the names of the columns in a dataset with the `rename()` command. For example, let's make the GDP variable name more explicit:

```
rename(approval, gdp_ppp_c2011 = gdp)
## # A tibble: 1,020 x 11
##   country  year quarter president  president_gender net_approval
##   <chr>   <dbl>   <dbl> <chr>      <chr>                   <dbl>
## 1 Argent~  2000       1 Fernando~  Male                     40.1
## 2 Argent~  2000       2 Fernando~  Male                     16.4
## 3 Argent~  2000       3 Fernando~  Male                     24.0
## # ... with 1,017 more rows, and 5 more variables
```

It is also possible to change multiple names at once. Note how we modify three names with just one command:

```
rename(approval,
       gdp_ppp_c2011 = gdp,
       unemployment_percentage = unemployment,
       gdp_percent_growth = gdp_growth)
## # A tibble: 1,020 x 11
##   country  year quarter president  president_gender net_approval
##   <chr>   <dbl>   <dbl> <chr>      <chr>                   <dbl>
## 1 Argent~  2000       1 Fernando~  Male                     40.1
## 2 Argent~  2000       2 Fernando~  Male                     16.4
## 3 Argent~  2000       3 Fernando~  Male                     24.0
## # ... with 1,017 more rows, and 5 more variables
```

2.3.3 Filter observations

Often, we want to keep only some observations from our dataset, filtering according to specific characteristics. We can do this thanks to the `filter()` function and logical operators. To start, let's keep only the observations for Chile:

```
filter(approval, country == "Chile")
## # A tibble: 60 x 11
##   country  year quarter president president_gender net_approval    gdp
##   <chr>   <dbl>   <dbl> <chr>     <chr>                   <dbl>  <dbl>
## 1 Chile    2000       1 Eduardo ~ Male                     6.22   3.63
## 2 Chile    2000       2 Ricardo ~ Male                    19.8    3.63
## 3 Chile    2000       3 Ricardo ~ Male                    19.5    3.63
## # ... with 57 more rows, and 4 more variables
```

We are telling `filter()`, through the second argument, only to retain the observations in which the variable country *is equal to* "Chile". This *is equal to* is a logical operator, which is written as "==" in R [3]. Here is a list of common logical operators:

TABLE 2.3: List of common logical operators.

Operator	Description
==	is equal to
!=	is different from
>	is greater than
<	is less than
>=	is greater or equal to
<=	is less or equal to
&	intersection
|	union
%in%	is contained in

For example, we can obtain all the observations (country-year-quarter) in which net presidential approval is positive:

```
filter(approval, net_approval> 0)
## # A tibble: 709 x 11
##   country  year quarter president president_gender net_approval    gdp
##   <chr>   <dbl>   <dbl> <chr>     <chr>                   <dbl>  <dbl>
## 1 Argent~  2000       1 Fernando~ Male                    40.1   14.0
## 2 Argent~  2000       2 Fernando~ Male                    16.4   14.0
## 3 Argent~  2000       3 Fernando~ Male                    24.0   14.0
## # ... with 706 more rows, and 4 more variables
```

---

[3] Stata users will find these familiar.

2.3 Basic operations

It is also possible to execute more complex filters. Let's obtain only the observations for the Southern Cone:

```
filter(approval,
       country == "Argentina" | country == "Chile" | country == "Uruguay")
## # A tibble: 180 x 11
##    country year quarter president president_gender net_approval    gdp
##    <chr>   <dbl>  <dbl>  <chr>    <chr>                   <dbl>  <dbl>
## 1 Argent~  2000      1  Fernando~ Male                     40.1   14.0
## 2 Argent~  2000      2  Fernando~ Male                     16.4   14.0
## 3 Argent~  2000      3  Fernando~ Male                     24.0   14.0
## # ... with 177 more rows, and 4 more variables
# The same, but with another logical operator:
filter(approval, country %in% c("Argentina", "Chile", "Uruguay"))
## # A tibble: 180 x 11
##    country year quarter president president_gender net_approval    gdp
##    <chr>   <dbl>  <dbl>  <chr>    <chr>                   <dbl>  <dbl>
## 1 Argent~  2000      1  Fernando~ Male                     40.1   14.0
## 2 Argent~  2000      2  Fernando~ Male                     16.4   14.0
## 3 Argent~  2000      3  Fernando~ Male                     24.0   14.0
## # ... with 177 more rows, and 4 more variables
```

We can also include small operations in our filters. Let's obtain all the observations in which the executive's corruption is greater than that of the dataset's mean:

```
filter(approval, corruption > mean(corruption))
## # A tibble: 252 x 11
##    country year quarter president president_gender net_approval    gdp
##    <chr>   <dbl>  <dbl>  <chr>    <chr>                   <dbl>  <dbl>
## 1 Argent~  2000      1  Fernando~ Male                     40.1   14.0
## 2 Argent~  2000      2  Fernando~ Male                     16.4   14.0
## 3 Argent~  2000      3  Fernando~ Male                     24.0   14.0
## # ... with 249 more rows, and 4 more variables
```

One practical caveat: you cannot search for missing values (NAs) with the intuitive `== NA` operator. You need to use the `is.na()` function. Our dataset does not have any missing values, and therefore a filter like the following will return no rows:

```
filter(approval, is.na(corruption))
## # A tibble: 0 x 11
## # ... with 11 variables
```

Exercise 2A. Select only the two columns that register the president's gender in the dataset.

Exercise 2B. Filter the dataset in order for it to contain only observations of the year 2000.

2.3.4 Change the order of a dataset

One of the most common operations with datasets is sorting them according to one variable. This can give us clear hints about our observations. We can do this thanks to the `arrange()` function. For example, let's sort the observations from the least corrupt country-year-quarter to the most corrupt:

```
arrange(approval, corruption)
## # A tibble: 1,020 x 11
##   country  year quarter president  president_gender net_approval    gdp
##   <chr>   <dbl>   <dbl> <chr>      <chr>                   <dbl>  <dbl>
## 1 Nicara~  2000       1 Arnoldo ~  Male                     7.60   85.7
## 2 Nicara~  2000       2 Arnoldo ~  Male                     7.57   85.7
## 3 Nicara~  2000       3 Arnoldo ~  Male                     3.87   85.7
## # ... with 1,017 more rows, and 4 more variables
```

If we wanted to order them inversely, we would have to add a - (minus sign) before the variable:

```
arrange(approval, -corruption)
## # A tibble: 1,020 x 11
##   country year quarter president  president_gender net_approval    gdp
##   <chr>  <dbl>   <dbl> <chr>      <chr>                   <dbl>  <dbl>
## 1 Brazil  2014       1 Dilma Va~  Female                   22.6   27.3
## 2 Brazil  2014       2 Dilma Va~  Female                   12.6   27.3
## 3 Brazil  2014       3 Dilma Va~  Female                   16.2   27.3
## # ... with 1,017 more rows, and 4 more variables
```

For utilizing an inverse alphabetic order (from Z to A), we have to use the `desc()` help function.

```
arrange(approval, desc(president))
## # A tibble: 1,020 x 11
##   country year quarter president president_gender net_approval    gdp
##   <chr>  <dbl>   <dbl> <chr>     <chr>                   <dbl>  <dbl>
## 1 Mexico  2001       1 Vicente ~ Male                     53.8   37.6
## 2 Mexico  2001       2 Vicente ~ Male                     40.7   37.6
## 3 Mexico  2001       3 Vicente ~ Male                     40.1   37.6
## # ... with 1,017 more rows, and 4 more variables
```

2.3 Basic operations

Lastly, we can sort the dataset by more than one variable. This is, sorting the dataset according to the first variable, and then sorting the draws according to a second variable. Let's examine the following example:

```
arrange(approval, president_gender, -net_approval)
## # A tibble: 1,020 x 11
##   country year quarter president   president_gender net_approval  gdp
##   <chr>   <dbl>  <dbl>  <chr>      <chr>                  <dbl> <dbl>
## 1 Brazil  2013       1  Dilma Va~  Female                  62.5  27.3
## 2 Brazil  2012       4  Dilma Va~  Female                  60.9  33.4
## 3 Brazil  2012       2  Dilma Va~  Female                  60.5  33.4
## # ... with 1,017 more rows, and 4 more variables
```

2.3.5 Transform and create variables

Most of the time we want to create new variables from the ones we already have. Let's suppose that we wanted to transform the population (`population`) scale into millions:

```
mutate(approval, population_mill = population / 1000000)
## # A tibble: 1,020 x 12
##   country year quarter president   president_gender net_approval  gdp
##   <chr>   <dbl>  <dbl>  <chr>      <chr>                  <dbl> <dbl>
## 1 Argent~ 2000       1  Fernando~  Male                    40.1  14.0
## 2 Argent~ 2000       2  Fernando~  Male                    16.4  14.0
## 3 Argent~ 2000       3  Fernando~  Male                    24.0  14.0
## # ... with 1,017 more rows, and 5 more variables
```

The previous command generates a new variable in the dataset, `population_mill`, which is `population` but on the scale of millions. We can execute all types of operations in our columns, like creating a GDP variable in a logarithmic scale:

```
mutate(approval, log_gdp = log(gdp))
## # A tibble: 1,020 x 12
##   country year quarter president   president_gender net_approval  gdp
##   <chr>   <dbl>  <dbl>  <chr>      <chr>                  <dbl> <dbl>
## 1 Argent~ 2000       1  Fernando~  Male                    40.1  14.0
## 2 Argent~ 2000       2  Fernando~  Male                    16.4  14.0
## 3 Argent~ 2000       3  Fernando~  Male                    24.0  14.0
## # ... with 1,017 more rows, and 5 more variables
```

We can also create new variables from operations between variables. For example, let's calculate the GDP *per capita*, which allows us to compare countries with different populations in a better way:

```
mutate(approval, gdp_pc = gdp / population)
## # A tibble: 1,020 x 12
```

```
##   country year quarter president    president_gender net_approval    gdp
##   <chr>   <dbl>  <dbl>  <chr>       <chr>                   <dbl>  <dbl>
## 1 Argent~  2000      1  Fernando~   Male                     40.1   14.0
## 2 Argent~  2000      2  Fernando~   Male                     16.4   14.0
## 3 Argent~  2000      3  Fernando~   Male                     24.0   14.0
## # ... with 1,017 more rows, and 5 more variables
```

Lastly, we can also generate more than one transformation at a time with `mutate()`, utilizing multiple arguments:

```
mutate(approval,
       population_mill = population / 1000000,
       gdp_pc          = gdp / population)
## # A tibble: 1,020 x 13
##   country year quarter president    president_gender net_approval    gdp
##   <chr>   <dbl>  <dbl>  <chr>       <chr>                   <dbl>  <dbl>
## 1 Argent~  2000      1  Fernando~   Male                     40.1   14.0
## 2 Argent~  2000      2  Fernando~   Male                     16.4   14.0
## 3 Argent~  2000      3  Fernando~   Male                     24.0   14.0
## # ... with 1,017 more rows, and 6 more variables
```

> **Exercise 2C.** Create a new *data frame*, that is sort from the country-year-quarter with the least presidential approval to the one with the highest (remember to *create* a new object and give it a descriptive name). In your new object, retain only the observations with women as presidents.
>
> **Exercise 2D.** Create a new variable, which registers unemployment as a proportion instead of a percentage.

2.3.6 Summaries

We can make summaries for our datasets with `summarize()`:

```
summarize(approval, unemployment_mean = mean(unemployment))
## # A tibble: 1 x 1
##   unemployment_mean
##               <dbl>
## 1              7.04
```

This process is often called *collapsing the dataset*: we are compressing the information from the rows to generate a single summary row. In this case, the summary function

2.3 Basic operations

`mean()` operates in the `unemployment` vector to obtain its mean. Like with the other operations, we can do multiple summaries at once:

```
summarize(approval,
          unemployment_mean = mean(unemployment),
          growth_mean = mean(gdp_growth),
          approv_mean = mean(net_approval))
## # A tibble: 1 x 3
##   unemployment_mean growth_mean approv_mean
##               <dbl>       <dbl>       <dbl>
## 1              7.04        3.77        15.3
```

2.3.7 Grouped summaries

This is a very fun function! This task consists in collapsing rows until obtaining one row per observation that summarizes the information of different groups in the dataset.

For doing this, first we need to have variables that group our observations (party, country, region, etc.). We will let R know what is the variable we are grouping our oberrvations, and this new dataset will be the same as the original dataset, but R will know that the next operations we make need to be grouped.

```
approval_by_country <- group_by(approval, country)
```

Let's do a summary operation in this new dataset:

```
summarize(approval_by_country,
          unemployment_mean = mean(unemployment),
          gdp_growth_mean   = mean(gdp_growth),
          approv_mean       = mean(net_approval))
## # A tibble: 17 x 4
##   country   unemployment_mean gdp_growth_mean approv_mean
##   <chr>                 <dbl>           <dbl>       <dbl>
## 1 Argentina              11.0            2.72        16.7
## 2 Bolivia                 3.70           4.24        11.3
## 3 Brazil                  8.35           3.4         34.2
## # ... with 14 more rows
```

The groups can also be combinations of variables. For example, our dataset in country-year-quarter level can be grouped into country-year, and then we can obtain the same previous calculus:

```
approval_by_country_year <- group_by(approval, country, year)
```

```
summarize(approval_by_country,
          unemployment_mean = mean(unemployment),
          gdp_growth_mean   = mean(gdp_growth),
          approv_mean       = mean(net_approval))
## # A tibble: 17 x 4
##   country   unemployment_mean gdp_growth_mean approv_mean
##   <chr>                 <dbl>           <dbl>       <dbl>
## 1 Argentina              11.0            2.72        16.7
## 2 Bolivia                 3.70           4.24        11.3
## 3 Brazil                  8.35           3.4         34.2
## # ... with 14 more rows
```

By the way, we can ungroup a dataset with **ungroup()**. This is a great idea if we no longer want to run grouped operations, avoiding mistakes:

```
approval_by_country_year %>%
  ungroup() # note that there's no longer "groups" in the data summary
## # A tibble: 1,020 x 11
##   country  year quarter president president_gender net_approval   gdp
##   <chr>   <dbl>   <dbl> <chr>     <chr>                   <dbl> <dbl>
## 1 Argent~  2000       1 Fernando~ Male                     40.1  14.0
## 2 Argent~  2000       2 Fernando~ Male                     16.4  14.0
## 3 Argent~  2000       3 Fernando~ Male                     24.0  14.0
## # ... with 1,017 more rows, and 4 more variables
```

2.4 Chain commands

Most of the time we want to make **more than one operation** in a dataset. For example, we could want (1) to create a new GDP per capita variable, and then (2) filter the observations with values equal or greater than the dataset's mean GDP per capita:

```
approval_with_gdp_pc <- mutate(approval,
                               gdp_pc = gdp / population)

filter(approval_with_gdp_pc, gdp_pc > mean(gdp_pc))
## # A tibble: 344 x 12
##   country  year quarter president president_gender net_approval   gdp
##   <chr>   <dbl>   <dbl> <chr>     <chr>                   <dbl> <dbl>
## 1 Bolivia  2000       1 Hugo Ban~ Male                    -19.0  60.5
## 2 Bolivia  2000       2 Hugo Ban~ Male                    -19.0  60.5
## 3 Bolivia  2000       3 Hugo Ban~ Male                    -23.8  60.5
## # ... with 341 more rows, and 5 more variables
```

2.4 Chain commands

The same chain of operations can be written in the following way:

```
approval %>%
  mutate(gdp_pc= gdp / population) %>%
  filter(gdp_pc > mean(gdp_pc))
## # A tibble: 344 x 12
##    country year quarter president president_gender net_approval   gdp
##    <chr>   <dbl>  <dbl>  <chr>    <chr>                  <dbl> <dbl>
## 1  Bolivia 2000      1  Hugo Ban~ Male                   -19.0  60.5
## 2  Bolivia 2000      2  Hugo Ban~ Male                   -19.0  60.5
## 3  Bolivia 2000      3  Hugo Ban~ Male                   -23.8  60.5
## # ... with 341 more rows, and 5 more variables
```

This code is surprisingly readable! Pipes (%>%) are read as "then" (or "but then"), and can be inserted with Ctrl/Cmd + Shift + M in RStudio[4]. The following reproduces our previous code in English:

> Take the `approval` dataset, *then* generate a new variable called gdp_pc (the division between `gdp` and `population`), *then* filter the observations so only the ones where `gdp_pc` is greater than then mean remain.

One of the most common uses of pipes is the `group_by()` + `summarize()` combo. Let's repeat our previous operations to make a grouped summary:

```
approval%>%
  group_by(country) %>%
  summarize(unemployment_mean = mean(unemployment),
            growth_mean       = mean(gdp_growth),
            approv_mean       = mean(net_approval))
## # A tibble: 17 x 4
##    country    unemployment_mean growth_mean approv_mean
##    <chr>                  <dbl>       <dbl>       <dbl>
## 1  Argentina               11.0        2.72        16.7
## 2  Bolivia                  3.70       4.24        11.3
## 3  Brazil                   8.35       3.4         34.2
## # ... with 14 more rows
```

> **Exercise 2E.** Calculate, with the help of pipes, the median executive's corruption and GDP by country. Remember that you can insert pipes with Ctrl/Cmd + Shift + M.

---

[4] You can look at all the keyboard shortcuts in RStudio in Help > Keyboard Shortcuts Help.

Exercise 2F. Again, using pipes, sort the countries in the dataset from the one that obtained the highest average GDP per capita in the 2010-2014 period to the lowest.

Exercise 2G. Which country-year-quarter, between the ones governed by women presidents, had the highest executive's corruption? And the highest net approval?

2.5 Recode values

A common exercise in dataset management is generating variables (or editing existing ones) based on certain logical conditions. We already constructed logical conditions before using `filter()`, so the general syntax should be familiar. For example, we could want to register the values of a binary categorical variable as zeros and ones, creating a *dummy* variable. This is easy thanks to the `if_else()` command. We can specify the logical condition (`president_gender==female`), and then the values to be assigned when this condition is fulfilled (1) or not (0):

```
approval %>%
  mutate(d_woman_pres = if_else(condition = president_gender == "female",
                                true      = 1,
                                false     = 0)) %>%
  select(country:president, president_gender, d_woman_pres) # for legibility
## # A tibble: 1,020 x 6
##   country   year  quarter president           president_gender d_woman_pres
##   <chr>     <dbl>   <dbl> <chr>               <chr>                   <dbl>
## 1 Argentina 2000        1 Fernando de la R~   Male                        0
## 2 Argentina 2000        2 Fernando de la R~   Male                        0
## 3 Argentina 2000        3 Fernando de la R~   Male                        0
## # ... with 1,017 more rows
```

It is possible to specify more complex logical conditions, just like in `filter()`. For example, let's generate a *dummy* variable for the countries-years-quarters in economic crisis, defined as in **Exercise 2A**: when the GDP growth is negative and/or the unemployment rate is greater than 20%. Under this simple classification, Argentina would be in crisis in 2001 and 2010:

```
approval %>%
  # we do not explicit the arguments to make the code concise:
  mutate(d_ec_crisis = if_else(gdp_growth < 0 | unemployment > 20, 1, 0)) %>%
  # the following is just to show the results more clearly:
```

2.5 Recode values

```
  select(country:quarter, gdp_growth, unemployment, d_ec_crisis) %>%
  filter(country == "Argentina" & year %in% c(2001, 2013))
## # A tibble: 8 x 6
##   country   year quarter gdp_growth unemployment d_ec_crisis
##   <chr>    <dbl>   <dbl>      <dbl>        <dbl>       <dbl>
## 1 Argentina 2001       1       -4.4         18.3           1
## 2 Argentina 2001       2       -4.4         18.3           1
## 3 Argentina 2001       3       -4.4         18.3           1
## # ... with 5 more rows
```

Nonetheless, `if_else()` is often not flexible enough, since it only allows to assign two values based on a logical condition. What happens if the variable that we want to create can assume more than two values? For example, we could want a variable that divides our observations into three categories, according to the country: "Southern Cone" (Argentina, Chile, Uruguay), "Central America" and "Rest of LA". To start, let's examine the values that the variable "country" can assume:

```
unique(approval$country)
## [1] "Argentina"    "Bolivia"     "Brazil"       "Chile"       "Colombia"
## [6] "Costa Rica"   "Ecuador"     "El Salvador"
## [ reached getOption("max.print") -- omitted 9 entries ]
```

`if_else()` would not allow us to generate this new variable, but its sibling function `case_when()` would.

```
approval %>%
  mutate(country_group= case_when(
    country %in% c("Argentina", "Chile", "Uruguay") ~ "Southern Cone",
    country %in% c("Costa Rica", "El Salvador", "Guatemala", "Honduras",
                   "Nicaragua", "Panama") ~ "Central America",
    TRUE ~ "Rest of LA"
  )) %>%
  # we will shrink the dataset to see the results better:
  filter(year == 2000 & quarter == 1) %>%
  select(country, country_group)
## # A tibble: 17 x 2
##   country   country_group
##   <chr>     <chr>
## 1 Argentina Southern Cone
## 2 Bolivia   Rest of LA
## 3 Brazil    Rest of LA
## # ... with 14 more rows
```

The new variable ("country group") is constructed based on multiple logical conditions, which are evaluated in order. If the first condition is fulfilled (`country %in% c("Argentina", "Chile", "Uruguay")`), the value "Southern Cone" is assigned to the new variable. The logical condition and the value assigned are separated by a "~"[5], that can be read as "therefore". The same will occur with the following condition, which will assign "Central America" if it is fulfilled. Our last argument for `case_when()` has a logical condition with great scope: *in all other cases*, the value "Rest of LA" will be applied.

Exercise 2H. `if_else()` can be thought of as a reduced version of `case_when()`: everything that we do with the first function could be converted into the syntax of the second. Translate one of the previous examples with `if_else()` into the `case_when()` syntax.

Exercise 2I. Generate a new variable that separates countries into three groups: "North America", "Central America" and "South America".

2.5.1 Data pivoting

The structure of the previous dataset, where the rows are the observations, the variables are the columns, and the dataset has only one observational unit, is the *tidy structure* of presenting data (Wickham, 2014). In general, R and the `tidyverse` works very well under this format, so we want to use it when possible.

Nevertheless, the data we work with in the real world is not always available in this format. Often, other formats are more suitable in contexts different than data analysis, for example, the manual recordings of public administration. To start with, let's create a dataset at country-year level with the mean levels of approval:

```
approval_annual <- approval %>%
  group_by(country, year) %>%
  summarize(net_approval = mean(net_approval)) %>%
  ungroup()

approval_annual
## # A tibble: 255 x 3
##    country    year net_approval
##    <chr>     <dbl>        <dbl>
## 1 Argentina  2000         15.6
## 2 Argentina  2001        -17.4
## 3 Argentina  2002        -16.0
## # ... with 252 more rows
```

---

[5] I learned some R users name this symbol a "pigtail"!

2.5 Recode values

This *tidy* dataset can be presented in different formats. The most common of them is the *wide* format[6], in which one of the id variables is distributed among the columns (in this case, "year"). Now we will load the dataset in *wide* format, from the book's package:

```
data("approval_wide1")
```

```
approval_wide1
## # A tibble: 17 x 16
##   country  `2000` `2001` `2002` `2003` `2004` `2005` `2006` `2007` `2008`
##   <chr>     <dbl>  <dbl>  <dbl>  <dbl>  <dbl>  <dbl>  <dbl>  <dbl>  <dbl>
## 1 Argent~    15.6  -17.4  -16.0   32.6   48.5   43.8   45.9   34.3   9.52
## 2 Bolivia   -18.8  -14.1  -5.77  -16.8 -0.301   24.5   34.7   28.1   16.9
## 3 Brazil    -8.72  -2.87   3.51   45.8   26.3   21.3   30.8   40.1   58.3
## # ... with 14 more rows, and 6 more variables
```

This dataset is the same as the one we created manually, it only changes its presentation form. This *wide* structure has some benefits, the most outstanding of which is its briefness: years are not repeated in multiple cells, as it happens in a *tidy* dataset. For a manual encoder, this saving of space (and time) is attractive. However, the *wide* format has a mayor disadvantage in comparison to the *tidy* format: in its traditional form, it is only possible to register information for one variable per dataset. In the example case, there is no easy way of adding, for instance, information about corruption of the country-years. As we have seen, this exercise is trivial in a *tidy* dataset, where variables can be added as columns. Having multiple variables in our dataset is exactly what we need for generating social data analysis, where we explore the different dimensions of our study phenomena.

Luckily, the `tidyr` package, which is loaded automatically with `tidyverse`, provides functions that rapidly convert data from a *wide* format into a friendlier version for analysis. This type of structure transformation is often called *pivot*. The key function here is `pivot_longer()`, which allows a dataset to pivot into a "longer" format. The dataset that we will obtain is equal to the *tidy* dataset we created before:

```
approval_wide1 %>%
  pivot_longer(cols = -country,
               names_to = "year", values_to = "net_approval")
## # A tibble: 255 x 3
##   country   year  net_approval
##   <chr>     <chr>        <dbl>
## 1 Argentina 2000          15.6
## 2 Argentina 2001         -17.4
## 3 Argentina 2002         -16.0
## # ... with 252 more rows
```

---

[6]Sometimes the term *long* (opposite of wide) is used instead of what we are calling *tidy*.

The first argument in `pivot_longer()`, `cols=`, asks us to select the columns to transform into an identification variable, utilizing the `select()` syntax that we learned earlier in this chapter. In this case, we are indicating `pivot_longer()` to transform all variables, except `country`, an identification variable. Then, the `names_to()` argument asks us how we want to call the new identification variable, which is created when the dataset is transformed. Lastly, `values_to=` requires naming the new variable that is created, based on the values of the cells of the original dataset.

In some cases it is also useful doing the inverse operation, to transform a *tidy* dataset into a *wide* format. For this, we can use another important function in `tidyr`, called `pivot_wider()`. Let's see an example, starting from a *tidy* dataset format:

```
approval_annual %>%
  pivot_wider(names_from = "year", values_from = "net_approval")
## # A tibble: 17 x 16
##   country  `2000` `2001` `2002` `2003` `2004` `2005` `2006` `2007` `2008`
##   <chr>    <dbl>  <dbl>  <dbl>  <dbl>  <dbl>  <dbl>  <dbl>  <dbl>  <dbl>
## 1 Argent~   15.6  -17.4  -16.0   32.6   48.5   43.8   45.9   34.3   9.52
## 2 Bolivia  -18.8  -14.1  -5.77  -16.8  -0.301  24.5   34.7   28.1  16.9
## 3 Brazil   -8.72  -2.87   3.51   45.8   26.3   21.3   30.8   40.1  58.3
## # ... with 14 more rows, and 6 more variables
```

The arguments, in this case, are practically mirrors of the previous. Here what we want is that the new dataset takes from "year" its column names across the width (`names_from="year"`), while the values are taken from our variable of interest "net_approval" (`values_from = "net_approval"`).

Thus, these commands are perfectly symmetrical. For example, the next chain of commands is harmless, since `pivot_wider()` will revert the transformation applied by `pivot_longer()`:

```
approval_wide1 %>%
  pivot_longer(cols = -country,
               names_to = "year", values_to = "net_approval") %>%
  pivot_wider(names_from = "year", values_from = "net_approval")
## # A tibble: 17 x 16
##   country  `2000` `2001` `2002` `2003` `2004` `2005` `2006` `2007` `2008`
##   <chr>    <dbl>  <dbl>  <dbl>  <dbl>  <dbl>  <dbl>  <dbl>  <dbl>  <dbl>
## 1 Argent~   15.6  -17.4  -16.0   32.6   48.5   43.8   45.9   34.3   9.52
## 2 Bolivia  -18.8  -14.1  -5.77  -16.8  -0.301  24.5   34.7   28.1  16.9
## 3 Brazil   -8.72  -2.87   3.51   45.8   26.3   21.3   30.8   40.1  58.3
## # ... with 14 more rows, and 6 more variables
```

Exercise 2J. Generate a *tidy* dataset with the mean GDP growth by country-year. Convert this dataset into a *wide* format, moving the years to the columns.

2.5.2 *Wide* datasets with more than one variable of interest

Previously, we mention that it is not possible to register, in a simple way, information for more than one variable of interest in a *wide* structure. Nonetheless, our data sources often will contain surprises that are unkind for us, like the next example:

```
data("approval_wide2")
```

```
approval_wide2
## # A tibble: 17 x 31
##   country gdp_2000 gdp_2001 gdp_2002 gdp_2003 gdp_2004 gdp_2005 gdp_2006
##   <chr>      <dbl>    <dbl>    <dbl>    <dbl>    <dbl>    <dbl>    <dbl>
## 1 Argent~  5.52e11  5.28e11  4.70e11  5.12e11  5.58e11  6.07e11  6.56e11
## 2 Bolivia  3.68e10  3.74e10  3.83e10  3.94e10  4.10e10  4.28e10  4.49e10
## 3 Brazil   1.99e12  2.02e12  2.08e12  2.11e12  2.23e12  2.30e12  2.39e12
## # ... with 14 more rows, and 23 more variables
```

Note that in this dataset the columns register information over time for two variables, `gdp` and `population`. What we want is to *extend* this information into the rows, reconstructing our country-years pair and the two variables of interest. First, we can pivot the dataset for leaving them in the country-year-variable level. In `pivot_longer()`, we can indicate that the names of the columns contain information of more than one variable. First, the argument `names_to = c("variable", "year")` takes two values in this occasion, the names of new variables after the pivot. Secondly, `names_sep= "_"` indicates that in the columns of the original dataset the information of the two variables is separated by an underscore (this could be another character, like a dash or a vertical bar[7]).

```
approval_wide2 %>%
  pivot_longer(cols = -country,
               names_to = c("variable", "year"), names_sep = "_")
## # A tibble: 510 x 4
##   country   variable year   value
##   <chr>     <chr>    <chr>  <dbl>
## 1 Argentina gdp      2000   552151219031.
## 2 Argentina gdp      2001   527807756979.
## 3 Argentina gdp      2002   470305820970.
## # ... with 507 more rows
```

Then, we can pivot the variables across the width to get our target dataset, just as we did before, with `pivot_wider()`. Let's do everything in a chain:

---

[7] If the separation between your variables is less clear, we can use the `names_pattern=` argument instead of the `names_sep=`. For this you will need to use regular expressions, a subject dealt with in the Chapter 13 of text analysis. For example, we could write the same operation here with following argument: `name_pattern = "(\\D+)_(\\d+)"`.

```
approval_wide2 %>%
  pivot_longer(cols = -country,
               names_to = c("variable", "year"), names_sep = "_") %>%
  pivot_wider(names_from = "variable", values_from = "value")
## # A tibble: 255 x 4
##   country   year            gdp population
##   <chr>     <chr>          <dbl>      <dbl>
## 1 Argentina 2000    552151219031.   37057452
## 2 Argentina 2001    527807756979.   37471509
## 3 Argentina 2002    470305820970.   37889370
## # ... with 252 more rows
```

3
Data Visualization

Soledad Araya[1]

Suggested readings

- Henshaw, A. L. and Meinke, S. R. (2018). Data Analysis and Data Visualization as Active Learning in Political Science. *Journal of Political Science Education,* 14(4):423–439.

- Kastellec, J. P. and Leoni, E. L. (2007). Using Graphs Instead of Tables in Political Science. *Perspectives on Politics,* 5(4):755–771.

- Tufte, E. R. (2006). *Beautiful Evidence.* Graphics Press, Cheshire, Conn.

Packages you need to install

- `tidyverse` (Wickham, 2019), `politicalds` (Urdinez and Cruz, 2020), `ggrepel` (Slowikowski, 2020).

3.1 Why visualize my data?

You have already learned to use the commands from `tidyverse`, and you probably want to dive in the world of graphs, and apply all you have been learning to your own dataset. With `tidyverse` and `ggplot2`, data management becomes an easy task, but there are a few steps you need to go through before writing your code. For example, knowing your variables. Are they continuous or categorical variables? When they are categorical, do they have two or more levels? Moreover, those levels, are they in order or not? These are not the only questions you have to consider. It appears an easy task, but if you do not consider this step in your work with `ggplot2` things can turn bad pretty quickly. Fun examples of this can be found in accidental aRt[2].

Quick question: Why represent our data graphically?

---

[1] Institute of Political Science, Pontificia Universidad Católica de Chile. E-mail: snaraya@uc.cl.
[2] See https://twitter.com/accidental__aRt

First, I know that many of us are interested in representing our data graphically because it is an attractive way of doing it. However, having a good or bad sense of aesthetics does not mean a a lot if our data is not clear. Thus, it is necessary to *understand* what we want to express, which can be a hard task if we do not reflect on why are we doing such types of representation. Sometimes, we can use tables to summarize quantities and/or patterns, but big data management nowadays makes this a complex and inefficient task. Therefore, let's return to the main question: why visualize? Why not simply make tables that express what we want to say? Through data visualization we can understand other types of problems that numbers on their own cannot show. By visualizing, we want to explore and *comprehend* our data. Also, graphing can help us with the interpretation of patterns, tendencies, distributions, and better communicating these to our readers.

FIGURE 3.1: Statistician Florence Nightingale (1820-1910).

Florence Nightgale (1820-1910) was a nurse and statistician who helped reorganize the administration of civic and military hospitals in Great Britain (Figure 3.1). She, with the help of a team, managed to make a record of the deaths and diseases in military hospitals during the Crimean War. For her surprise, the vast majority of deaths were avoidable, and the main reason behind them were the poor conditions of the health system. One of her reports for the British government was the diagram in Figure 3.2.

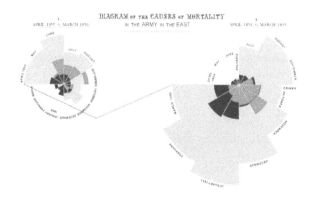

FIGURE 3.2: Diagram of the causes of death in the British army.

In dark gray are highlighted the death from war injuries, in light gray those deaths due to avoidable diseases, and in black, deaths caused by other type of causes. This chart does not only deliver quantitative information about deaths, it also signals a substantial problem about the military's health system at that time.

Nightingale's diagram revealed the problem, which was the initial step for a series of reforms. Thus, visualization becomes a tool that can be applied in all stages of investigation. At an initial stage, it is important for the exploration of data, and to understand how variables relate to each other, their distributions and frequencies. When interpreting the data, visualization is useful for displaying possible tendencies or patterns in the data. Finally, visualization is a great tool for the dissemination of knowledge. But remember, with great power comes great responsibility, and spurious relations stop being funny when people take them too seriously[3].

Monogan (2015, ch. 3) had already explained in a simple way for social scientists, why data visualization is important when working with quantitative data. In the introduction of the chapter, Monogan states the importance and the advantages of working with figures, from the simple distribution of variables, outliers or biases, to the trends over time. For this reason, data visualization is a crucial tool for anyone who works with data. It is not, for any reason, an "aesthetic movement"; graphing is extremely useful.

Nonetheless, for some people, data visualization is both a functional element for analysis and an aesthetic element for excellence. For Edward Tufte (2006), visualizing data in an effective way has an inevitable artistic component. With statistician formation and a PhD in Political Science from Yale University, Edward Tufte was dedicated to understanding and explaining how science and art have in common an *open-eye observation that generates empirical information*. His book *Beautiful Evidence* (Tufte, 2006) describes the process of how *watching* is transformed into *showing*, and how empirical observation becomes explanations and evidence.

We need to understand that data visualization is a language just like any other. As broadcasters, we need to know our audience: who are the receivers of our message, if it is an expert audience or just the general public. In any circumstance, we would adjust our message to the type of audience. The same occurs when we visualize data. The graphs that we make must be adapted to our audience. Yet, even with the most knowledgeable people we should not get too excited. It is not about applying all we know immediately, but to understand what we are trying to communicate. Understanding the functions of this language is essential.

In the following subsection we will talk about how `ggplot2` works. From now on, we will start with applied examples. The most common types of visual representation are the histogram, the bar graph, the density chart and the line graph. Also, we will introduce other utility packages for making more sophisticated graphs. Finally, we will

---

[3]Still, it is always entertaining to watch how a correlation exists between per capita cheese consumption and the number of people strangled to death by their sheets in the United States[4]!

learn about other packages that can be useful within social sciences, and in particular, political science, as are `sf` and `ggparliament`.

Tip: After this chapter, if you want to learn more about data visualization, check Data Visualization: A Practical introduction[5] by Kieran Healy, a book available for free that is fun and helpful for learning `ggplot2` step by step. In this book you will not only find a theoretical part, but also a practical one. On the other hand, the web page From Data to Viz[6] can help you learn how to present your data, but not only that: whether you work with R or Python, you can find the packages and codes for its application.

3.2 First steps

Now that we understand the process before constructing a graph, we have to get familiarized with `ggplot2`, the package to create graphs that is part of the `tidyverse`. A Layered Grammar of Graphics[7], from Hadley Wickham, explains in detail how this new "grammar" works for making graphs. We recommend that you read from the original source how this package was created to later understand the use of layers in the construction of the graphs.

Although the use of `ggplot2` was rapidly expanded, within the R community there are constant discussions about the teaching of `ggplot2` as a first option over the base R graphs. For example, David Robinson[8] has in his blog different entries about this topic, where he explains in detail the advantages of `ggplot2` over other options. If you are a beginner in R, starting with `ggplot2` will give you a powerful tool, and its learning curve is not as steep as base R's.

Some advantages that David Robinson mentions in "Why I use `ggplot2`"[9] are:

- Captions! Base R requires more knowledge from the users in order to add captions in graphs. Our friend `ggplot2` does it automatically.
- Faceting! Basically, we can create sub-graphs with a third or fourth variable and overlay them, which will allow us a better understanding of the behavior of our data.

---

[5]See http://socviz.co
[6]See https://www.data-to-viz.com/
[7]See https://byrneslab.net/classes/biol607/readings/wickham_layered-grammar.pdf
[8]See http://varianceexplained.org/r/why-I-use-ggplot2/
[9]See http://varianceexplained.org/r/why-I-use-ggplot2/

3.2 First steps

- It works together with `tidyverse`. This means that we can make more with less. At the end of this chapter you will understand what I mean. There are shortcuts for everything.
- Aesthetically, it is better. There are thousands of options for chromatic palettes, themes and fonts. If you do not like it, there is a way of changing it.

With this in consideration, let's start with the practical.

3.2.1 The layers of the "ggplotian multiverse"

Let's start with our topic of interest: How does `ggplot2` works? This package is included in the `tidyverse`, so it is not necessary to load it separately. Also, we will use tools from both packages throughout the entire chapter. Then, the first step is loading the package:

```
library(tidyverse)
```

The intuition behind `ggplot2` is straightforward. The construction of the data is based on layers that contain a certain type of information.

3.2.1.1 Data

The first layer corresponds to the data we will use. To make this more demonstrative, we will load the dataset that will be used throughout the chapter.

```
library(politicalds)
data("municipal_data")
```

The dataset should now be in our environment. This dataset corresponds to information about Chilean municipalities. Some are from the Electoral Service[10] and others from the National System of Municipal Information[11] of Chile. In the first dataset, we find electoral results of local, regional and national elections of the country; while in the second we find economic, social and demographic characteristics of the Chilean municipalities. In this case, we have the communal electoral data from 1992 to 2012, with descriptive data such as population, total income of the municipality, social assistance expenditure and the percentage of people living in poverty based on the communal total of the National Socio-economic Characterization Survey (CASEN).

```
glimpse(municipal_data)
## Rows: 1,011
## Columns: 6
## $ year         <chr> "2004", "2004", "2004", "2004", "2004", "2004",...
```

---

[10]See http://www.servel.cl
[11]See http://datos.sinim.gov.cl/datos_municipales.php

```
## $ zone         <chr> "Upper North", "Upper North", "Upper North", "U...
## $ municipality <chr> "Alto Hospicio", "Arica", "Camarones", "Camina"...
## $ gender       <chr> "0", "0", "1", "0", "0", "0", "0", "0", "0", "0...
## $ income       <int> 1908611, 12041351, 723407, 981023, 768355, 5580...
## $ poverty      <dbl> NA, 23.5, 10.6, 37.3, 58.3, 38.8, 31.3, 7.7, 23...
```

When looking at the dataset, we find that there are continuous (numeric) and categorical (character) variables. Knowing what type of variable we are working with is essential for the next step.

3.2.1.2 Aesthetics

The second layer corresponds to the mapping of the variables within the space. In this step, we use `mapping = aes()`, which will contain the variable we will have on our x-axis and y-axis (Figure 3.3). For `aes()`, there are many options we will see throughout the chapter: some of them are, for example, `fill =`, `color =`, `shape =`, and `alpha =`. All these options are a set of signs that will allow us to better translate what we want to say through our graphic. Normally, these options are called *aesthetics* or `aes()`.

```
ggplot(data    = municipal_data,
       mapping = aes(x = year, y = poverty))
```

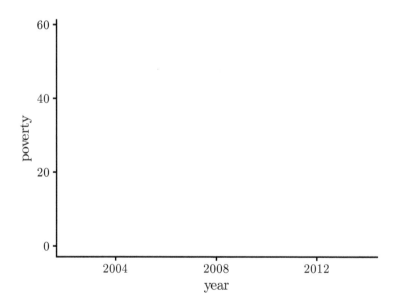

FIGURE 3.3: Empty frame plot.

The result shows an empty frame. This is because we have not told R which geometrical object to use.

3.2 First steps

3.2.1.3 Geometrical Object

It sounds strange, but when we talk about the geometrical object or "geom", we are referring to the type of graph we want to make, either a line graph, a bar graph, a histogram, a density graph, or a dot graph, or if we want to make a boxplot. This corresponds to the third layer. In this case, since we have data from the CASEN survey, we will make a boxplot to look at how municipalities are distributed in our sample (Figure 3.4).

```
ggplot(data    = municipal_data,
       mapping = aes(x = year, y = poverty)) +
  geom_boxplot()
```

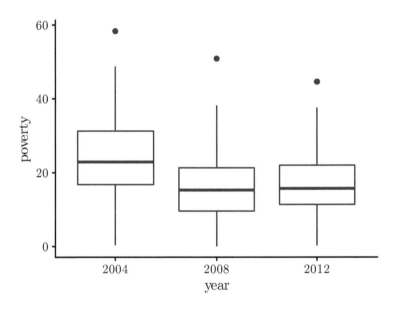

FIGURE 3.4: Adding a geometric object to your plot.

The first thing we notice is the absence of data for three periods. Unfortunately, there is no data before 2002, so no entries are found for those years. Because of this, it is a great idea to filter the data and leave only those years which contain data about the CASEN survey. Besides that, our graph does not tell us much about the percentage of poverty and its distribution. Considering Chile's geography, it is a great idea to see the distribution of poverty by geographical region zone.

3.2.1.4 Faceting

Now, we will use our new skills for doing two things: first, we will use `filter()` to keep only those years we are interested in. Second, we will divide the results by zones using `facet_wrap()`, which corresponds to the fourth layer we can use for building a graph with ggplot2. When we use this layer, what we want is to organize the geoms we are using as a function of a categorical variable. In this case, zone. However, *faceting*, as an action, is much more than that. `facet_wrap()` and `facet_grid()` can adopt a series of arguments, the first one being the most important. In this case, the syntax we use is the same used for formulas in R, and we denote the first argument with a "~" sign. With the arguments `nrow =` and `ncol =` we can specify how we want to order our graph.

Finally, we add two lines of code, one for filtering and the other one for subdividing our information. This is what we accomplish (Figure 3.5):

```
ggplot(data    = municipal_data %>% filter(year == c(2004, 2008, 2012)),
       mapping = aes(x = year, y = poverty)) +
  geom_boxplot() +
  facet_wrap(~ zone, nrow = 1)
```

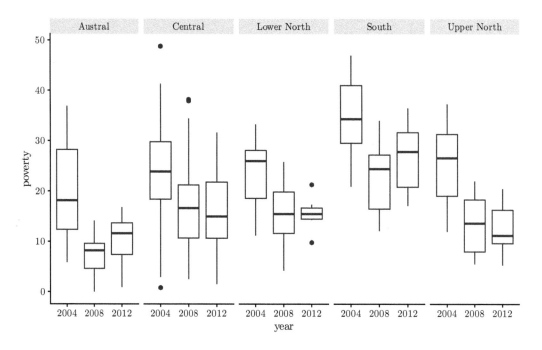

FIGURE 3.5: Adding a facet to your plot.

Both with `facet_wrap()` and `facet_grid()` we can use more than one argument, but the results are different. `facet_wrap()` does not only order the geoms, but is capable of crossing them, creating graphs with two or more dimensions using categorical variables. Look at the next examples in Figures 3.6 and 3.7:

```
ggplot(data    = municipal_data %>% filter(year == c(2004, 2008, 2012)),
       mapping = aes(x = year, y = poverty)) +
  geom_boxplot() +
  facet_wrap(zone ~ gender)
```

3.2 First steps

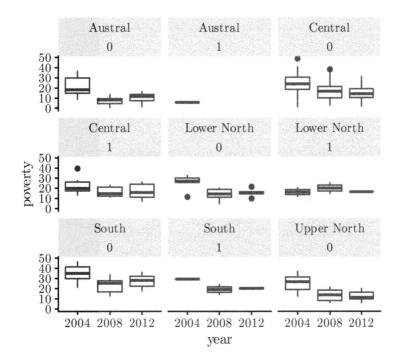

FIGURE 3.6: Comparing wraps and grids, example A.

```
ggplot(data    = municipal_data %>% filter(year == c(2004, 2008, 2012)),
       mapping = aes(x = year, y = poverty)) +
  geom_boxplot() +
  facet_grid(zone ~ gender)
```

This graph shows that, by zone, the percentage of poverty has varied considerably from 2004 to 2012, and that there is a high interregional variability. Furthermore, it shows us how `ggplot2` delivers high-quality results without much complexity. The `facet_wrap()` function is an optional layer within the multiple layers of "A Layered Grammar of Graphics", but it is important to remember that the other three must be present for any type of results.

3.2.1.5 Transformations

Another layer you can use is one that allows us to do scale transformations in the variables (Figure 3.8). Normally, it will appear with the name `scale_x_discrete()`, which will vary depending on the aesthetic used within our mapping. Thus, we can encounter `scale_fill_continous()` or `scale_y_log10()`. For example, we can see how municipalities' income is distributed according to the poverty rate of our sample. Usually, we would do this as follows:

```
ggplot(data    = municipal_data %>% filter(year == c(2004, 2008, 2012)),
       mapping = aes(x = poverty, y = income)) +
  geom_point()
```

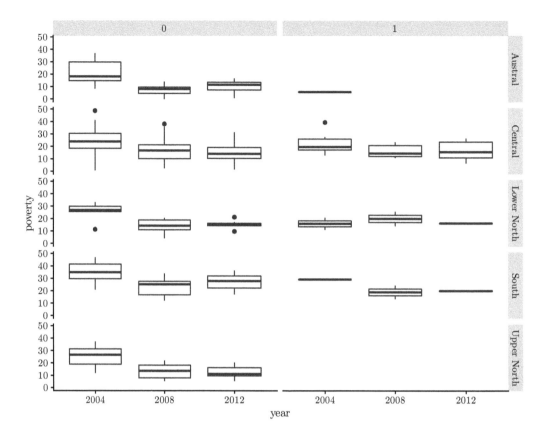

FIGURE 3.7: Comparing wraps and grids, example B.

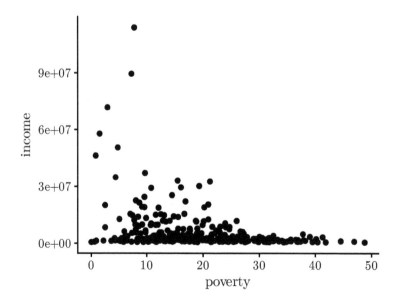

FIGURE 3.8: Example of a plot in which we do not use scaling.

3.2 First steps

Most frequently, when we use a variable related to money, we apply a logarithmic transformation (Figure 3.9). However, how does this translate into our figure?

```
ggplot(data    = municipal_data %>% filter(year == c(2004, 2008, 2012)),
       mapping = aes(x = poverty, y = income)) +
  geom_point() +
  scale_y_log10()
```

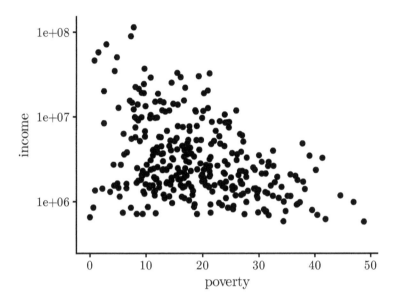

FIGURE 3.9: Example of a plot in which we rescaled the y axis.

3.2.1.6 Coordinate system

Usually, we will work with an x-axis and a y-axis. There are functions in `ggplot2`, such as `coord_flip`, that allow us to change the direction of our graph. However, we can also use this type of layer when working with geographical data, or when, for example, we want to make a pie chart. Although, normally, we do not want to make pie charts[12]. The more you use `ggplot2`, the more you will learn about each option.

3.2.1.7 Themes

When mapping data, we use aesthetic options. When we want to change how a graph looks, we change the theme. You can do this through `theme()`, which allows modifying things that are not related to the content of the graph. For example, the background colors or the font of the letters in the axes. You can also change where the caption or the title will be located. Lastly, you can also change the title, the name on the axes,

---
[12] See https://www.datapine.com/blog/common-data-visualization-mistakes/

add annotations, etc. You just need to know `labs()` and `annotate()`. Now, it is time to apply everything we "apparently" already understand.

3.3 Applied example: Local elections and data visualization

As we previously mentioned, the primary issue is to understand that visualization enables us to explore our data and answer substantive questions of our investigation. Usually, means, standard deviations or another type of parameters do not tell us much. We can express the same data by visualizing it. For example, a boxplot can be useful for representing the distribution of the data and see its possible outliers, while a bar graph can help us to look at the frequency of our categorical data, and a line graph is practical for understanding change over time. These are just some examples within a variety of possibilities.

In this third section, we will learn how to visualize different types of graphs with data from municipal re-election in Chile. To contextualize, the smallest political-administrative division in Chile is the commune or municipality, which every four years chooses its local authorities: a mayor and a municipal council. Since 1992 to 2000, mayors were indirectly elected, and since 2004 they started being elected directly by citizens.

Since we already know our data, we can start with the simplest. A good idea, for example, is to see the number of women elected as mayors compared to the number of men elected. For that, we can use a bar graph (Figure 3.10). As we learned in the previous section, for constructing any type of graph we need to know the variable(s) that we want to use and what geometry or "geom" allows us to represent it. In this case, we will use `geom_bar()` to see how many men and women were elected since 1992.

3.3.1 Graph bar

```
plot_a <- ggplot(municipal_data, mapping = aes(x = gender))

plot_a +
  geom_bar()
```

As we can see, constructing a bar graph is an easy task. We see that, from 2004, over 800 men were elected as mayors, a number that exceeds by far the number of women elected for the same charge in the same period.

Perhaps, this number has changed over time, and we cannot see it in this type of graph? This appears to be a good reason for using `facet_wrap` (Figure 3.11).

3.3 Applied example: Local elections and data visualization

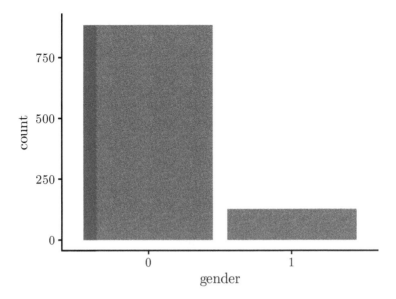

FIGURE 3.10: Simple graph bar.

```
plot_a +
  geom_bar() +
  facet_wrap(~year, nrow = 1)
```

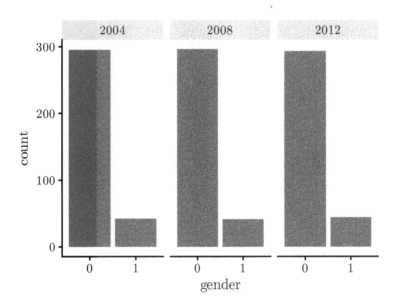

FIGURE 3.11: Bar graph with a facet by year.

As we can see, the number of women mayors appears to increase, although it is a much smaller increase than would be expected. This could be a substantive problem for making an analysis of the local government in Chile.

Geometries like `geom_bar()`, `geom_col()`, `geom_density()` and `geom_histogram()` tend not to carry a y-axis explicit in its aesthetics, since they are a count on the x-axis. Nonetheless, one can modify the y-axis in these geometries by applying some type of transformation. For example, when specifying `y = ..prop..` as an aesthetic within the geometrical object, we are commanding the calculus of the proportion, not the count. Normally, we will use `aes()` in addition to the data in `ggplot()`, but depending on your preferences, it is also possible to use it with geoms. The latter is more common when we occupy more than one dataset or when we want to do a transformation.

For example, we could be interested in the number of local authorities per geographical zone. For that, it would be helpful to use a proportion, since every geographical zone is made up of a different number of municipalities. By doing this, comparing the situation between zones will be easier (Figure 3.12).

```
plot_a +
  geom_bar(mapping = aes(y = ..prop.., group = 1)) +
  facet_wrap(~zone, nrow = 1)
```

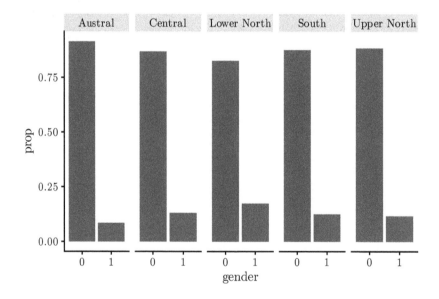

FIGURE 3.12: Bar graph with a facet by zone.

But why do we use `group = 1`? When we want to calculate a proportion with `y = ..prop..`, we have to take some precautions if we are using `facet_wrap`. This function does not calculate the proportion based on the sum of both genders by zone. For example, this function registers that there are 89 elected men and 13 elected women in the Upper North zone. It concludes that "in the Upper North, the 89 men correspond to 100% of elected men, and the 13 women correspond to 100% of elected

3.3 Applied example: Local elections and data visualization

women". Clearly, this is not what we are trying to represent in the graph. This is why we use `group=1`. Try to see the result without `group=1` to check out what happens.

We have done it! We see that there are no big differences, where the "Lower North" zone is the one with more women in the mayor office than men. Nonetheless, there are no major differences between zones, and the results from the first bar graph are replicated in this one.

Now, we can change the graph's presentation. Every good graph must contain, for example, an clear title, the source of the data and the detail of the axes (Figure 3.13).

Suggestion. The *Chicago Guide to Writing about Multivariate Analysis* (Miller, 2013) has lots of good advice on how to create effective charts.

```
plot_a +
  geom_bar(mapping = aes(y = ..prop.., group = 1)) +
  facet_wrap(~zone, nrow = 1) +
  labs(title = "Proportion of men and women elected as mayors (2004-2012)",
       subtitle = "By economic zones of Chile",
       x = "Gender", y = "Proportion",
       caption = "Source: Based on data from SERVEL and SINIM (2018)")
```

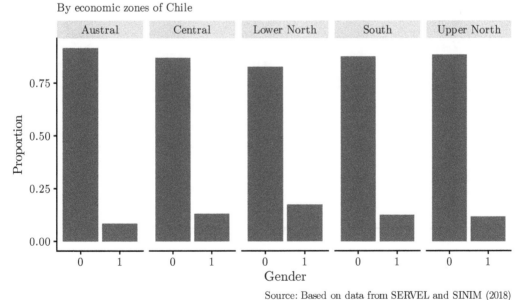

FIGURE 3.13: Bar graph with title and sources.

Now, we only need to add labels for the x-axis (Figure 3.14). We can easily do this with `scale_x_discrete()`. You have to consider which aesthetic from `aes()` you will modify, since this will change the `scale =` you need. If we were to examine labels from `fill =`, for example, we would have to use `scale_fill_discrete()`. You also have to take into account the type of variable you are using. `scale_x_discrete()` does not have "discrete" at the end for no reason. As you will comprehend, it depends totally on the type of variable we are using.

```
plot_a +
  geom_bar(mapping = aes(y = ..prop.., group = 1)) +
  facet_wrap(~zone, nrow = 1) +
  scale_x_discrete(labels = c("Men", "Women")) +
  labs(title = "Proportion of men and women elected as mayors (2004-2012)",
       subtitle = "By economic zones of Chile",
       x = "Gender", y = "Proportion",
       caption = "Source: Based on data from SERVEL and SINIM (2018)")
```

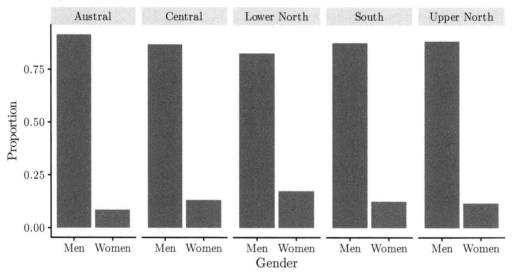

FIGURE 3.14: Plot with group labels.

Tip. With `labels =` we can change the labels. Consider the number of values in your categorical variable so that they match the variable, and you do not miss any category.

3.3.2 Line graph

In the last graph of the previous section we saw that, although the election of women as mayors in Chile has risen, this rise appears to not be significant: in 2012, only 13% of elected mayors were women. Perhaps this may be because socio-economic changes have not affected perceptions of gender roles in society. Looking at the economic data from municipal income or the percentage of poverty according to the CASEN might help us understand why women's election in municipal instances has not increased substantially. For this, we can use `geom_line()`, the geometric object that permits watching the evolution over time of our subject of interest (Figure 3.15). The intuition would be to make the figure this way:

```
plot_b <- ggplot(data    = municipal_data,
                 mapping = aes(x = year, y = income))

plot_b +
  geom_line()
```

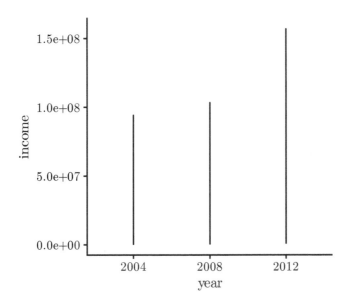

FIGURE 3.15: A wrong specification for a line graph.

The problem is that it does not deliver the expected result. The intuition is correct, but we have to help `geom_line()` with some specifications. In this case, it groups by what it makes the most sense: by year. For this reason, we have to specify which is the variable that groups all the information and, as we know, the information we possess is grouped by municipality. When we add this information, the result changes and it resembles what we are looking for (Figure 3.16):

```
plot_b +
  geom_line(mapping = aes(group = municipality))
```

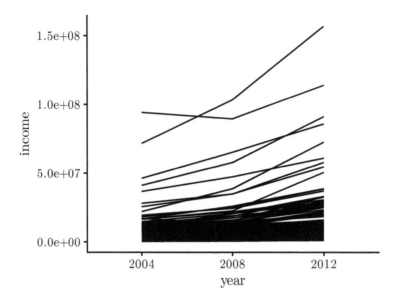

FIGURE 3.16: Yearly evolution of income by municipality.

One of the issues that emerges at first glance is that, considering that Chile has 345 municipalities, it appears impossible to display them all in just one graph.

Now, we can separate the graph as we have done before. It can be done by zones or regions, considering your interests (Figure 3.17). We have already seen different results by zone, so it would be worthwhile to see the income in the same way:

```
plot_b +
  geom_line(aes(group = municipality)) +
  facet_wrap(~zone, nrow = 2)
```

Since our sample is composed of a small number of years, we are not able to see much variability and, at first glance, the income of all municipalities has incremented considerably. Perhaps, we can still make some adjustments to our graph. Most likely, you are not familiar with scientific notation and you are better off reading big numbers. Perhaps you know that it is better to work with a monetary variable in its logarithmic transformation, as we have been taught in different courses of methodology. In addition, you may want to add another type of information into this graph, for example, the means (Figure 3.18).

What do you think about this graph?

3.3 Applied example: Local elections and data visualization

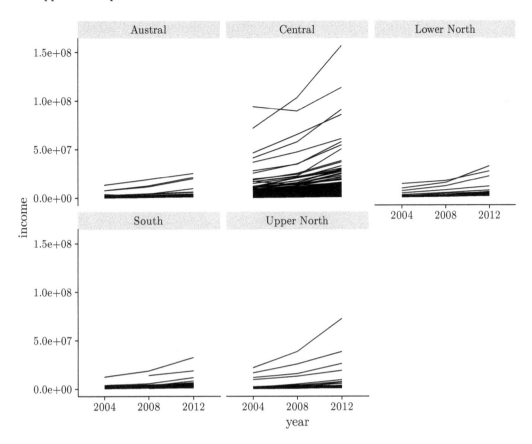

FIGURE 3.17: Yearly evolution of income by municipality faceted by zone.

```
means <- municipal_data %>%
  group_by(zone) %>%
  summarize(mean = mean(income, na.rm = T))

plot_b +
  geom_line(color = "lightgray", aes(group = municipality)) +
  geom_hline(aes(yintercept = mean), data = means, color = "black") +
  scale_x_discrete(expand = c(0,0)) +
  scale_y_log10(labels = scales::dollar) +
  facet_wrap(~ zone, nrow = 2) +
  labs(title = "Municipal income in electoral years (2004-2012)",
       y = "Income",
       x = "Years") +
  theme(panel.spacing = unit(2, "lines"))
```

What did we specify?

1. First, we created a dataset ("mean") that contains the mean income by every zone. We did this by using group_by() and summarize() of the tidyverse.

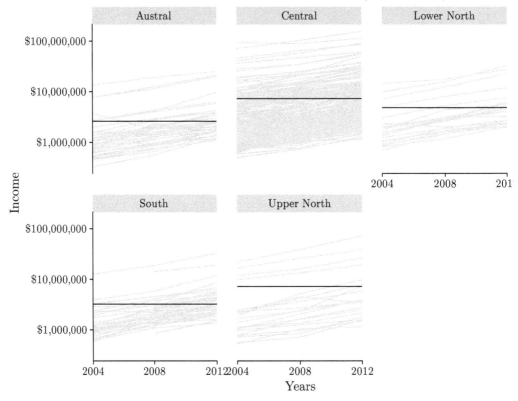

FIGURE 3.18: Complete version of our line graph for municipality income in electoral years.

```
municipal_data %>%
  group_by(zone) %>%
  summarize(mean = mean(income, na.rm = T))
## # A tibble: 5 x 2
##   zone          mean
##   <chr>         <dbl>
## 1 Austral     2609648.
## 2 Central     7302625.
## 3 Lower North 4816249.
## # ... with 2 more rows
```

2. Then, we specify the color of the `geom_line()`.

3. After that, we add to our code `geom_hline()`. This geometrical object, like `geom_vline()` or `geom_abline()`, allows us to add lines with information. In this case, I used it to add the mean income of every zone. We specify the variable that contains the means `yintercept = mean`, the dataset `means`, and the color with `color = "black"`.

3.3 Applied example: Local elections and data visualization   57

4. Following that, we use `scale_x_discrete()` to specify the expansion of the panels. If before we previously saw a grey space without information, we removed it. This is aesthetic.

5. Then, we use `scale_x_discrete()` to scale our data. This is a logarithmic transformation that is usually done when we work with linear models that contain monetary data. Also, we changed the labels of the y-axis: it no longer appears with scientific notation. This was done with a package called `scales`. Here we call the function directly with `scales::dollar`.

6. We add the title and names of the x and y axes with `labs()`.

7. Finally, we specify information about the theme. Without it, the years between one panel and another would crash. For that, we specify it with `panel.spacing = unit(2, "lines")` in the `theme()` layer.

3.3.3 Boxplot

We already saw that the income of the municipalities in Chile increased between 2004 and 2012. While we looked at the graph without functional transformations, we noted that some municipalities had incomes well above the average and stood out within their zones. The intuition is that they are probably *outliers*. We could see this clearly with a boxplot, which allows us to graph diverse descriptive data in our variables such as the median, minimum and maximum. In this case, we will use it to observe if our intuition is correct or not[13].

Let's start by filtering the data as we did in the previous graph. In our x-axis we will place the zones of Chile and in the y-axis the income. This is the result we will work with (Figure 3.19):

```
plot_c <- ggplot(data    = municipal_data %>%
                   filter(year %in% c(2004, 2008, 2012)),
                 mapping = aes(x = income, y = zone)) +
  geom_boxplot() +
  facet_wrap(~year, ncol = 1)

plot_c
```

We can clearly observe some *outliers*. Perhaps, after looking at these results, we would like to identify which municipalities receive the highest total income. For this, we can use the aesthetic `label =`, which is included in `geom_text()`. For naming just the outliers, we have to make a subset of data (Figure 3.20):

```
plot_c +
  geom_text(data    = municipal_data %>% filter(income > 50000000),
            mapping = aes(label = municipality))
```

---

[13]Chapter 16 will be very useful if you wish to detect outliers through maps.

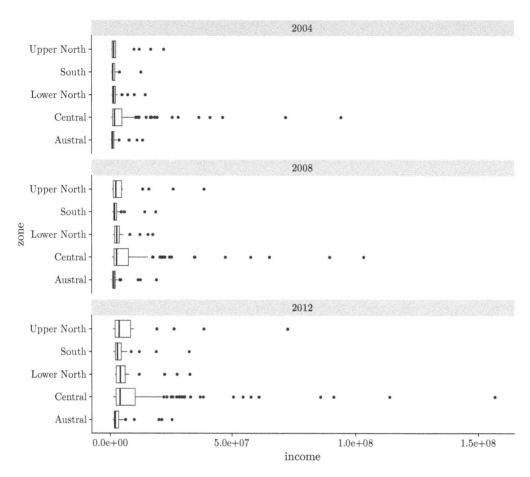

FIGURE 3.19: Boxplot of municipality income per zone, faceted by year.

Unfortunately, the labels are over the points, and, in some cases, these overlap when these are close to each other. We can solve this with the package `ggrepel`, which has an "improved" `geom_text()` geometric element that avoids the coalition of the labels.

```
library(ggrepel)

plot_c +
  geom_text_repel(data    = municipal_data %>%
                    filter(income > 50000000),
                  mapping = aes(label = municipality))
```

The cut-off can be at $50,000,000 or at greater or smaller numbers (Figure 3.21). It depends entirely on what we want to observe. Furthermore, with `geom_text()` or `geom_text_repel()` we can not only change the color, but also the type of font of the text, or if it should be in bold, italic or underlined. To see more options, you can type ?geom_text or call a `help("geom_text")`.

3.3 Applied example: Local elections and data visualization

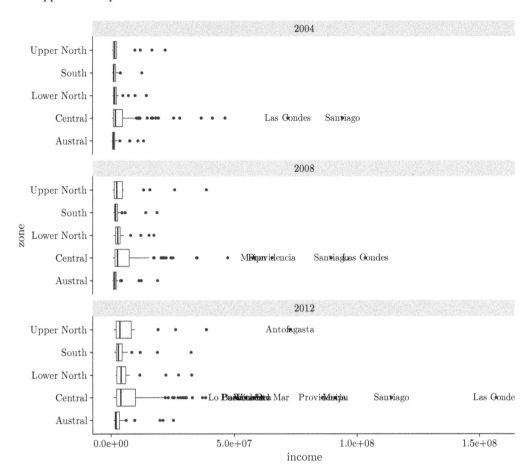

FIGURE 3.20: Through the use of labels a boxplot can be usefull to identify outliers.

We could also add other information or change the current way the graph is presented (Figure 3.22).

```
plot_c +
  geom_text_repel(data = municipal_data %>%
                  filter(income > 50000000),
                  mapping = aes(label = municipality)) +
  scale_x_continuous(labels = scales::dollar) +
  labs(title = "Municipal income by zone (2004-2012)",
       x = "Income", y = "Zone",
       caption = "Source: Based on data from SINIM (2018)")
```

Some other specifications: 1. We added the descriptive information into the graph. 2. We changed the font size. This was important because of the number of municipalities that are above $50,000,000 in income. 3. Again, we changed the y-axis labels with `scales::dollar`. 4. Lastly, with `guides`, and specifying the `aes()` we wanted to address, we wrote the code `color=F` to remove the label, since it was repeated information within the graph.

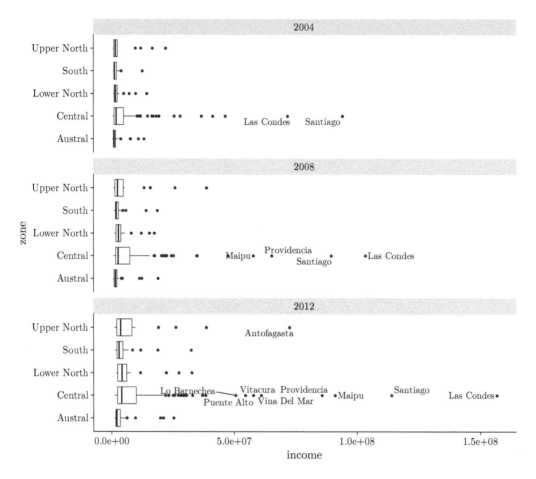

FIGURE 3.21: We can fix overlapping labels using ggrepel.

Exercise 3A. We invite you to play with `geom_text()`: change the colors, size, fonts, etc. We also encourage you to install packages that will allow you to further customize your graphics: `ggthemes` from jrnorl[14] has themes for graphs from programs and well-known magazines such as Excel or The Economist. On the other hand, `hrbrthemes` from hrbrmstr[15] has elaborated some minimalist and elegant themes that will make all your graphs look better. If you are into colors, you can check the `wespalette` package from karthik[16], a chromatic palette based on Wes Anderson movies, or create your own palettes based on images with `colorfindr`. You can find more about the latter in the following link[17].

---

[14]See https://github.com/jrnold/ggthemes
[15]See https://github.com/hrbrmstr/hrbrthemes
[16]See https://github.com/karthik/wesanderson
[17]See https://github.com/zumbov2/colorfindr

3.3 Applied example: Local elections and data visualization

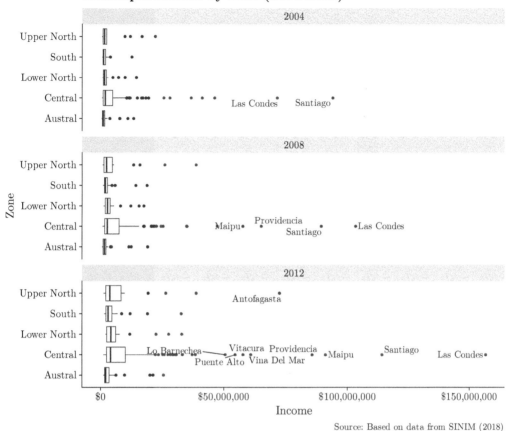

FIGURE 3.22: Polished version of our boxplot.

3.3.4 Histogram

As we observed in our boxplot, many municipalities–especially those in the central zone–are well above the median income per zone. We can see the distribution of this data through a histogram. Constructing a histogram is an easy task, and as previously mentioned, `geom_histogram()` does not have an explicit y-axis, since it counts the frequency of an event within an interval.

When creating the histogram according to our intuition, the result is as follows (Figure 3.23):

```
ggplot(data    = municipal_data,
       mapping = aes(x = income)) +
  geom_histogram()
## Warning: Removed 7 rows containing non-finite values (stat_bin).
```

As we can see, the graph gives a "Warning" that indicates the existence of "rows that contain non-finite values". This warning has been present throughout this chapter, and it means nothing more than "There are unknown values within this variable" and

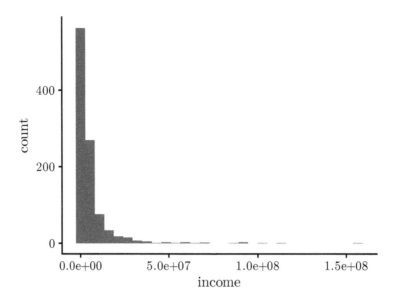

FIGURE 3.23: Simplest version of a histogram of the municipality's tax income.

it is because there is no data for the first years. So do not worry, if we filter the data with `filter(!is.na(income))`, this warning will surely disappear.

Also, the console gives the following message: `stat_bin()` using `bins = 30`. Pick better values with `binwidth`. Simply, it says that it is possible to modify the intervals for the histogram.

The following step is to modify the x-axis. Personally, I have never been good at reading numbers with scientific notation. On the other side, we will try changing the number of intervals with `bins` (Figure 3.24).

```
ggplot(data    = municipal_data,
       mapping = aes(x = income)) +
  geom_histogram(bins = 50) +
  scale_x_continuous(labels = scales::dollar)
## Warning: Removed 7 rows containing non-finite values (stat_bin).
```

Exercise 3B. What happens if we put `bins = 15` intervals?

Next we will make a subset of the data. Considering the number of outliers we found, we will eliminate the municipalities that have incomes above $50,000,000. Also, we will use another type of aesthetic: `fill =`, which will help us identify each zone by color[18]. Aesthetic properties such as `fill =`, `color =`, `size =`, change when they are used with discrete or continuous variables (Figure 3.25).

---

[18]The colors don't translate well to grayscale, of course. You should see them if you are running the code yourself!

3.3 Applied example: Local elections and data visualization   63

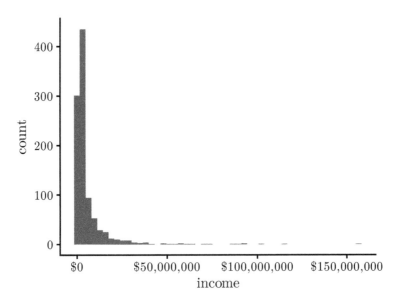

FIGURE 3.24: Histogram of municipality's tax income with a corrected scale in x.

```
ggplot(data    = municipal_data %>% filter(income < 50000000),
       mapping = aes(x = income, fill = zone)) +
  geom_histogram(alpha = 0.5, bins = 50) +
  scale_x_continuous(labels = scales::dollar) +
  labs(title = "Number of municipalities according to their annual income",
       subtitle = "Chile (2004-2012)",
       x = "Income", y = "Number of municipalities",
       caption = "Source: Based on data from SINIM (2018)")
```

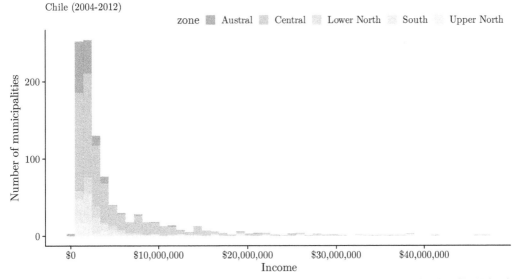

FIGURE 3.25: Polished version of our histogram in which we 'fill' by zone

3.3.5 Relation between variables

It is likely that one of your biggest concerns is whether the two variables you are studying are related in any way. With `ggplot2` this is easy to verify. In this case, we have two continuous variables: the poverty rate, from the CASEN dataset, and the municipal income. Following the theory, there should be some kind of correlation: the higher the municipal income, the lower the poverty rate in the municipality. We create our empty plot:

```
plot_f <- ggplot(data    = municipal_data,
                 mapping = aes(x = poverty, y = log(income)))
```

For this type of graph, we will use `geom_smooth()`. With this object, you can modify the way in which the variables relate with `method`. You can also introduce your own formulas. By default, a linear relation between variables comes specified, so it is not necessary to write it (Figure 3.26).

```
plot_f +
  geom_smooth(method = "lm", color = "black")
```

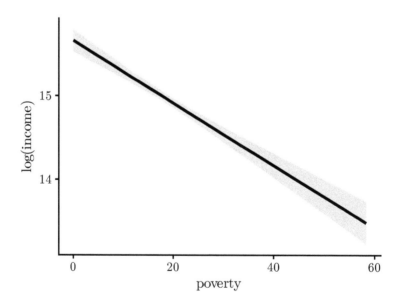

FIGURE 3.26: Linear fit of povery on the log of income.

It looks empty, doesn't it? Normally, we use `geom_smooth()` with other geometric figures, as `geom_point()`, to indicate the position of the columns within space. We use `alpha` to see the overlapping of the points. As they are not too many, there are no problems in seeing how they are distributed (Figure 3.27).

3.3 Applied example: Local elections and data visualization

```
plot_f +
  geom_point(alpha = 0.3) +
  geom_smooth(method = "lm", color = "black")
```

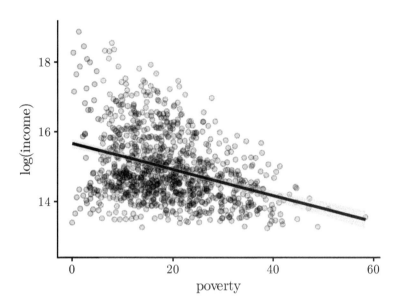

FIGURE 3.27: Linear fit plus the scattered observations.

Now we can make two improvements. First, we will insert the title and the name of the axes. Second, in `scale_x_continuous()` we will specify where our graph starts and ends. We had already used this with `geom_line()` (Figure 3.28).

```
plot_f +
  geom_point(alpha = 0.3) +
  geom_smooth(method = "lm", color = "black") +
  scale_x_continuous(expand = c(0,0)) +
  labs(title = "Relation between municipality income and its poverty rate",
       subtitle = "Chile (2004-2012)",
       x = "CASEN's poverty rate", y = "Income",
       caption = "Source: Based on data from SINIM (2018)")
```

Clearly, there is a negative correlation between both variables. This is what we expected! Now, we can calculate the correlation between both variables, to be more certain of the results obtained visually:

```
cor(municipal_data $poverty, municipal_data$income,
    use = "pairwise.complete.obs")
## [1] -0.27
```

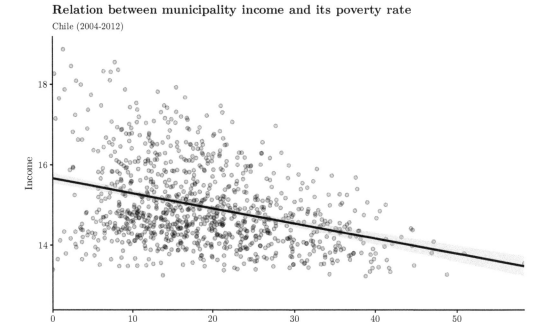

FIGURE 3.28: Polished version of our linear fit plot.

The correlation between both variables is -0.27. It would be interesting to add this information into the graph. We can do this with `annotate()`. We just need to specify the type of geometric object we want to generate. In this case, what we want to create is text `geom = "text"`, but it could be a box highlighting a specific point in the graph `geom = "rect"` or a line `geom = "segment"`. We specify where we want to locate it and, finally, annotate what we want to annotate (Figure 3.29).

```
plot_f +
  geom_point(alpha = 0.3) +
  geom_smooth(method = "lm", color = "black") +
  scale_x_continuous(expand = c(0, 0)) +
  labs(
    title = "Relation between municipality income and CASEN's poverty rate",
    subtitle = "Chile (2004-2012)",
    x = "CASEN's poverty rate ", y = "Income",
    caption = "Source: Based on data from SINIM (2018)") +
  annotate(geom = "text", x = 50, y = 15, label = "Correlation: -0.27")
```

3.4 To continue learning

There are many roads for visualizing your data. In this entry, you learned the main functions from `ggplot2`, a package within `tidyverse`, but there are many more

3.4 To continue learning

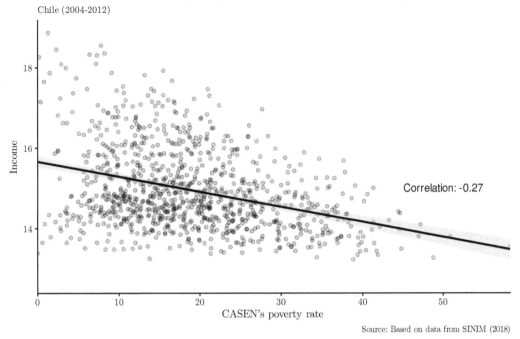

FIGURE 3.29: We add the correlation coeficient using 'annotate'.

packages that can be of great help for other types of visualizations. Although `ggplot2` may not have all the geometrical objects you need, there are packages for visualizing other types of data that work under `ggplot2` and the layers that constitute its "grammatical" form.

3.4.1 Other packages

3.4.1.1 sf

It allows to visualize spatial elements. For `ggplot2` it works with `geom_sf()`. It enables the creation of geometric figures with different types of spatial data. In Chapter 16 of spatial data, Andrea and Gabriel deliver the tools to work with `sf`, its main functions and directrices. Here[19] you can find more details about how to install it and its performance depending on your computer.

3.4.1.2 ggparliament

All political scientists should know this package. It allows you to visualize the composition of the legislative power. It is a dream for those who work with this type of information. It enables you to specify the number of seats, the color of each party, and

---

[19] See https://github.com/r-spatial/sf

add different characteristics to your graph. Here[20] you can find more details about the tools from `ggparliament` (Figure 3.30).

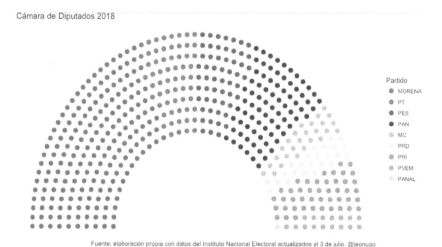

FIGURE 3.30: Example of a plot built using ggparliament with data from the Mexican Parliament. Source: @leonugo, Twitter.

3.4.1.3 ggraph

If you study networks and you know how `ggplot2` works, this package can become your new best friend. It is made for all types of relational data, and even though it works under the `ggplot2` logic, it has its own geometric objects, *facets*, among others. Here[21] you can find more information. In Chapter 14 Andres will show you how to use this package in depth.

3.4.1.4 patchwork

This is a great tool to combine separate ggplots into the same graphic. You will use +, | and / to arrange them (Figure 3.31):

```
plot_a2 <- plot_a +
  geom_bar()

plot_b2 <- plot_b +
  geom_line(mapping = aes(group = municipality))

library(patchwork)
(plot_a2 | plot_b2) / plot_c
```

---

[20] See https://github.com/RobWHickman/ggparliament
[21] See https://github.com/thomasp85/ggraph

3.4 To continue learning

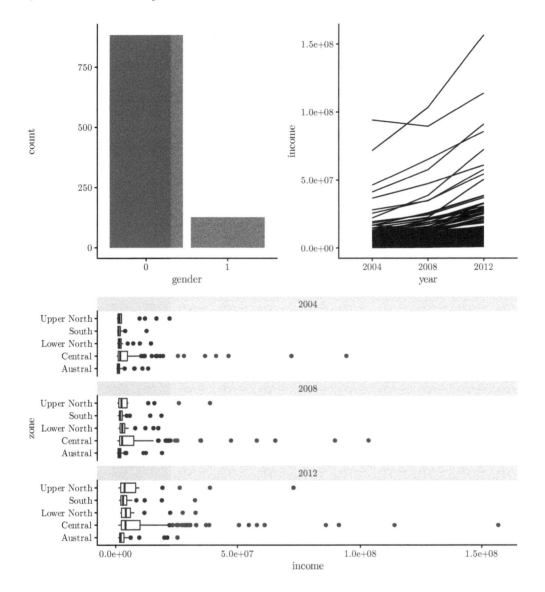

FIGURE 3.31: Patchwork example.

You already know how to make graphs and, as basic as it may appear, you now have numerous tools to keep working. However, there are some things we did not addressed:

Exercise 3C. We already learned how to make a histogram, yet, density charts tend to be more used for looking at the distribution of one variable. Using the same variables, make a density chart with `geom_density()`.

Exercise 3D. Usually, bar graphs are presented with the frequency or proportion within the bar. We can also do this with `ggplot2`. Using `geom_bar()` and `geom_text()`, note the number of mayors by geographical area. A tip:

you have to do some calculus with `tidyverse` before adding that information into the plot.

Exercise 3E. Choosing only one year, make a line graph with `geom_smooth()` that indicates the relation between income and poverty rate. Now, with `annotate()`, make a box that encloses the municipalities with the highest poverty rate and, above it, write down the corresponding municipality.

4
Data Loading

Soledad Araya and Andrés Cruz[1]

Packages you need to install

- `tidyverse` (Wickham, 2019), `politicalds` (Urdinez and Cruz, 2020), `haven` (Wickham and Miller, 2020), `readxl` (Wickham and Bryan, 2019), `data.table` (Dowle and Srinivasan, 2019), `ff` (Adler et al., 2020).

4.1 Introduction

Loading datasets is not always an easy task. People who create datasets use different file formats, trying to optimize diverse parameters (usability for office automation users, size, etc.). Sometimes, the information you need is distributed in multiple small datasets. In other cases, the problem relies on the magnitude of the data, with datasets that, employing the usual data loading methods, will affect the performance of your computer. The next three subsections will help you sort your way through these challenges.

Unlike the other chapters in this book, here you will need to download the datasets directly to your computer without using the book's package. The idea is that you learn how to load data in the real world! Start by creating a new folder in your computer, which will be the folder of your project. Then, download from this link[2] the .zip file with the datasets for this chapter, and save it in the folder you created. In Windows, for example, this should look similar to this image:

FIGURE 4.1: Folder with .zip file

---

[1] Institute of Political Science, Pontificia Universidad Católica de Chile. E-mails: snaraya@uc.cl and arcruz@uc.cl.
[2] See https://arcruz0.github.io/politicalds/data_load_politicalds.zip

Afterwards, you need to decompress the contents of the .zip file. The exact way of doing this will depend on your operating system and its programs, but it tends to be something similar to `Right click > Decompress here`. Now you should have a subfolder called "files":

FIGURE 4.2: Folder with subfolder already decompressed.

Check the new subfolder. Here are the datasets that are going to be used in this chapter. As you will see, we will learn how to load and utilize multiple common formats:

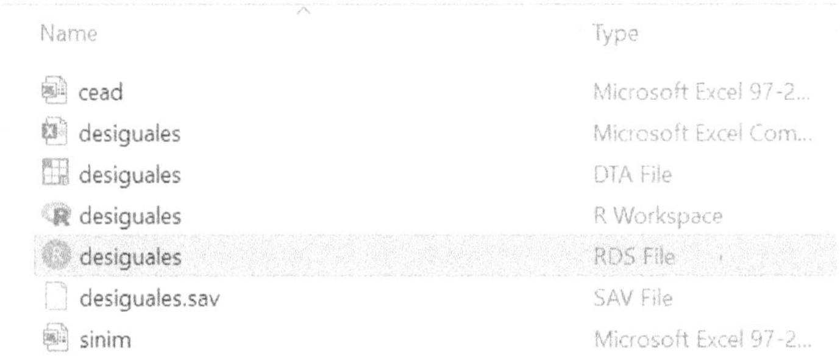

FIGURE 4.3: Subfolder with datafiles.

We are almost done! As justified and explained with greater detail in Chapter 2, we will use *RStudio Projects* in order to sort our work. Lets create a RStudio project clicking the top right corner of RStudio, "Project...", and then "New Project". Choosing "Existing Directory" in the next window you can link the project to the folder you already created (in this case, `load-dataset`):

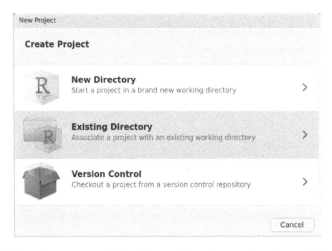

FIGURE 4.4: Creating a new RStudio project in an already existing folder.

Perfect! Now you should have a new project to start working on in a tidy way with the data of this chapter.

4.2 Different dataset formats

In this subsection, you will learn how to load different dataset formats in R: separated by delimiters, created in R, with labels in Stata or SPSS, or from Excel.

4.3 Files separated by delimiters (.csv and .tsv)

A plain text file (*flat-file dataset*) is a file constituted only by text or characters. It is a format supported by different programs, and is effortless to work with. The only difference relies on the delimiters it uses: (1) in the .csv case, the delimiter is a comma (,) and, (2) in the case of .tsv, it is delimited by a tabulation (or bumpers) that are expressed with `<TAB>` or `\t`.

4.3.1 Base R: `utils`

Base R has functions that read these kinds of files. You probably have encountered `read.table` or `read.csv`, which are specific functions that reads csv files with the integrated `utils` R package. The first one should be sufficient to open any of these files. We just need to focus on the specific characteristics of each one. As we stated before, the separations of these files are often different, and this function allows us to specify the type of separation ("," , ";" , "|" or "^") with the argument `sep =`.

Nonetheless, there are better options for loading these types of datasets. Although `read.table()` can be friendly when learning R, it will not be sufficient for working with bigger and larger files. The second disadvantage of `read.table()` is that, usually, it will not work properly with files that contain special characters, a great disadvantage for non-English speakers!

For those cases, we have different alternatives that we introduce below.

4.3.1.1 `readr`

The `readr` package is in charge of introducing a series of functions that can read different types of files. At first glance, the difference between the `utils` and `readr` functions is an underscore. Nevertheless, the `readr` functions are better when working with big datasets, and it is much faster when loading them. `readr`, as many of the packages already presented, is part the main `tidyverse` packages. You can decide to call it by itself or through `tidyverse`.

```
library(tidyverse)

library(readr) # alternative
```

Let's start by presenting one of the datasets that will be used in this chapter, which comes from the *Desiguales*[3] project by UNDP —United Nations Development Programme. This project sought to map the complexity of inequalities in Chile with the help of a survey applied in 2016. We will use a small subsection of this dataset, which contains information for 300 respondents through 20 variables. As you may have noticed, we have the dataset in six different formats for the development of this chapter:

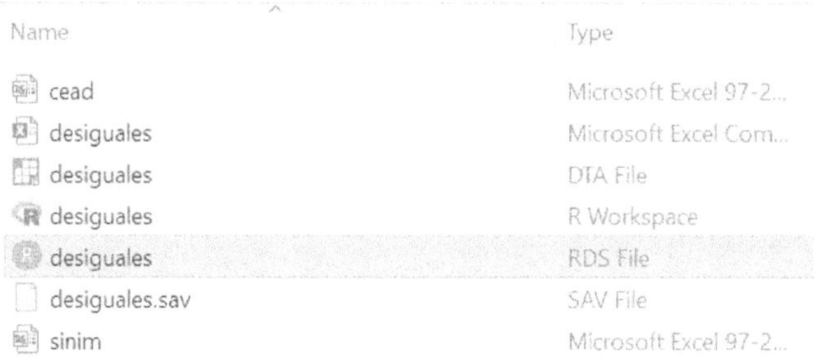

Name	Type
cead	Microsoft Excel 97-2...
desiguales	Microsoft Excel Com...
desiguales	DTA File
desiguales	R Workspace
desiguales	RDS File
desiguales.sav	SAV File
sinim	Microsoft Excel 97-2...

FIGURE 4.5: Subfolder with datafiles.

How can we load the dataset in the .csv format? As you might suspect, we will start with the simplest dataset to load. Thanks to the `read_csv()` function in `readr`, we will only need to type down the file path inside our project:

```
df_desiguales_csv <- read_csv("data/desiguales.csv")
```

We start the name of the object with "df" as short for *dataframe*. In order to confirm that our file is in the environment as an object, we will use `ls()`. With this function, you can see the list of the created or loaded objects. You can also check this in the "Environrment" panel of RStudio.

```
ls()
## [1] "df_desiguales_csv"
```

---

[3]See (https://www.desiguales.org/)

FIGURE 4.6: Environment tab, top right corner.

Now, we confirm with `class()` that this file is in the *dataframe* and *tibble* format. Because we will keep on using the `tidyverse` tools, it is useful to work with this format.

```
class(df_desiguales_csv)
## [1] "spec_tbl_df" "tbl_df"      "tbl"         "data.frame"
```

It is important to know more about the function so we can sort our way through the obstacles that arise when loading datasets in this format. To start, it may be that your file is separated by a different type of separator. You can define this with `delim=`, which fulfils the same function as `sep=` in `read.table`. Also, when managing files with a certain type of digitation, it is useful to know `locale` for specifying decimal and thousand separators. The English-speaking countries use a point to separate decimals, but continental Europe and South American countries use a comma. A second note: You probably saw the function `read_csv2()` suggested by RStudio when you wrote `read_csv` in R. The only difference between the two is the character we are using to delimit our data: `read_csv2()` has ";" as the default delimiter, while `read_csv()` has ",". For .tsv files you can use the `read_tsv()` function, which is not included by default in the `utils` package.

Let's briefly explore the data we are working on with `glimpse()`, as we did in the previous chapter. The tools from the `dplyr` package should already be available, since you previously loaded `tidyverse`. We will use `glimpse()` to see a brief panorama of the first 10 columns of our data.

```
df_desiguales_csv %>% select(1:10) %>% glimpse()
## Rows: 300
## Columns: 10
## $ id        <dbl> 34, 36, 70, 75, 99, 121, 122, 128, 160, 163, 166, ...
## $ sexo      <dbl> 1, 1, 1, 2, 2, 2, 2, 2, 1, 1, 2, 2, 2, 1, 2, 2, 1,...
## $ zona      <dbl> 2, 2, 2, 2, 2, 2, 2, 2, 2, 2, 2, 2, 1, 2, 2, 2, 2,...
## $ macrozona <dbl> 4, 4, 2, 2, 4, 2, 2, 4, 2, 2, 2, 2, 2, 4, 4, 4, 4,...
## $ region    <dbl> 13, 13, 7, 7, 13, 7, 7, 13, 7, 7, 7, 5, 7, 13, 13,...
## $ edad      <dbl> 63, 52, 73, 78, 22, 51, 18, 21, 57, 41, 55, 64, 26...
## $ p1_anyo   <dbl> 1952, 1963, 1943, 1938, 1993, 1964, 1997, 1995, 19...
## $ p1_mes    <dbl> 8, 7, 2, 2, 12, 11, 10, 1, 12, 4, 10, 12, 10, ...
## $ p2        <dbl> 1, 1, 4, 7, 8, 1, 5, 5, 3, 3, 7, 4, 5, 1, 5, 1, 1,...
## $ p3        <dbl> 2, 3, 2, 3, 3, 2, 3, 3, 2, 2, 3, 1, 1, 1, 3, 1, 2,...
```

It all looks great! Let's keep on exploring other ways of loading data into R.

4.3.2 Files created with R (.Rdata and .rds)

We will now focus on files created with R, which have a .Rdata (.rda is also valid) or .rds extension. These file formats allow us to store *any* object in R, either a vector, *dataframe*, matrix, list, etc. After reading the file, R will load the object just as it was saved. Thus, files in this format stand out for (a) their flexibility when storing, not being limited to their dataset, and (b) for their perfect compatibility with R: for example, you can be sure that every variable in a .rds dataset will be loaded in the correct format (numerical vectors as numeric, factors as factors, and so on.[4])

The difference between a .Rdata (or .Rda) and a .rds file is simple but important: while the first can contain any number of objects, the second is limited to one object. Now you will learn how to load objects created in R, in either of both formats.

4.3.2.1 .Rdata Files (one or more objects)

.Rdata files, crucially, can contain more than one object. Although this sounds convenient, it includes a limitation: when loading the objects, they will automatically adopt the name under which they were created. As an example, in the .Rdata file for the *Desiguales* dataset we have saved two objects: the *data frame* we saw earlier (now called `df_desiguales_rdata`) and a numeric vector for the ages of the surveyed (called `age_vector`). For loading the file you just need to use the `load()` function.

```
load("data/desiguales.Rdata")
```

As previously stated, you can use the `ls()` command to check if the objects were correctly loaded into the R session. If the two objects were loaded properly, then both should be named in addition to the previous data frame originated from a .csv file.

```
ls()
## [1] "df_desiguales_csv"   "df_desiguales_rdata" "vector_edades"
```

4.3.2.2 .rds Files (just one object)

.rds files are limited to storing just one object. Although it sounds less appealing than the flexibility of .Rdta, this format stands out for its modularity, which helps keeping your files in order. Another positive trait of .rds is that the syntax used for loading these files is familiar, very similar to the one you used for loading .csv files. Now you can name the object when creating it! The command is as follows, using the `read_rds()` function (which we loaded before with the `tidyverse`):

```
df_desiguales_rds <- read_rds("data/desiguales.rds")
```

---

[4] This is not always true in a .csv file, for example, that does not contain this stored information. What R does for that certain format—and others—is to infer the type of format.

Again, you can use the `ls()` command or RStudio to ensure that the object was created without any problems in the session:

```
ls()
## [1] "df_desiguales_csv"   "df_desiguales_rdata" "df_desiguales_rds"
## [4] "vector_edades"
```

4.3.3 Data sets with labels (from Stata or SPSS)

Data sets with labels are of common usage in social sciences, especially in SPSS (.sav) and Stata (.dta) files. The main idea is relatively simple: explanatory information is saved into the file in addition to the values of the dataset. For example, the variable "p2" in the *Desiguales* dataset, which corresponds to the second question of the survey, could have a label that describes the question itself ("Civil Status") or the explicit question ("Regardless of whether or not you have a partner right now, what is your current marital o civil status?"). We will call this type of label *variable label*.

Moreover, labels can register information on the values of the variables. For example, question "p2" had the following possible answers:

Value	Label
1	Married for the first time
2	Married for the second time or more
3	Legally married, but single in fact
4	Divorced
5	Single, has never married
6	Single, but with a marriage legally annulled
7	Widower
8	Couple
88	Doesn't know
99	Doesn't answer

A dataset can store the numerical values registered by the pollster, while explanatory information can be saved in labels. Although a dataset might only register a "4" in variable "p2" for a specific case, its label helps us identify the civil status answered by the surveyed ("Divorced"). We will call this type of label *value label*.

In the following section, you will learn how to load SPSS (.sav) or Stata (.dta) datasets with labels into R. This will allow you to report your analysis by labels, both for variables and values. You will need the package **haven** to use this function. If you installed **tidyverse** in the past you already have access to **haven**. Nevertheless, `library(tidyverse)` is not enough to install it, so you will need to load it separately:[5]

```
library(haven)
```

The commands used for loading datasets with labels are similar to those you used previously in order to load .csv and .rds files. The function used for reading SPSS files is `read_spss()`, while the one for Stata is `read_stata()`. Let's load both datasets:

```
df_desiguales_spss <- read_spss("data/desiguales.sav")
```

```
df_desiguales_stata <- read_stata("data/desiguales.dta",
                                  # handling special characters
                                  encoding = "UTF-8")
```

You can check if the objects were created by using the `ls()` command or in the "Environment" panel in RStudio.

```
ls()
## [1] "df_desiguales_csv"   "df_desiguales_rdata" "df_desiguales_rds"
## [4] "df_desiguales_spss"  "df_desiguales_stata" "vector_edades"
class(df_desiguales_spss)
## [1] "tbl_df"     "tbl"        "data.frame"
class(df_desiguales_stata)
## [1] "tbl_df"     "tbl"        "data.frame"
```

Datasets with labels are different from others not by the class of the object, but by its variables. Using the tools learned in Chapter 2, you can explore the first ten variables of any of the datasets (both will give the same results, so from now on we will use the SPSS dataset).

```
df_desiguales_spss %>% select(1:10) %>% glimpse()
## Rows: 300
## Columns: 10
## $ id        <dbl> 34, 36, 70, 75, 99, 121, 122, 128, 160, 163, 166, ...
## $ sexo      <dbl+lbl> 1, 1, 1, 2, 2, 2, 2, 2, 1, 1, 2, 2, 2, 1, 2, 2...
## $ zona      <dbl+lbl> 2, 2, 2, 2, 2, 2, 2, 2, 2, 2, 2, 2, 1, 2, 2, 2...
## $ macrozona <dbl+lbl> 4, 4, 2, 2, 4, 2, 2, 4, 2, 2, 2, 2, 2, 4, 4, 4...
## $ region    <dbl+lbl> 13, 13, 7, 7, 13, 7, 7, 13, 7, 7, 7, 5...
## $ edad      <dbl> 63, 52, 73, 78, 22, 51, 18, 21, 57, 41, 55, 64, 26...
## $ p1_anyo   <dbl+lbl> 1952, 1963, 1943, 1938, 1993, 1964, 1997, 1995...
```

---

[5] Although `haven` is a part of `tidyverse`, it is not included in the "core" of it. The core of `tidyverse` includes a few packages, which are the most used ones. The rest of the packages, for example `haven`, need to be loaded (but not installed) separately.

```
## $ p1_mes     <dbl+lbl> 8, 7, 2, 2, 12, 11, 10, 1, 12, 4, 10, 12...
## $ p2         <dbl+lbl> 1, 1, 4, 7, 8, 1, 5, 5, 3, 3, 7, 4, 5, 1, 5, 1...
## $ p3         <dbl+lbl> 2, 3, 2, 3, 3, 2, 3, 3, 2, 2, 3, 1, 1, 1, 3, 1...
```

As you can see, the majority of variables are not just numerical vectors ("dbl" or "double" [6]), but they also include labels (+ "lbl"). Let's obtain the first six values and the information of the labelled variable p2 by utilizing the head() command:

head(df_desiguales_stata$p2)

```
## <labelled<double>[6]>: P2 - ¿podria decirme cual es su estado
    conyugal...?
## [1] 1 1 4 7 8 1
##
## Labels:
##  value                                         label
##      1                        Casado(a) por primera vez
##      2                   Casado(a) por segunda vez o más
##      3 Casado(a) legalmente, pero separado de hecho
##      4                                      Divorciado
## [ reached 'max' / getOption("max.print") -- omitted 6 rows ]
```

Let's briefly analyse the result shown in the console. First, we can see that the variable at issue is a numerical vector with a label ("Labelled double"). Then, after colons, it registers the * label variable: *"P2 - what is your current marital o civil status?"*. Afterwards, the first six values of the variable are displayed, just as we requested through the head() command. At last, value labels are found, which give us information over the meaning of each number in the context of a specific variable.

Thus, we obtained all the information registered by labels in this dataset. With the commands learned you can be certain of the meaning of each value in the labelled dataset. Lastly, note that R, with the help of haven, will also show the labels in other cases when convenient. For example, let's see a simple summary of the first two variables in the dataset. Look how easy it is to read it!

df_desiguales_stata %>% select(region, p2)
```
## # A tibble: 300 x 2
##                              region                        p2
##                           <dbl+lbl>                 <dbl+lbl>
## 1 13 [Metropolitana de Santiago] 1 [Casado(a) por primera vez]
## 2 13 [Metropolitana de Santiago] 1 [Casado(a) por primera vez]
## 3  7 [Del Maule]                 4 [Divorciado]
## # ... with 297 more rows
```

---

[6]This is a way of naming real numbers in computing.

4.3.4 Excel Files

While most of the time we will use datasets that come in the previous formats, it is important to acknowledge that not all institutions present their information in that way. Often you will face formats that bring headaches. For example, Excel. In Chile, a substantial amount of governmental organizations still work with Excel, and the problem is not the format but the structure of the datasets.

Most of the time we will face something like this:

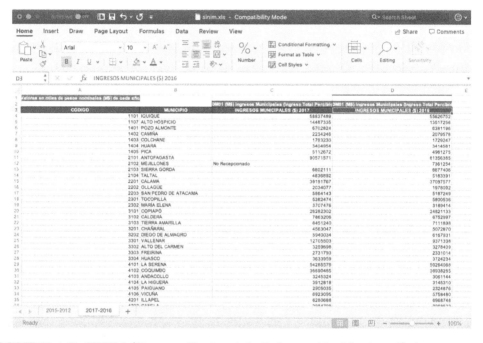

FIGURE 4.7: SINIM (Sistema Nacional de Información Municipal) dataset.

Or like this:

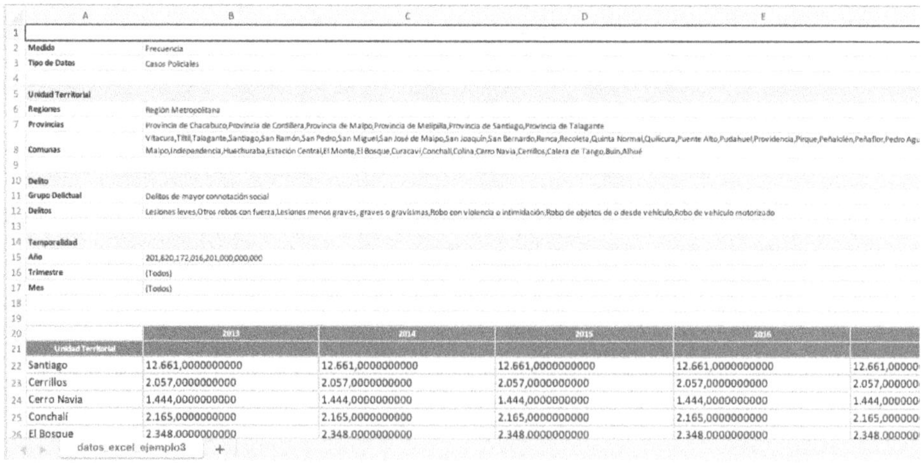

FIGURE 4.8: CEAD's dataset.

4.3 Files separated by delimiters (.csv and .tsv)

Here is where problems arise.

To begin, let's load one of the most used packages for reading Excel files (.xls or .xlsx), `readxl`, a package from `tidyverse`[7]:

```
library(readxl)
```

For the next example, we will use the CEAD (Centro de Estudios y Análisis del Delito)[8] dataset. CEAD is a Chilean institution which focuses on monitoring and studying crime in that country. Their objective is to create and analyze crime information in order to aid public policy formation. This dataset is the one that can be downloaded by default in their webpage.

For loading the file, we will use `read_excel()` from `readxl`. It is only necessary to put the .xls or .xlsx file path:

```
df_cead_excel <- read_excel("data/cead.xls")
```

First, let's check that the data was correctly loaded into the Environment with `ls()`. Then, let's use `glimpse` to examine the first observations in our dataset:

```
ls()
## [1] "df_cead_excel"       "df_desiguales_csv"   "df_desiguales_rdata"
## [4] "df_desiguales_rds"   "df_desiguales_spss"  "df_desiguales_stata"
## [7] "vector_edades"
df_cead_excel %>% glimpse()
## Rows: 79
## Columns: 7
## $ Medida     <chr> "Tipo de Datos", NA, "Unidad Territorial", "Regio...
## $ Frecuencia <chr> "Casos Policiales", NA, NA, "Región Metropolitana...
## $ ...3       <chr> NA, NA, NA, NA, NA, NA, NA, NA, NA, NA, NA, NA, N...
## $ ...4       <chr> NA, NA, NA, NA, NA, NA, NA, NA, NA, NA, NA, NA, N...
## $ ...5       <chr> NA, NA, NA, NA, NA, NA, NA, NA, NA, NA, NA, NA, N...
## $ ...6       <chr> NA, NA, NA, NA, NA, NA, NA, NA, NA, NA, NA, NA, N...
## $ ...7       <chr> NA, NA, NA, NA, NA, NA, NA, NA, NA, NA, NA, NA, N...
```

Something looks wrong. If we look again at the picture of the CEAD dataset, we will notice what's going on. First, the Excel file has in its initial rows information of the dataset, but these are not observations.

To solve this, it is necessary to use `skip =`. This argument will help us omit rows we are not interested in, in this case, the first 18:

```
df_cead_excel_v2 <- read_excel("data/cead.xls", skip = 18)
```

---

[7] Just as `haven`, `readxl` is a part of `tidyverse`, but not a main member of its nucleus. Thus, it is necessary to load it separately.
[8] See http://cead.spd.gov.cl/estadisticas-delictuales/

FIGURE 4.9: CEAD dataset (Centro de Estudios y Análisis del Delito).

Another way of doing the same this is by delimiting the information we want to load through a range. Thus, we can delimit range with `range`, obtaining only the information we need inside the specified rectangle. The range can be delimited with the letter of the column and the number of the row in the .xls or .xlsx file.

```
df_cead_excel_v3 <- read_excel("data/cead.xls", range = "A20:G81")
```

We check that the data was loaded correctly with `ls()`, and then we can see the file type with `class()`. For the purposes of this chapter, we will only use the latest .xls file uploaded (`df_cead_excel_v3`).

```
ls()
## [1] "df_cead_excel"        "df_cead_excel_v2"     "df_cead_excel_v3"
## [4] "df_desiguales_csv"    "df_desiguales_rdata"  "df_desiguales_rds"
## [7] "df_desiguales_spss"   "df_desiguales_stata"  "vector_edades"
class(df_cead_excel_v3)
## [1] "tbl_df"     "tbl"        "data.frame"
```

As you can observe, it is very easy to load this types of files after practicing for a while. For now, that's what you need to know for loading .xls or .xlsx files. You might have to deal with digitation problems, with impossible variable names or with typing errors that can cause more than one problem when working with `skip`. But you already know how to use some `tidyverse` tools that can help you with that.

Let's try to make some basic changes into the CEAD dataset:

```
names(df_cead_excel_v3)
## [1] "...1"   "2013"   "2014"   "2015"   "2016"   "2017"   "2018"
```

The first problem is the name of the column that indicates the county. The second problem is that we want a *wide* dataset, so it does not help us to have the years as

4.3 Files separated by delimiters (.csv and .tsv)

columns! For the first problem, we use `rename`, and for the second `pivot_longer`. By doing this, we end up with a *long* dataset.

```
df_cead_excel_v4 <- df_cead_excel_v3 %>%
  rename(county = `...1`) %>%
  pivot_longer(cols = -county, names_to = "year", values_to = "n_crime") %>%
  filter(county != "Unidad Territorial")
```

Another problem that can arise when using Excel files is that information might be divided into different sheets. As you will see, `read_excel()` resolves those problems in a very simple way. To prove this, we will use the SINIM (*Sistema Nacional de Información Municipal*)[9] dataset. SINIM is a Chilean information system through which citizens can access a variety of data regarding county administration:

```
df_sinim_excel <- read_excel("data/sinim.xls",
                     sheet = 2, skip = 2, na = "Not received")
```

With the `sheet=` argument we can select the sheet we want to load into R. It can be loaded with both the position or the name of the sheet. As we already learned, with `skip` we select the number of rows we want to skip when loading the dataset, and with `na` we can let the program know what other phrase, word or situation has to be listed as NA besides the blank cells.

One of the main tasks when working on Excel is dataset cleaning. Unfortunately, these two examples are widespread, and for dealing with some of these problems there is `janitor`.

Usually, variable names in Excel tend to come in a very detailed and/or descriptive way: they can have spaces, special characters and capital letters that hinder working easily with our variables of interest. For example, the names in the columns of this dataset have all of the above:

```
names(df_sinim_excel)
## [1] "CODIGO"                    "MUNICIPIO"
## [3] "INGRESOS MUNICIPALES ($) 2017" "INGRESOS MUNICIPALES ($) 2016"
```

Can you imagine having to write those names every time we wanted to do some type of analysis?

For that we have the `clean_names()` function from the `janitor` package. This package was created to ease data cleaning, and, without being a `tidyverse` package, it can be used without problems with pipes. And, for our luck, it is optimized to work with `readr` and `readxl`.

---

[9]See http://datos.sinim.gov.cl/index.php

The `clean_names()` function works in a simple way after installing `janitor` (with `install.packages("janitor")`):

```
df_sinim_excel_v2 <- df_sinim_excel %>% janitor::clean_names()
```

Now, let's look again at the variable names:

```
names(df_sinim_excel_v2)
## [1] "codigo"                    "municipio"
## [3] "ingresos_municipales_2017" "ingresos_municipales_2016"
```

It looks much better! These functions make working in R a much simpler and friendly experience, regardless of your level of level of experience and what you are looking to do. Working with variable names can be a headache, and this tool will help you avoid overworking, allowing you to focus on the important parts of the analysis (which, of course, should not be column names).

With these tools, you should be more than prepared to face the hideous world of working with Excel.

4.4 Large tabular datasets

Along with technological advancements and faster internet connections, large datasets are interestingly available to social scientists. Nevertheless, the usual tabular dataset managing tools tend to not work properly in large datasets, and it is necessary to find alternatives. The complexity of facing the size of the dataset depends on multiple factors (for example, the nature of the data and the characteristics of the computer), but, in general, a tabular dataset of more than 1GB in size will generate problems in R in an average personal computer. What can we do to handle this type of dataset?

In the following section, you will work with the *Desiguales* dataset in a .csv format, just as you did in the 4.2 subsection of this chapter. We will pretend that it is a case of a "large dataset", even though its only 15KB in size. By the way, if you want to try doing the following analysis with real-world datasets that are of political science interest, you can find some for download in the *Observatorio de Complejidad Económica*[10], which registers the bilateral trade of countries by the different productive categories[11]

So, how can we manage a large dataset? A first alternative is to evaluate the necessity of using the entire dataset for the analysis, or, if it can be shortened before using it.

---

[10]See https://atlas.media.mit.edu/es/

[11]These are found in a .tsv format, so you will have to make slight alterations to the syntax, just as you learned in the subsection for .csv/.tsv files (4.2).

4.4 Large tabular datasets

To start an exploratory analysis that can clarify this point, we recommend using the n_max= argument in `read_csv()` and its siblings functions (for example, `read_tsv()`). By doing this we can examine the first one hundred observations of the dataset, making the computational process less demanding:

```
df_desiguales_large_100 <- read_csv("data/desiguales.csv", n_max = 100)
```

Surely your computer will handle with ease this new-reduced dataset. Now, what happens if, after checking the data, you found out you only need a couple of variables for the analysis? Trimming the dataset will allow an efficient computational resource usage. Note that the `n_max` argument is no longer useful, since it trims rows instead of columns. For example, let's suppose that from the *Desiguales* dataset (the "large" one) you only need the `age` and `p2` variables. With an assistance function from our `politicalds` package (`cols_only_chr()`), you can make `read_rds()` load the dataset with only those two variables, omitting all the others, and as well as avoiding the computational cost that they imply:

```
library(politicalds)
```

```
## Warning: The following named parsers don't match the column names: age
```

```
df_desiguales_laod_2vars <- read_csv("data/desiguales.csv",
                            col_types = cols_only_chr(c("age",
                                                        "p2")))
```

However, it is possible that some analysis will not allow the omission of variables, or that even after all the procedures to reduce its size, the dataset remains too large. In this case, the R ecosystem also provides alternatives, that we will briefly present below.

A first option is the `fread()` functions from the `data.table` package. This function, optimized for speed, is usually faster than `read_csv()` and its associates, although it does not have the same variety of options and is not as friendly to use. Once `data.table` is installed in our system –that is to say, after `install.packages("data.table")`– the next command will allow us to load the dataset:

```
library(data.table)
df_desiguales_large_fread <- fread("data/desiguales.csv")
```

Note that, although the created object is a *data frame*, it is also a special type called `data.table`:

```
class(df_desiguales_large_fread)
## [1] "data.table" "data.frame"
```

The `data.table` package has various functions for dealing with large types of objects, and tends to be more computationally efficient –however, it is often less intuitive and legible. If you want to learn more about `data.table`, you can check the multiple vignettes[12] available on the package's web page.

Finally, even if `data.table` is not efficient enough for dealing with your data, we have another option available to you. The `ff` package provides an interesting solution for the large dataset problem: instead of loading them into the RAM, like other packages, it occupies the hard drive directly (which usually has more available space). Although this makes most of the traditional R functions not work, `ff` opens the gate for using giant datasets: the package brings a whole new family of *ad hoc* analysis functions, inspired by R.

You can find more information about `ff`, if necesary, on their web page[13] and their official R help file[14]. If you want first try to load a dataset with this package, you can use the following command[15]–assuming the package is installed, with `install.packages("ff")`:

```
library(ff)
df_desiguales_large_ff <- read.csv.ffdf(file = "data/desiguales.csv")
```

Exercise 4A. From the Latinobarómetro[16] web page, download the 2017 edition in SPSS (.sav) format and import it to R. Be careful with the labels.

Exercise 4B. Now, repeat the process downloading the Stata (.dta) dataset.

---

[12]See https://cran.r-project.org/web/packages/data.table/index.html
[13]See http://ff.r-forge.r-project.org/
[14]See https://cran.r%20project.org/web/packages/ff/ff.pdf
[15]It is important to add the `file =` argument, since this is not the first argument the function in question receives.
[16]See http://www.latinobarometro.org/latContents.jsp

Part II

Models

5

Linear Models

Inés Fynn[1] and *Lihuen Nocetto*[2]

Suggested readings

- Angrist, J. D. and Pischke, J. S. (2008). *Mostly Harmless Econometrics: An Empiricist's Companion*. Princeton University Press, Princeton, NJ.

- Dunning, T. (2012). *Natural Experiments in the Social Sciences: A Design-Based Approach*. Cambridge University Press, Cambridge.

- Lewis-Beck, C. and Lewis-Beck, M. (2016). *Applied Regression: An Introduction*. SAGE, Thousand Oaks, CA.

- Wooldridge, J. M. (2016). *Introductory Econometrics: A Modern Approach*. Cengage Learning, Boston, MA, 6th edition.

Packages you need to install

- `tidyverse` (Wickham, 2019), `politicalds` (Urdinez and Cruz, 2020), `skimr` (Waring et al., 2020), `car` (Fox et al., 2020), `ggcorrplot` (Kassambara, 2019), `texreg` (Leifeld, 2020), `prediction` (Leeper, 2019), `lmtest` (Hothorn et al., 2019), `sandwich` (Zeileis and Lumley, 2019), `miceadds` (Robitzsch et al., 2020).

5.0.1 Introduction

In this chapter, we will learn how to do linear regressions. Here the function is linear, that is, it is estimated by two parameters: the slope and the intercept. When we face a multivariate analysis, the estimation gets more complex. We will cover how to interpret the different coefficients, how to create regression tables, how to visualize predicted values, and we will go further into evaluating the Ordinary Least Squares (OLS) assumptions, so that you can evaluate how well your models fit.

---

[1] Institute of Political Science, Pontificia Universidad Católica de Chile. E-mail: ifynn@uc.cl.
[2] Institute of Political Science, Pontificia Universidad Católica de Chile. E-mail: lnocetto@uc.cl.

5.1 OLS in R

In this chapter, the dataset we will work is a merge of two datasets constructed by Evelyne Huber and John D. Stephens[3]. These datasets are:

- Latin America Welfare Dataset, 1960-2014 (Evelyne Huber and John D. Stephens, Latin American Welfare Dataset, 1960-2014, University of North Carolina at Chapel Hill, 2014.): it contains variables on Welfare States in all Latin American and Caribbean countries between 1960 and 2014.

- Latin America and Caribbean Political Data Set, 1945-2012 (Evelyne Huber and John D. Stephens, Latin America and Caribbean Political Dataset, 1945-2012, University of North Carolina at Chapel Hill, 2012): it contains political variables for all Latin American and Caribbean countries between 1945 and 2012.

The resulting dataset contains 1074 observations for 25 countries between 1970 and 2012 (data from the 1960s was excluded since it contained many missing values).

First, we load the `tidyverse` package.

```
library(tidyverse)
```

We will import the dataset from the book's package:

```
library(politicalds)
data("welfare")
```

Now, the dataset has been loaded into our R session

```
ls()
## [1] "welfare"
```

In the chapter, we will use the paper of Huber et al. (2006) as the example for analysis. In this article, they estimate the determinants of inequality in Latin America and Caribbean. Working from this article allows us to estimate a model with multiple control variables that have already been identified as relevant for explaining the variation of inequality in the region. Thus, the dependent variable we are interested in explaining is income inequality in Latin American and Caribbean countries, operationalized

---

[3] See http://huberandstephens.web.unc.edu/common-works/data/

according to the Gini Index (`gini`). The control variables that we will incorporate into the model are the following:

- Sectorial dualism (it refers to the coexistence of a traditional low-productivity sector and a modern high-productivity sector) - `sector_dualism`
- GDP - `gdp`
- Foreign Direct Investment (net income as % of the GDP) - `foreign_inv`
- Ethnic diversity (dummy variable coded as 1 when at least the 20% but no further than the 80% of the population is ethnically diverse) - `ethnic_diversity`
- Democracy (type of regime) - `regime_type`
- Education expenditure (as percentage of the GDP) - `education_budget`
- Health expenditure (as percentage of the GDP) - `health_budget`
- Social security expenditure (as percentage of the GDP) - `socialsec_budget`
- Legislative balance - `legislative_bal`

During this chapter, we will try to estimate what is the effect of education expenditure in the levels of inequality in Latin American and Caribbean countries. Thus, our independent variable of interest will be `education_budget`.

5.1.1 Descriptive Statistics

Before estimating a linear model with Ordinary Least Squares (OLS) it is recommended you first identify the distribution of the variables you are interested in: the dependent variable y (also called response variable) and the independent variable of interest x (also called explanatory variable or regressor). In general, our models will have, besides the independent variable of interest, other independent (or explanatory) variables that we will call "controls", since their task is to make the *ceteris paribus* scenario as credible as possible. That is, "keeping all other factors constant" to get as close as possible to an experimental scenario where we can control all the variables that affect y, and to observe how the variation of a single independent variable x affects the variation of the dependent variable (y). Then, before estimating the model, we will observe the descriptive statistics of the variables that will be included in the model (both the dependent and the independent ones). The goal is to pay attention to the following issues:

1. Variation in x: that the independent variables (but especially the one of interest) have variation in our sample. If there is no variation in x, we cannot estimate how this variation affects the variation of y.

2. Variation in y: if the dependent variable does not vary, we cannot explain its variation according to the variation of x.

3. Unit of measurement of the variables: here is where we evaluate how our variables are measured (in addition to reviewing the codebooks that usually come with the datasets we are working with), so we can understand what types of variables they are (nominal, ordinal, continuous), and also to correctly interpret the results later on.

4. Type of variables: in OLS estimation, the dependent variable must be, in general, *continuous*, (although it is possible to work with dichotomous dependent variables). Thus, we need to be sure that the dependent variable is continuous and numerical. Also, it is important to know the type of independent variable and check that its type is coherent with its coding format (i.e. if we have an independent variable of "age ranges", the variable must be categorical or factorial, not numerical), so that our interpretations of the results are correct. Problems with coding format are very common, be careful!

5. Identify missing values: if our variables have too many missing values, we need to check where are our missings, and, eventually, decide if an imputation is desireable (as explained in Chapter 11).

5.1.2 Descriptive statistics and distribution of the variables in the model

A starting visualization of our variables of interest can be done by using the `skimr::skim()` command which gives us some descriptive statistics, as it is shown in Figure 5.1:

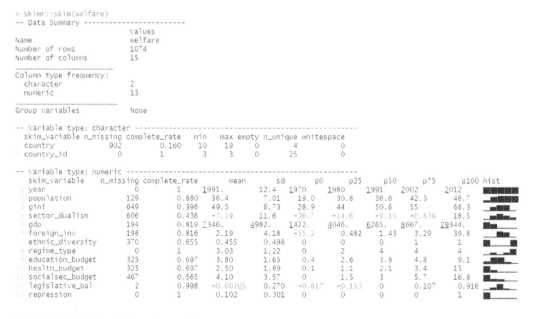

FIGURE 5.1: Skim of our dataset.

5.1 OLS in R

In the output, the results are ordered by the type of variable, and it indicates the number of missing values for each of them, their respective mean, standard deviation, the corresponding values of percentiles and a small histogram that shows how they are distributed.

5.1.3 Correlation matrix of independent variables

After identifying all the variables that we will incorporate into the model, it is recommended to observe how they are related to each other. To do this, we will create a correlation matrix of the independent variables with the `ggcorrplot()` function of the homonymous package, with which we can evaluate Pearson's correlation between all variables (Figure 5.2).

Anyhow, it is important to remember that **correlation does not imply causation**. Here, we simply want to understand if the variables of the model are related in a certain way. This step is not only important for recognizing our data and variables, but also because we want to avoid having perfect multicollinearity (that there are independent variables that are perfectly correlated) in our model, since this is a central assumption of OLS.

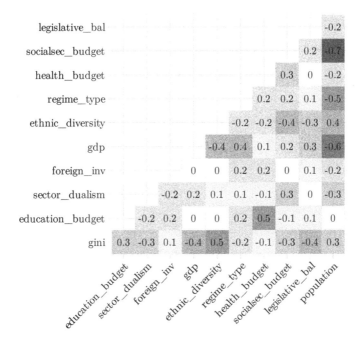

FIGURE 5.2: Correlation matrix among selected variables.

```
library(ggcorrplot)

corr_selected <- welfare %>%
  select(gini, education_budget, sector_dualism, foreign_inv, gdp,
```

```
            ethnic_diversity, regime_type, health_budget, socialsec_budget,
            legislative_bal, population) %>%
  # calculate correlation matrix and round to 1 decimal place:
  cor(use = "pairwise") %>%
  round(1)

ggcorrplot(corr_selected, type = "lower", lab = T, show.legend = F)
```

Now that we know all the variables that will be incorporated into the model, and how they correlate to each other, we will go further into the key variables of interest: the dependent and the independent.

5.1.4 Distribution of the variables of interest

As we previously mentioned, we want to estimate how the change of an independent variable (its variation) affects the variation of a dependent variable. That is, how y changes when x changes. In this case, let's suppose that we are interested in estimating how the levels of inequality of a country vary (measured through Gini Index), related to the variation of the budget on public education). Thus, our independent variable of interest is education expenditure, while the dependent variable is inequality.

Let's observe how these variables are distributed in our dataset (Figures 5.3 and 5.4):

```
ggplot(welfare, aes(x = gini, na.rm = T)) +
  geom_histogram(binwidth = 1) +
  labs(x = "Gini Index", y = "Frequency",
       caption = "Source: Huber et al (2012)")
## Warning: Removed 649 rows containing non-finite values (stat_bin).
```

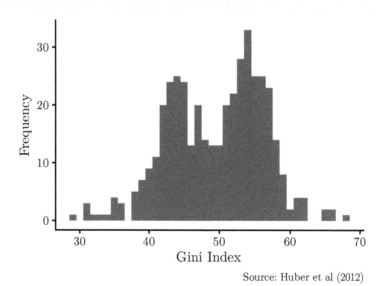

FIGURE 5.3: Histogram of our dependent variable.

5.1 OLS in R

The independent variable: Education Expenditure (% of GDP)

```
ggplot(welfare, aes(x = education_budget, na.rm = T))+
  geom_histogram(binwidth = 1) +
  labs(x = "Education Expenditure", y = "Frequency",
       caption = "Source: Huber et al (2012)")
## Warning: Removed 325 rows containing non-finite values (stat_bin).
```

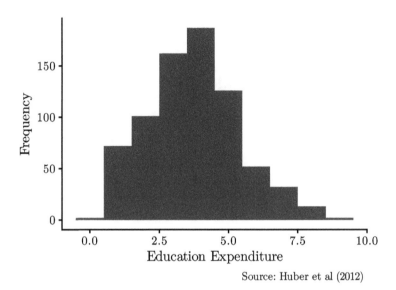

Source: Huber et al (2012)

FIGURE 5.4: Histogram of our independent variable: Education Expenditure as a Percentage of GDP.

5.1.5 Relation between the dependent and independent variable

After observing how the variables of interest are distributed, we can graphically see how they are related. That is, we graph the correlation between these two variables: in the x (horizontal) axis we locate the independent variable, while in the y (vertical) axis the dependent variable. As a result, each "dot" of the Figure 5.5 represents an observation of our sample with a particular value of education expenditure (x) and a particular value in the Gini index (y).

```
ggplot(welfare, aes(education_budget, gini)) +
  geom_point() +
  labs(x = "Education Expenditure (% of GDP)", y = "Gini",
       caption = "Source: Huber and Stephens (2012)")
## Warning: Removed 718 rows containing missing values (geom_point).
```

This is a first visualization of the relationship between our variables that allows us to observe if there is any kind of relationship between them. Here, we clearly see a positive relation (the higher the education expenditure, the higher the Gini). Anyway, until now we cannot say anything concluding about the effect of education expenditure on the levels of inequality. For this, it is necessary to estimate a model. Until now we have only explored our data. Let's move on into regressions!

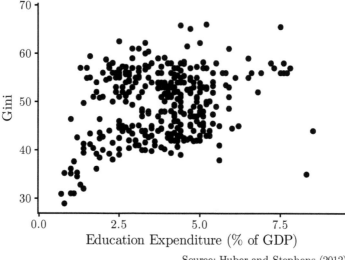

Source: Huber and Stephens (2012)

FIGURE 5.5: Scatterplot of Education Expenditure and Gini Index.

Exercise 5A. Imagine that now we are interested in the effect of Foreign Direct Investment `foreign_inv` on economic inequality (Gini). Analyze the distribution of this variable and make a graph to evaluate the relation between this variable and our independent variable (`gini`), Are there any signs of correlation between the variables? What is the direction (positive/negative) of the relation?

5.2 Bivariate model: simple linear regression

The simple linear model assumes that a random response variable y is a linear function of an independent variable x plus an error term u. We also say that the dependent variable y is the result of a data generating process (DGP) that can be written as

$$Y = \beta_0 + \beta_1 x + u$$

Thus, the linear model involves defining that Y is generated by a linear function of x_1, in addition to a constant term β_0 an the unobserved u variable.

Another two assumptions are necessary for deriving the estimators of OLS. The first one refers to the expectation of u being equal to 0

$$E(u) = 0$$

This implies that, at a population level, all the unobserved factors average zero. The most important assumption is that the average of u for each value of x is zero:

$$E(u|x) = 0$$

This assumption is known in the econometric literature as the zero conditional mean, or conditional independence. In the experimentalist literature, it is known as exogeneity of x, or that x and u are orthogonal. All these terms imply the assumption that, for each value of the independent variable of interest x, the unobserved factors will average zero. In other words, if the expectation of the unobserved is zero for each x value, the expected value of Y only depends on the x and the constant. When these assumptions are met, we can identify the effect of the x on y, keeping everything else constant. Also, when $E(u|x) = 0$, it is also true that

$$cov(x, u) = 0$$

In summary, under the conditional independence assumption, x and y do not correlate and this allows deriving the estimators of OLS through the *first order conditions*. The first two order conditions are that

$$E(u) = 0$$

and

$$E(u|x) = 0$$

Understanding that these are the conditions that allow deriving the OLS *estimator* is key for understanding why we cannot test the error independence starting from an estimation. This means that, by construction, the residuals ($\hat{u}$) of the regression will always average zero in the sample and at every value of x. Demanding exogeneity involves being able to argue that u is indeed orthogonal to the $x's$, something more credible for experiments and quasi-experiments than observational data (see Gerber and Green, 2012; Dunning, 2012).

5.2.1 Estimating a linear model in R

The function `lm()` that is part of base R is the main tool for estimating linear models. The general form that the function takes is

```
lm(Y ~ 1 + X)
```

From which it is understood that a *linear model* (lm) is estimated for a regressed (~) dependent variable Y into an independent variable X. The "1" is not usually included, but we added it to denote the intercept (β_0). Based on the investigation from Huber et al. (2006), our model poses that inequality is a linear function of the education expenditure in addition to an unobserved error term u and a constant β_0. Formally:

$$Inequality = \beta_0 + \beta_1 EducationExpenditure + u$$

For the moment, we will assume that the dataset contains 1074 independent observations.[4] The independent and identically distributed observations assumption is what allows us to write the model for a i individual picked randomly as

$$Inequality_i = \beta_0 + \beta_1 EducationExpenditure_i + u_i$$

Remember that dependent variable "Inequality" is proxied with the Gini index and that the name of the variable is `gini`, while the independent variable "Education Expenditure" is `education_budget`. Since data is stored in a data.frame, we need to indicate its name in the function to bring the data from the corresponding dataset. That is what occurs after the comma in the following command

```
# after the comma we indicate the data.frame that contains the data
model_1 <- lm(gini ~ 1 + education_budget, data = welfare)

class(model_1) # we verify that the object class is "lm"
## [1] "lm"
```

In the first line of code we created an object (`model_1`) which saves the results of the `lm()` function. This function creates `lm()` class objects, which are vectors that include the estimated coefficients, the standard errors, residuals, fit, among other results of the estimation. To see the components of the object, a quick way to do so is to use the `summary()` function

```
summary(model_1)
##
## Call:
## lm(formula = gini ~ 1 + education_budget, data = welfare)
##
## Residuals:
##     Min      1Q  Median      3Q     Max
## -20.04   -5.62    1.09    4.99   15.57
##
## Coefficients:
```

---

[4]In reality, the structure is panel data: the same country has different observations through time. Nonetheless, we are not currently capable of approaching these data. We will cover this in the chapter of panel data.

5.2 Bivariate model: simple linear regression

```
##                  Estimate Std. Error t value Pr(>|t|)
## (Intercept)         44.80       1.02   43.80  < 2e-16 ***
## education_budget     1.23       0.25    4.93 1.3e-06 ***
## ---
## Signif. codes:  0 '***' 0.001 '**' 0.01 '*' 0.05 '.' 0.1 ' ' 1
##
## Residual standard error: 6.7 on 354 degrees of freedom
##   (718 observations deleted due to missingness)
## Multiple R-squared:  0.0642, Adjusted R-squared:  0.0615
## F-statistic: 24.3 on 1 and 354 DF,  p-value: 1.29e-06
```

More elegant presentations can be achieved with the function `screenreg()` from the `texreg` package. This will be used in all chapters of econometric models. Let's see the presentation of the results with the `screenreg()` function

```
library(texreg)
screenreg(model_1)
##
## ============================
##                   Model 1
## ----------------------------
## (Intercept)       44.81 ***
##                    (1.02)
## education_budget   1.23 ***
##                    (0.25)
## ----------------------------
## R^2                0.06
## Adj. R^2           0.06
## Num. obs.        356
## ============================
## *** p < 0.001; ** p < 0.01; * p < 0.05
```

We can add names to the variables with `custom.coef.names =`:

```
library(texreg)
screenreg(model_1,
         custom.model.names = "Model 1",
         custom.coef.names = c("Constant", "Education expenditure"))
##
## =================================
##                         Model 1
## ---------------------------------
## Constant                44.81 ***
##                          (1.02)
## Education expenditure    1.23 ***
##                          (0.25)
## ---------------------------------
```

```
## R^2                       0.06
## Adj. R^2                  0.06
## Num. obs.                 356
## =====================================
## *** p < 0.001; ** p < 0.01; * p < 0.05
```

And we can export the table in a .doc format for adding it into our manuscript. The file will be saved in our working folder.

```
htmlreg(list(model_1), file = "model_1.doc",
        custom.model.names = "Model 1",
        custom.coef.names = c("Constant", "Education expenditure"),
        inline.css = FALSE, doctype = T, html.tag = T,
        head.tag = T, body.tag = T)
```

As you can notice, the results are better displayed with `texreg` than with `summary()`. It is clearly seen that `education_budget`, education expenditure, has a positive effect of magnitude 1.233. In particular, when education expenditure as a percentage of the GDP increases by one unit, inequality increases by 1.23 percentage points. The effect of Education Expenditure is significant at a 99.9% confidence level. We know this because besides the coefficient three stars appear, which refer to a significance level of 0.01%. Statistical significance is the result of a t-test. This test indicates the standardized distance, where the estimated beta is found in the distribution under the null hypothesis that $\beta_1 = 0$. The estimator has a t-Student distribution with degrees of freedom equal to $n - k - 1$, where k is the number of independent variables and 1 is added for the estimation of the constant β_0. A manual approximation of the estimated beta distance in the estimator distribution under the null hypothesis $\beta_1 = 0$ is obtained when we divide the estimation by its standard error:

```
1.233 / 0.25
## [1] 4.9
```

The same value is delivered by the third column of the "Coefficients" section in the `summary()` of `model_1`. The t-value is always interpreted as the distance of the $\hat{\beta}_1$ estimate from the mean of the estimator distribution under $H_0 = \beta_1 = 0$. In this case, the value 1.233 is at 4.93 standard deviations from the estimator distribution when H0 is true (the mean of the distribution is 0).

Since t-distributions overlap the normal distribution as the degrees of freedom increase, and we know that approximately until 2 standard deviations the 95% of the probability is found in a normal distribution, we can state that the probability of observing our estimate, if H0 was true, is less than 0.05 when the t-statistic exceeds the value of 2. When this occurs, we reject the null hypothesis with a confidence level of 95%.

5.2 Bivariate model: simple linear regression

In this case, the estimated $\hat{\beta}_1$ is more than 4.93 standard deviations from the mean of the distribution under $H_0 = \beta_1 = 0$, so it is unlikely to have observed an effect of 1.23 if H_0 was true. The precise probability can be observed in the fourth column of the `summary()` of the model, which can be requested to R with the following command.

```
coef(summary(model_1))[, "Pr(>|t|)"]
##     (Intercept) education_budget
##        5.5e-145          1.3e-06
```

The probability of observing an estimation of 1.23 if the H0 was true is of 0.00000128. Thus, we can reject H0 even at a 99.9% level of confidence.

Exercise 5B. Using the same data, estimate a model where the independent variable is Foreign Direct Investment (`foreign_inv`) and the dependent variable is Inequality (`gini`) and export it into a .doc file. Is the effect statistically significant?

5.2.2 Graphic representation

As we saw earlier, one of the easiest ways of presenting the relationship between two variables is through graphs. You already learned that `ggplot2` is a convenient tool for generating various types of plots. Let's explore its use for OLS. In the first code, all the observations are plotted according to their independent and dependent variable values (Figure 5.6).

```
ggplot(data = welfare, # the dataset is selected
       aes(x = education_budget, y = gini))+ # indep. and dep. variables
    geom_point() + # the observed values are plotted
    geom_smooth(method = "lm", # the regression line is overlapped
                se = F, # the error area is not plotted at a 95% CI
                color = "black") + # color line
    labs (x = "Education Expenditure", y = "Inequality")
## Warning: Removed 718 rows containing non-finite values (stat_smooth).
## Warning: Removed 718 rows containing missing values (geom_point).
```

Usually, it is also helpful to display a graphic representation of the prediction error of the line. `ggplot2` allows us to edit a shaded area where the predicted values with a certain significance level could have been located. Although the 95% of confidence is the default value, we can also edit that value. The first chunk shows the regression line and its error for a level of statistical significance of 95% (Figure 5.7). Take into account that this line will not represent the coefficient you obtained through `lm` after you include controls to your regression.

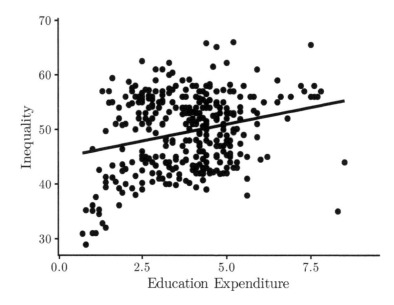

FIGURE 5.6: Linear fit between Education Expenditure and Inequality.

```
ggplot(data = welfare, aes(x = education_budget, y = gini)) +
  geom_point() +
  geom_smooth(method = "lm", color = "black",
              se = T) + # we add the prediction
  labs(x = "Education Expenditure", y = "Inequality")
## Warning: Removed 718 rows containing non-finite values (stat_smooth).
## Warning: Removed 718 rows containing missing values (geom_point).
```

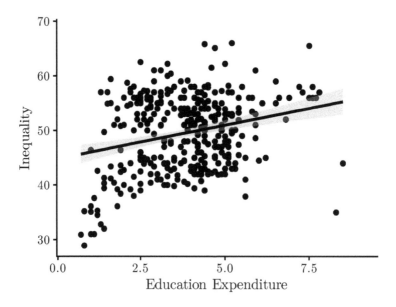

FIGURE 5.7: Linear fit in which we added a 95% confidence interval.

5.3 Multivariate model: multiple regression

Although we tend to be interested in the effect of a single independent variable over a dependent one, it is common to estimate models where Y is both the result of an independent variable of interest and of a set of control variables. Formally,

$$Y = \beta_0 + \beta_1 x_1 + \beta_1 x_2 + ... + \beta_j x_j + u$$

In contrast to a simple regression, now the random variable Y is a function of multiple variables plus an error term u. As in simple regressions, the expectancy of the conditional error in the values of x_j has to be equal to zero. Formally, $E(u|x_1, x_2, ..., x_j) = 0$. To estimate an unbiased multiple linear model, the zero conditional mean assumption is not the only requirement, but all the needed assumptions will be presented in a later section.

For the moment, we will estimate a population model where social inequality (`gini`) is a linear function of the education expenditure as percentage of the GDP (`education_budget`), of the foreign direct investment (`foreign_inv`), of the health expenditure as percentage of the GDP (`health_budget`), of the social security expenditure as percentage of the GDP. (`socialsec_budget`), of the young population (`population`), of the structural dualism of the economy (`sector_dualism`), of the ethnic division (`ethnic_diversity`), of the real GDP per cápita (`gdp`), of the type of regime (`regime_type`), and of the balance between State powers (`legislative_bal`).

As it can be seen, a variety of variables have been added that are suspected to predict inequality (Huber et al., 2006). The multiple regression analysis will allow us to estimate the extent to which our model is correct. First of all, we have to run the OLS estimation. The `lm()` function also estimates multiple models, and the only difference is that the independent variables must be added. Before estimating the model, we will filter the dataset by eliminating all the cases with missing values (NA).[5] Here, for practical purposes, we will only consider those cases (country-year) that are complete in all the variables of our model, dropping missing values[6]:

```
welfare_no_na <- welfare %>%
  drop_na(gini, education_budget, foreign_inv, health_budget,
          socialsec_budget, population, sector_dualism, ethnic_diversity,
          gdp, regime_type, legislative_bal)
```

We are now able to estimate model 2:

```
model_2 <- lm(gini ~ 1 + education_budget + foreign_inv + health_budget +
                socialsec_budget + population + sector_dualism +
```

---

[5] For this, you can check Chapter 11 about imputation of missing values.
[6] We do this to make the example easier, with your data you should not do this as it might create undesired biases.

```
                    ethnic_diversity + gdp + factor(regime_type) +
                    legislative_bal,
              data = welfare_no_na)
```

We have indicated that the the variable `regime_type` is categorical using `as.factor`. By doing this, each regime category is measured by a "dummy" coefficient.

Exercise 5C. Remember: including the 1 is not necessary for estimating the model (we place it there just to remind you that we are also estimating the intercept). Try testing the model without it, and you will see that results do not change.

Like the simple model, we can display and plot the results of the estimation with `summary()` or `screenreg()`.

```
screenreg(model_2)
##
## ==================================
##                       Model 1
## ----------------------------------
## (Intercept)            85.94 ***
##                        (8.73)
## education_budget        1.59 ***
##                        (0.45)
## foreign_inv             0.24
##                        (0.18)
## health_budget          -0.83 **
##                        (0.26)
## socialsec_budget       -0.83 ***
##                        (0.20)
## population             -0.93 ***
##                        (0.17)
## sector_dualism         -0.17 ***
##                        (0.03)
## ethnic_diversity        3.68 ***
##                        (1.04)
## gdp                    -0.00 **
##                        (0.00)
## factor(regime_type)2   -2.29
##                        (4.75)
## factor(regime_type)3   -2.90
##                        (4.70)
## factor(regime_type)4   -5.14
##                        (4.62)
## legislative_bal       -10.40 ***
```

5.3 Multivariate model: multiple regression

```
##                                (2.22)
## -----------------------------------
## R^2                           0.59
## Adj. R^2                      0.56
## Num. obs.                     167
## ==================================
## *** p < 0.001; ** p < 0.01; * p < 0.05
```

These functions also allow us to compare two or more models. When presenting the conclusions of an investigation, it is often recommended to show how the results change (or not) under different specifications. First, we save the models into a list. To the **screenreg** command we add the names of the variables, as we previously learned. In this case, the comparison of the models is as follows:

```
models <- list(model_1, model_2)

screenreg(models,
          custom.model.names = c("Model 1", "Model 2"),
          custom.coef.names = c(
             "Constant", "Education expenditure", "FDI",
             "Health expenditure", "Social sec. expenditure",
             "Young population", "Dualism in economy",
             "Ethnic division", "pc GDP", "Democratic.reg", "Mixed.reg",
             "Authoritarian.reg", "Balance between powers")
          )
##
## =====================================================
##                            Model 1      Model 2
## -----------------------------------------------------
## Constant                   44.81 ***    85.94 ***
##                            (1.02)       (8.73)
## Education expenditure       1.23 ***     1.59 ***
##                            (0.25)       (0.45)
## FDI                                      0.24
##                                         (0.18)
## Health expenditure                      -0.83 **
##                                         (0.26)
## Social sec. expenditure                 -0.83 ***
##                                         (0.20)
## Young population                        -0.93 ***
##                                         (0.17)
## Dualism in economy                      -0.17 ***
##                                         (0.03)
## Ethnic division                          3.68 ***
##                                         (1.04)
## pc GDP                                  -0.00 **
##                                         (0.00)
## Democratic.reg                          -2.29
```

```
##                                                  (4.75)
## Mixed.reg                                        -2.90
##                                                  (4.70)
## Authoritarian.reg                                -5.14
##                                                  (4.62)
## Balance between powers                          -10.40 ***
##                                                  (2.22)
## -----------------------------------------------------
## R^2                               0.06           0.59
## Adj. R^2                          0.06           0.56
## Num. obs.                          356            167
## =====================================================
## *** p < 0.001; ** p < 0.01; * p < 0.05
```

As you can see, the point estimation of the effect of education expenditure changed slightly. While in the simple model the effect is of 1.23, in the multiple model this effect changes to 1.59. In this case, the interpretation is that when education expenditure increases by one unit, inequality increases by an average of 1.59 percentage points, *keeping all other factors constant*. As in model 1, the variable remains significant at 99.9% confidence, so we say that the effect of education expenditure is *robust* at different specifications. When researchers include new controls to the model and the main variable of interest remains significant and with relatively stable magnitudes, we gain evidence in favor of the effect of the latter. In other words, it is increasingly unlikely that the effect observed in the first instance was spurious.

Another contribution of model 2 is the addition of nominal variables. Dichotomous and categorical variables pose a slight interpretation challenge. Observe the ethnic diversity variable, which is dichotomous, i.e. the value 1 implies that more than 20% of the population belongs to an ethnic minority and 0 that there is not such a relevant minority. The coefficient of `ethnic_diversity` is 3.7, significant at 99.9%. How do we interpret this coefficient? Simply, the predicted value of inequality is 3.7 points superior when an ethnic minority exists, for any value of the others $x's$. To interpret these coefficients, we always need to know the base category. Since `ethnic_diversity` is equal to 0 when there are no ethnic minorities, the coefficient is interpreted as the effect of having an ethnic minority.

In the case of variable `sector_dualism`, given that the base category is 0 for "no dualism", the coefficient is interpreted such that having a dual economy *decreases* (negative coefficient) inequality in approximately 0.17 points.

Exercise 5D. Export the table with both models (with and without controls) into a .doc file, we will wait for you.

5.3 Multivariate model: multiple regression

The following graph shows the difference in the effect of education expenditure on inequality according to the presence or absence or ethnic minorities in the country. For didactic purposes, we estimated a new model restricted to two variables of interest: education expenditure and ethnic minorities.

```
model_2_restricted <- lm(gini ~ 1 + education_budget + ethnic_diversity,
                         data = welfare_no_na)
```

```
screenreg(model_2_restricted) #we observe new coefficients
##
## ================================
##                   Model 1
## --------------------------------
## (Intercept)       45.40 ***
##                   (1.83)
## education_budget   0.67
##                   (0.44)
## ethnic_diversity   6.82 ***
##                   (0.99)
## --------------------------------
## R^2                0.24
## Adj. R^2           0.23
## Num. obs.          167
## ================================
## *** p < 0.001; ** p < 0.01; * p < 0.05
```

To make the graph, we need to load the `prediction` package that calculates the predicted values of a model and that will be used to draw the line predicted by the model (Figure 5.8).

```
library(prediction)
```

```
pred_model_2_restricted <- as_tibble(prediction(model_2_restricted))

ggplot(data = pred_model_2_restricted) + # the new predicted values
  geom_point(mapping = aes(x = education_budget, y = gini,
                           color = factor(ethnic_diversity))) +
  # the regression lines are drawn (differentiated by color):
  geom_line(mapping = aes(x = education_budget, y = fitted,
                          color = factor(ethnic_diversity),
                          group = factor(ethnic_diversity))) +
  labs(x = "Education Expenditure", y = "Inequality",
       color = "Ethnic division")
```

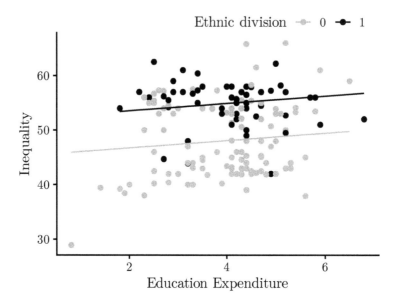

FIGURE 5.8: Predicted values by ethnic division.

As we clearly see in the graph, the effect of education expenditure on inequality is positive because, as expenditure increases, the expected value of the Gini coefficient also increases. However, countries with ethnic minorities (value of 1 in `ethnic_diversity`) have almost 7 more percentage points in their Gini coefficient at any value of education expenditure. It can also be stated that the average effect of having ethnic minorities on inequality is 6.82. Models with more regressors (as the previously estimated model 2) are more complex to graph because the predicted value does not only depend on two variables (as in this graph), but on all variables present in the model. In any case, as we previously saw, the interpretation is the same.

5.4 Model adjustment

Goodness of fit is defined as the explanatory capacity of the model. Intuitively, it refers to the portion of the variation of the dependent variable y is explained by the specified model. The measurement of the goodness of fit is R^2 and is defined as 1-RSS/TSS, where RSS is the Residual Sum of Squares and TSS the Total Sum of Squares. In simple terms, RSS is a measure of everything that is *not* explained by the model, while TSS is the variability of y. A model that explains *all* the variation of y will have a R^2 of 1. A model that explains nothing of the variability of the dependent variable will have a value of 0.

As a general rule, as the number of independent variables increases, the R^2 never decreases, so the adjusted R^2 is often used as a measure that penalizes the inclusion of unsubstantiated variables. As we can observe in the comparison of the previously estimated models, the simple linear model has a R^2 of 0.06. This can be read as model 1 explaining 6% of the variability of inequality. The multiple model 2 increases its explanatory power to 59%.

Some researches try to increase the goodness of fit of the model. Nevertheless, for estimating the effect of a single variable it is not necessary to increase the goodness of fit but to simply meet the assumptions of the Classic Linear Model, such as the zero conditional mean, the linearity in parameters and the rest of assumptions that are described in the corresponding section.

5.5 Inference in multiple linear models

As in the simple linear regression, the estimators of each β_j parameter has a t-Student distribution, so it is possible to do inferences about the point estimations of each $\hat{\beta}_j$ through a t-test. However, we often want to impose multiple linear constrains to the model of the type $H_0 = \beta_1 = \beta_2 = 0$. Here we are stating that the effect of the two variables x_1 and x_2 is equal to zero. A typical case that requires this type of null hypothesis is that of categorical variables that enter the model as dichotomous dummy variables. The dichotomous "high-school education" variable and the dichotomous "superior education" variable are, in reality, categories of the single nominal variable "education level", which can only enter the regression analysis in a *dummy* format.

The test that allows us to make inference for multiple linear constrains is the F test. This means that the H_0 of a multiple constraint distributes as Fisher's F. Let's suppose that we want to test the null hypothesis $H_0 = \beta_1 = \beta_2 = \beta_3 = 0$. According to this hypothesis, variables x_1, x_2 and x_3 do not affect Y when considered together. The alternative hypothesis is that at least one of the betas is different from 0. If H_0 is true, then a model that excludes these variables should explain the same as a model that includes them, that is, these variables are redundant.

The way of testing this hypothesis is through a F-test, in which the sum of the squared residuals of the complete and restricted models are compared. In simple terms, if the variables do not explain the variability of Y, the Residual Sum of Squares of both models (another way is to regard the R^2) should not significantly change. In this case, it does not make any sense to keep the variable in the model.

It uses the fact that the comparison of the squared residuals distributes F

$$F = \frac{(RSS_r - RSS_c)/q}{RSS_c/(n-k-1)}$$

Where RSS_r is the Residual Sum of Squares of the restrained model, RSS_c is the Residual Sum of Squares of the complete model, q is the quantity of excluded variables and $n - k - 1$ are the degrees of freedom of the complete model. In R, we can use the anova function for comparing the models.

For example, let's suppose that a colleague states the legislative balance (`legislative_bal`), the type of regime (`regime_type`) and the ethnic diversity (`ethnic_diversity`) need to be excluded from the model. Then, we need to estimate a restricted model so that

```
model_2_restricted <- lm(gini ~ 1 + education_budget + foreign_inv +
                         health_budget + socialsec_budget +
                         population + sector_dualism + gdp,
                    data = welfare_no_na)
```

As we can see, the listed variables were excluded from the syntax. Now we need to compare the explanatory capacity of each model

```
anova(model_2, model_2_restricted)
## # A tibble: 2 x 6
##     Res.Df    RSS    Df `Sum of Sq`      F  `Pr(>F)`
##      <dbl>  <dbl> <dbl>       <dbl>  <dbl>     <dbl>
## 1      154  3148.    NA          NA     NA        NA
## 2      159  4737.    -5      -1588.   15.5  2.31e-12
```

The significance of the last column of the test (Pr(>F)) clearly shows that the null hypothesis is rejected as the p-value is below the 0.05 threshold, so these variables should *not* be excluded from the model.

Exercise 5E. Estimate a model where you exclude health expenditure (`health_budget`) and social security expenditure (`socialsec_budget`), and compare its explanatory capacity with the complete model. According to the results, should we exclude these variables from the model?

5.6 Testing OLS assumptions

The OLS estimator will be useful (it will unbiasedly estimate the population parameter) if the Gauss-Markov assumptions are met. This allows it to be the Best Linear Unbiased Parameter (BLUE). For more in-depth discussion of the assumptions, see Wooldridge (2016) and Stock and Watson (2019). It is important to evaluate whether these

5.6.1 Zero conditional mean

assumptions are met in our estimation. As we will see, this evaluation is theoretical and, in some cases, could be approached empirically. All these assessments are usually added to the articles as appendices or in the data replication files, and not necessarily in the body of the text.

5.6.1 Zero conditional mean

This is the main assumption for using the Ordinary Least Squares estimator. The central premise is the independence between the independent variables and the error term. This allows us to isolate the effect of the x from the unobservable factors (contained in the error term u). This assumption cannot be evaluated empirically because, by definition, we do not know the factors contained in the error term. Thus, the defense of this assumption will always be *theoretical*.

5.6.2 Random sample

This is an assumption about data generation. A random sample of size n is assumed, which implies that the sample was taken in such a way that all population units had the same probability of being selected. That is, there is no sampling bias.

5.6.3 Linearity in parameters

OLS assumes that the dependent variable (y) has a linear relationship with the independent variable(s) and the error term (u). That is, an increase by one unit of x results in a constant effect on the dependent variable y. Hence, the functional form of the regression equation:

$$Y = \beta_0 + \beta_1 x + u$$

If the relationship is not linear, then we face a problem of model specification. That is, the values predicted by our model do not fit into the reality of our data and, as a consequence, the estimations will be biased. Hence, it is crucial to evaluate if the relationship we want to estimate is linear or if the functional form that characterizes such a relationship is another one (for example, it could be quadratic, cubic, logarithmic, etc).

The good news is that if we have theoretical and empirical reasons to believe that the relationship is not linear, it is possible to make transformations to our variables in order to achieve a better specification of the model. A classic example is that of the parabolic relationship between age and income: as age increases, income increases until a turning point is met where an increase in age is related to lower levels of income, as an inverted U. In this case, it is recommended to make a quadratic transformation to the age variable for achieving a better model specification.

To evaluate linearity, we make a scatter plot of the predicted values against the u residuals (Figure 5.9). The objective is to assess whether the average residuals tends to be randomly located above and below zero. If the residuals present an increasing or decreasing pattern – or any other type – then the functional form of one of the variables in question is non-linear. For this we use `ggplot2`:

```
ggplot(mapping = aes(x = model_1$fitted.values, y = model_1$residuals)) +
  labs(x = "Predicted Values", y = "Residuals") +
  geom_point() +
  geom_hline(mapping = aes(yintercept = 0))
```

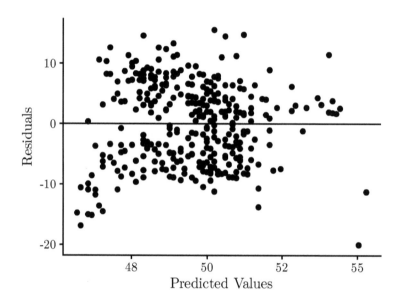

FIGURE 5.9: Linearity test in predicted values.

Also, we can make a partial residual plot where every independent variable of the model is graphed against the residuals (see Figure 5.10). The goal is to obtain a "partial" plot for observing the relationship between the independent variable(s) and the dependent variable by accounting for (by controlling) all the other variables of the model. A dotted line shows us the OLS prediction, and another (light gray) line shows us the "real" relationship. If we observe that one of our variables *does not have a linear relationship* we can make transformations (to the variables!) so that the functional form approximates the empirical. It should be noted that, besides the empirical justification, this linear transformation must *always* be supported by a theoretical argument as to why the relationship between the two variables takes such a form.

A common transformation you will regularly see in papers is the logarithmic transformations of variables. These are present both in dependent and independent variables.

5.6 Testing OLS assumptions

For this reason, we offer you a table that will be helpful (Table 5.1). This will allow you to know how the interpretation of results changes when one of the variables (or both) is transformed.

TABLE 5.1: Interpretation in the presence of logarithmic transformations

Case	Dependent variable (y)	Independent variable (x)	Interpretation of β_1
Level-Level Regression	y	x	$\Delta y = \beta_1 \Delta x$ If we increase x by 1 unit, we expect a change of β_1 units on y
Log-Level Regression	$ln(y)$	x	$\%\Delta y = 100\beta_1 \Delta x$ If we increase x by 1 unit, we expect y to change by $100\beta_1\%$
Level-Log Regression	y	$ln(x)$	$\Delta y = (\beta_1/100)\%\Delta x$ If we increase x by 1%, we expect a change of $(\beta 1 / 100)$ units on y
Log-Log Regression	$ln(y)$	$ln(x)$	$\%\Delta y = \beta_1 \%\Delta x$ If we increase x by 1%, we expect y to change by $\beta_1\%$

Note: Beware of observations with a value of 0 as they will be omitted (the logarithm of zero is not defined), creating bias.

For example, we decided to transform our dependent variable in such a way that:

```
model_1_log <- lm(log(gini) ~ 1 + education_budget, data = welfare)

screenreg(model_1_log)
##
## ==========================
##                   Model 1
## --------------------------
## (Intercept)       3.78 ***
##                  (0.02)
## education_budget  0.03 ***
##                  (0.01)
## --------------------------
## R^2               0.07
## Adj. R^2          0.07
## Num. obs.         356
## ==========================
## *** p < 0.001; ** p < 0.01; * p < 0.05
```

The interpretation would be as follows: if we increase health expenditure by one unit, we would expect that the Gini increases by 3%, *ceteris paribus*. In order to know when to transform our variables, we will see with an example how we can diagnose a measurement problem in our variables (Figure 5.10).

```
library(car)
crPlots(model_1)
```

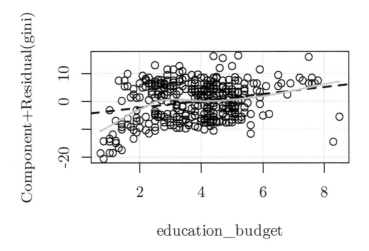

FIGURE 5.10: Linearity test.

The relationship of our variable of interest with the dependent variables seems to be increasing quadratic. Then, it is reasonable to make a quadratic transformation to the variable. Let's evaluate this graphically:

```
welfare_no_na <- welfare_no_na %>%
  mutate(cseduc2 = education_budget * education_budget)

model_1_quadratic <- lm(gini ~ 1 + cseduc2 + education_budget,
                        data = welfare_no_na)
```

```
crPlots(model_1_quadratic)
```

Based on a visual diagnosis (Figure 5.11), we observe an increasing tendency in the residuals as it advances in the predicted values. Also, we detected a non-linear relationship between education expenditure and the levels of inequality. We suspect that this relationship might be quadratic (increasing quadratic parabola) and, according

5.6 Testing OLS assumptions

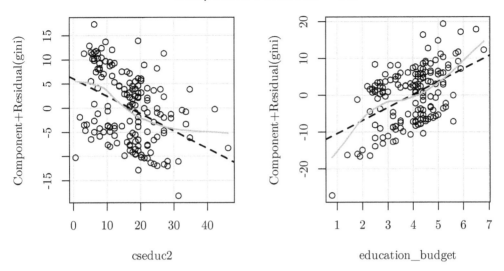

FIGURE 5.11: Alternative linearity test.

to the partial residual plot, it seems that the transformed variable is much closer to the linear relationship estimated by OLS (marked by the dotted line). Note that the scale in the left figure is from 0 to 15, while the one in the right is from 0 to 20, which denotes a steeper slope.

To confirm the visual observations, a statistical test is often used to diagnose a poor functional specification of the model: Ramsey's RESET Test. The idea is, precisely, to assess whether there is an error of specification in the regression equation. What this test does is to estimate again the model but adding the predicted values of the original model with some non-linear transformations of the variables. Then, starting from a F-Test, the model is evaluated with the non-linear specifications to check if it has a better fit than the original model without the non-linear transformation. The null hypothesis states that the new variables (in this case, `csdeuc2`) are not significant for explaining the variation of the dependent variable; that is, that their coefficient is equal to zero ($beta = 0$)

```
library(lmtest)
resettest(model_1, power = 2, type = "fitted", data = welfare_no_na)
##
##   RESET test
##
## data:  model_1
## RESET = 9, df1 = 1, df2 = 353, p-value = 0.004
```

According to the result of the F-Test, we confirm what we graphically observed: adding a quadratic term to education expenditure improves the fit of our estimation.

We reached this conclusion by observing the *p-value* of Ramsey's RESET Test: at a statistical significance level of 5%, the null hypothesis that the incorporation of the quadratic term does not improve the fit of the model is rejected.

- *Note*: This evaluation was made for a simple (bivariate) regression model, but the same procedure can be made for multivariate models.

5.6.4 Variation in independent variables and no perfect collinearity

We stated that there needs to be variation in the independent variable(s). A variable that does not vary, is not a variable! If we do not have variation, the estimation of the coefficients will be indeterminate. Additionally, greater variation in the independent variables will allow us to make more precise estimations. On the other hand, no perfect collinearity implies that the independent variables are not perfectly *linearly* correlated. That is, although independent variables tend to have some type of relationship between each other, we do not want them to measure the same phenomenon! This will be evaluated with multicollinearity tests.

5.6.4.1 Multicollinearity issues

A. *Loss of efficiency*, because their standard errors will be infinite. Even when multicollinearity is less than perfect, the regression coefficients have larger standard errors, which means that they cannot be estimated with great precision.

Let's review the standard error formula of the coefficients

$$\hat{\sigma}_{\hat{\beta}_1} = \frac{\hat{\sigma}}{\sqrt{\sum(X_j - \bar{X})^2(1 - R_j^2)}}$$

- $\hat{\sigma}$ Is the error term variance: $\frac{\sum \hat{u}}{n-k-1}$
- $\sum(X_j - \bar{X})^2$ Is the variability of x_j (SST)
- $1 - R_j^2$ Is the portion of x_j that is not explained by the rest of the x in the model (R_j^2 indicates the variance of x_j which is explained by the rest of the x's of the model). It is because of this term that non-perfect collinearity is so important!

B. *Coefficient estimates can vary widely* depending on what other independent variables are in the model. In an OLS estimation, the idea is that we can change the value of an independent variable and not the others (this is what *ceteris paribus* means, to keep the other co-variables constant). However, when independent variables are correlated, changes in one variable are linked to changes in another variable. The stronger the correlation, the more difficult it is to change one variable without changing another. It becomes difficult for the model to estimate the relationship between each independent variable and the dependent variable while keeping the rest constant,

5.6 Testing OLS assumptions

because independent variables tend to change simultaneously. Let's review the formula for estimating the coefficient in a multiple regression:

$$\hat{\beta}_1 = \frac{\sum(\hat{r}_{i1}\hat{y}_i)}{\sum(\hat{r}_{i1}^2)}$$

where:

- $\hat{r}_{i1}$ are the residuals of a x_1 regression over the rest of the x in the model (that is, the part of x_1 that cannot be explained –or that is not correlated– with the rest of the x)

Thus, $\hat{\beta}_1$ measures the sample relationship between y and x_1 after removing the partial effects of x_2, x_3...x_k. To evaluate multicollinearity, the first step is to observe the correlation matrix of the variables of our model (just as we did in the descriptive statistics analysis stage) (Figure 5.12).

```
library(ggcorrplot)

corr_selected <- welfare %>%
  select(gini, education_budget, sector_dualism, foreign_inv, gdp,
         ethnic_diversity, regime_type, health_budget, socialsec_budget,
         legislative_bal, population) %>%
  # calculate correlation matrix and round to 1 decimal place:
  cor(use = "pairwise") %>%
  round(1)

ggcorrplot(corr_selected, type = "lower", lab = T, show.legend = F)
```

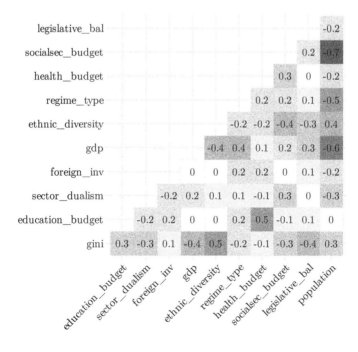

FIGURE 5.12: Correlation matrix, where we will diagnose multicollinearity issues.

We see that some of our variables have strong correlations, such as social security expenditure `socialsec_budget` and the number of young inhabitants in the country `pop014w`, which has a negative correlation of 0.7. In any case, to detect if multicollinearity is a problem, it is necessary to perform a VIF test (*variance inflation factors*), because looking at pairs of correlations does not help us to establish whether more than two variables have a linear correlation. What the VIF test reveals is how much the coefficient errors "grow" when the rest of the variables are present (how much error variance increases).

```
vif(model_2)
##                       GVIF Df GVIF^(1/(2*Df))
## education_budget       1.8  1             1.4
## foreign_inv            1.5  1             1.2
## health_budget          1.8  1             1.3
## socialsec_budget       4.8  1             2.2
## population             5.0  1             2.2
## sector_dualism         1.2  1             1.1
## ethnic_diversity       1.9  1             1.4
## gdp                    2.5  1             1.6
## factor(regime_type)    2.7  3             1.2
## legislative_bal        1.7  1             1.3
```

We then make a query as to whether the square root of VIF for each variable is less than 2 (the squared root because we are interested in the standard error and not in the variance). As a rule of thumb, the score should be less than 2. If it is greater than 2, it means the variance is high. Thus, there are multicollinearity issues.

```
sqrt(vif(model_2)) > 2
##                       GVIF     Df GVIF^(1/(2*Df))
## education_budget     FALSE FALSE           FALSE
## foreign_inv          FALSE FALSE           FALSE
## health_budget        FALSE FALSE           FALSE
## socialsec_budget      TRUE FALSE           FALSE
## population            TRUE FALSE           FALSE
## sector_dualism       FALSE FALSE           FALSE
## ethnic_diversity     FALSE FALSE           FALSE
## gdp                  FALSE FALSE           FALSE
## factor(regime_type)  FALSE FALSE           FALSE
## legislative_bal      FALSE FALSE           FALSE
```

According to the query, it seems that we do not have multicollinearity issues. However, if we have them, should we correct them? In general, the handbooks of econometrics agree on the necessity of reducing multicollinearity depending on its seriousness and on what is the main goal of the regression model. We have to consider the following three points:

1. Seriousness of the issues increases the degree of multicollinearity. Thus, if we have a moderate multicollinearity, it is plausible not to resolve it.

5.6 Testing OLS assumptions

2. Multicollinearity only affects the specific independent variables that are correlated. Thus, if multicollinearity is not present in the independent variables of interest, it is plausible not to resolve it.

3. Multicollinearity only affects the coefficients and the standard errors, but it does not directly influence the predicted values of the model, nor the precision of these predictions and the goodness of fit statistics. If our main goal is to make predictions, and we do not need to understand the role of each independent variable, we do not need to reduce multicollinearity.

5.6.4.2 Solutions to multicollinearity

1. Removing one of the independent variables that is strongly correlated. This constitutes a trade-off, and there needs to be a theoretical justification that explains why some variables were kept and others not, in addition to making evident the high degree of correlation.

2. By combining the variables that are strongly correlated, for example, by making an index (as we show in Chapter 15).

- Until now, we have reviewed four assumptions that enables us to state that our OLS estimators are not biased. That is, it allows us to trust that the expectation of the estimation made through OLS will be equal to the population's average: $E(\hat{\beta}) = \beta$

5.6.5 Homoscedasticity

The fifth assumption is related to efficiency. That is, with the error term variance of our estimation. The error term variance is a constant. That is to say, given any value for the explanatory variables, the error has the same variance. $Var(u \mid x) = \sigma^2$, that is $Var(u) = \sigma^2$

Thus, the variance of the unobservable error, u, conditional on the explanatory variables, is constant. As we previously stated, this assumption **does not affect the bias** of the estimator (that is, that the sample distribution of our estimation $\hat{\beta}_1$ is centered in β_1), but its *efficiency* (how much dispersion there is around the $\hat{\beta}_1$ estimate of the β_1 parameter)

This assumption is crucial for calculating the variance of OLS estimators, and is the one that allows it to be the lowest variance estimator among the unbiased linear estimators. If we evaluate the standard error formula of the coefficients, the need of this assumption becomes evident:

$$\hat{\sigma}_{\hat{\beta}_1} = \frac{\hat{\sigma}}{\sqrt{\sum(X_j - \bar{X})^2(1 - R_j^2)}}$$

- $\hat{\sigma}$ Is the error term variance: $\frac{\sum \hat{u}}{n-k-1}$

In order to implement this formula, we need σ^2 to be constant. When this assumption is not met, that is, the error term does not remain constant for different x values, we face a **heteroscedasticity** scenario. It is quite common to have heteroscedasticity. The good news is that this does not hinder the use of the OLS estimator: there is a solution!

5.6.5.1 Evaluating the assumption

To evaluate this assumption, two steps are usually followed:

1. Visual diagnosis: We want to observe if the residuals (the distance between the points and the regression line) are constant for different values of x. First, we make a simple scatter plot of the independent variables we are interested in and the dependent variable (Figure 5.13):

```
ggplot(welfare_no_na, aes(education_budget, gini)) +
  geom_point() +
  geom_smooth(method = "lm", color = "black")
```

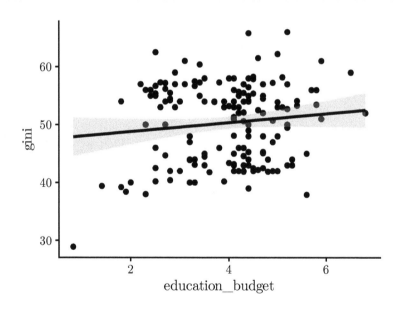

FIGURE 5.13: Visual evaluation of the homoscedasticity assumption.

It seems that at low levels of education expenditure, the variability of inequality levels is significantly higher than at higher levels of education expenditure. We can make a better visual diagnosis if we use the estimated model (and not only the relationship between the two variables) and we graph the residuals (Figure 5.14). First, we do it for the bivariate model, simply using `residualPlot()` from the `car` package:

```
residualPlot(model_1)
```

5.6 Testing OLS assumptions

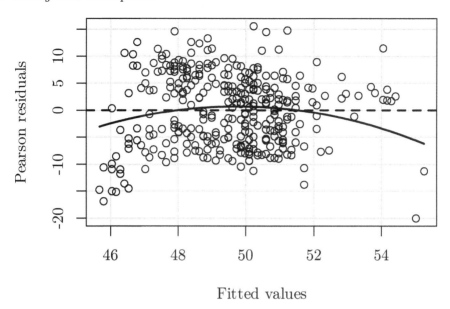

FIGURE 5.14: Evaluation of the homoscedasticity for the bivariate model.

Then, we do it for the multivariate model (Figure 5.15):

```
residualPlot(model_1)
```

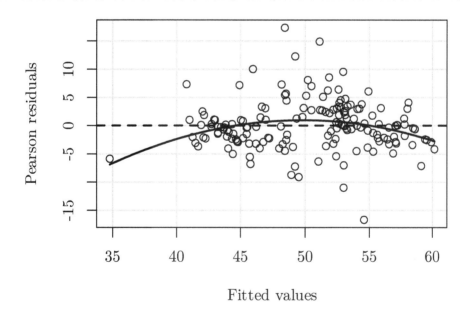

FIGURE 5.15: Evaluation of the homoscedasticity for the multivariate model.

These graphs plot the fitted values against the residuals. Let's remember that under the homoscedasticity assumption, since $Var(u \mid x) = \sigma^2$, then $Var(Y \mid x) = \sigma^2$. In other words, the variance of the residuals of the predicted values based on the x's **should**

be constant. Therefore, if there is absolutely no heteroscedasticity (that is, we are facing a homoscedasticity scenario), we should observe a completely random and equal distribution of points across the X-axis range and a constant solid line. Nonetheless, we clearly observe that the residuals are not constant for the different values of the education expenditure variable. We are facing a case of **heteroscedasticity**.

- We can also evaluate each variable in the model and thus identify in which specific variables heteroscedasticity is present. Again, what we expect is that the solid line matches the dashed line (at zero). We will use the `residualPlots()` fuction in `car` package (Figure 5.16):

```
residualPlots(model_2, layout = c(3, 4), tests = F, fitted = F)
```

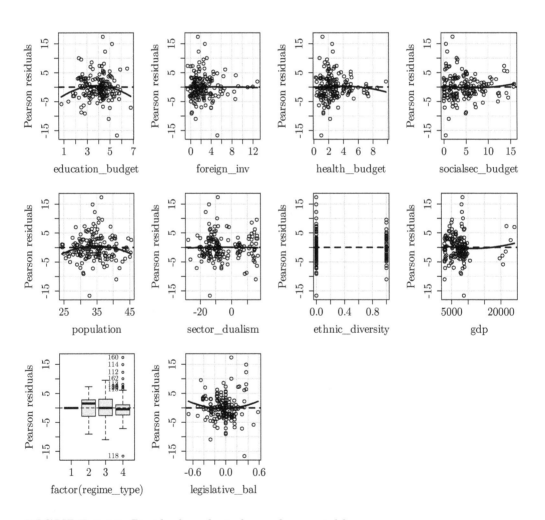

FIGURE 5.16: Residual analysis for each covariable.

5.6.5.2 Statistical diagnosis

In the second step we make a statistical diagnosis. There are different ways of evaluating homoscedasticity, but the *Breusch-Pagan* test is the most frequently used. The logic behind this test is as follows: a regression is made, where the dependent variable consists of the squared residuals as to assess whether the independent variables of the model have any relationship with u or not. What we expect is that the effect is 0, because if the error variance is constant, the error (residuals) should not vary in relation to the values of the $x's$. In short, we do not want to reject the null hypothesis!

```
bptest(model_2, studentize = T)
##
##   studentized Breusch-Pagan test
##
## data:  model_2
## BP = 32, df = 12, p-value = 0.001
```

Since the p-value is less than 0.05, the null hypothesis is rejected. Thus, we are in a scenario of heteroscedasticity.

5.6.5.3 Solutions to heteroscedasticity

Once we identify that we have heteroscedasticity, it is necessary to solve it. The first alternative is to correct the *functional form*. You can be facing a case where the non-constancy of the error term is the result of a non-linear relationship between the variables, a problem we have already learned how to solve, for example, by exponentiating the variables. The second alternative often occurs when the empirical nature of the relationship makes the error not to be constant. We know that we cannot calculate the standard errors of the estimators as we always do in OLS: since the error variance is **not constant**, it is necessary to modify the way in which we calculate the errors.

So, in order to make an inference, we need to fit the estimate of the error in such a way that we can make a valid estimation in the presence of **heteroscedasticity of the unknown form**. This is, although we might not know the type of heteroscedasticity we have, we can improve our precision and make valid statistical inference. The usual formula of the standard error of the estimator is:

$$\hat{\sigma}_{\hat{\beta}_1} = \frac{\sum_{i=1}^{n}(x_i - \overline{x})^2 \hat{\sigma}}{\sqrt{\sum(X_j - \bar{X})^2(1 - R_j^2)}}$$

When we have **homoscedasticity**, the dominator is as follows: $\sum_{i=1}^{n}(x_i - \overline{x})^2 \hat{\sigma} = \hat{\sigma}$. However, under homoscedasticity, since $\hat{\sigma}$ is no longer constant, the equality is no longer present! This is because the value assumed by $\hat{\sigma}$ depends on the different values of x. Also, remember that when estimating a **multiple** regression, it is necessary to subtract the variation of x_1 in the standard error estimation, which is explained by the rest of the x_k of the model.

Thus, in a multiple regression, a valid estimator of $\hat{\sigma}_{\hat{\beta}_1}$ under heteroscedasticity will be:

$$\hat{\sigma}_{\hat{\beta}_1} = \frac{\sum_{i=1}^{n} r_{ij}^2 \hat{u}^2}{\sqrt{\sum(X_j \check{} \bar{X})^2(1-R_j^2)}}$$

where:

r_{ij}^2 Represents the squared residuals of the regression of the rest of the independent variables on the independent variable j. $\sqrt{\sum(X_j \check{} \bar{X})^2(1-R_j^2)}$ Represents the Total Variance of x after subtracting the effect of the rest of the x's.

This way of estimating the standard errors is called **robust standard errors**, also known as **reinforcing** the error, which consists on addressing and allowing heteroscedasticity by making the errors more demanding.

5.6.5.3.1 Robust Standard Errors

While there are several ways to reinforce the errors (we can even do it by hand), R allows us to calculate them easily with the `coeftest` command of the `lmtest` package. Also, the `sandwich` package, with its `vcovHC` function, allows us to incorporate the specifications of the robust variance-covariance matrix.

- HC0 = is White's original (Wooldridge, 2016)
- HC1= Is the one used by the Stata software
- HC3 =Is the most conservative, thus, is highly recommended

```
library(lmtest)
library(sandwich)
model_2_robust_3 <- coeftest(model_2, vcov = vcovHC(model_2, "HC3"))
model_2_robust_1 <- coeftest(model_2, vcov = vcovHC(model_2, "HC1"))
model_2_robust_0 <- coeftest(model_2, vcov = vcovHC(model_2, "HC0"))

models_robust <- list(model_2, model_2_robust_0,
                      model_2_robust_1, model_2_robust_3)

screenreg(models_robust,
          custom.model.names = c("w/o robust SE",
                                 "robust HC0", "robust HC1", "robust HC3"))
##
## ===============================================================
##                    w/o robust SE   robust HC0   robust HC1   robust HC3
## ---------------------------------------------------------------
## (Intercept)           85.94 ***      85.94 ***    85.94 ***    85.94
##                       (8.73)         (8.77)       (9.14)
## education_budget       1.59 ***       1.59 **      1.59 **      1.59
```

5.6 Testing OLS assumptions

```
##                              (0.45)       (0.50)       (0.52)
## foreign_inv                   0.24         0.24         0.24         0.24
##                              (0.18)       (0.14)       (0.14)
## health_budget                -0.83 **    -0.83 ***   -0.83 ***    -0.83
##                              (0.26)       (0.22)       (0.23)
## socialsec_budget             -0.83 ***   -0.83 **    -0.83 **     -0.83
##                              (0.20)       (0.25)       (0.26)
## population                   -0.93 ***   -0.93 ***   -0.93 ***    -0.93
##                              (0.17)       (0.20)       (0.21)
## sector_dualism               -0.17 ***   -0.17 ***   -0.17 ***    -0.17
##                              (0.03)       (0.03)       (0.03)
## ethnic_diversity              3.68 ***    3.68 ***    3.68 ***     3.68
##                              (1.04)       (0.92)       (0.96)
## gdp                          -0.00 **    -0.00 *     -0.00 *      -0.00
##                              (0.00)       (0.00)       (0.00)
## factor(regime_type)2         -2.29        -2.29        -2.29       -2.29
##                              (4.75)       (1.36)       (1.41)
## factor(regime_type)3         -2.90        -2.90 *     -2.90 *     -2.90
##                              (4.70)       (1.16)       (1.20)
## factor(regime_type)4         -5.14        -5.14 ***   -5.14 ***   -5.14
##                              (4.62)       (0.84)       (0.88)
## legislative_bal             -10.40 ***  -10.40 ***  -10.40 ***  -10.40
##                              (2.22)       (2.29)       (2.38)
## ------------------------------------------------------------------------
## R^2                           0.59
## Adj. R^2                      0.56
## Num. obs.                   167
## ========================================================================
## *** p < 0.001; ** p < 0.01; * p < 0.05
```

All the alternatives deliver similar robust errors. The difference is given by the different specifications about the robust variance-covariance matrix (HC).

- Tip: To reproduce the Stata default behavior of using the robust option in a call to regress you need to request vcovHC to use the HC1 robust variance-covariance matrix.

5.6.5.4 A special case of heteroscedasticity: error variance associated to clusters

We know that there are observations that can be related with each other within specific groups (or clusters). For example, countries in Latin America could be related by belonging to similar regions (South America versus Central America or Caribbean, Andean regions versus non-Andean, etc.). Thus, their standard errors could be correlated based on the region they belong to. Then, we have that the error variance conditioned by the region is not constant.

While working with panel data, as is our case, this is much clearer. By having education expenditure by country for several years, *autocorrelation of the error* exists between observations of the same country. That is, errors are correlated between the observations of the same country for each year (unobserved factors that are correlated both with X and Y may be constant over time). Then, when our observations belong to clusters, the correction will be to **cluster the standard errors**.

What we are doing when we cluster the standard errors is to allow the existence of an error correlation within the cluster (the assumption of homoscedasticity is loosened). Thus, we allow the error variance not to be constant, but to be different according to the clusters. The selection of the relevant clusters will be theoretically defined. In this case, it makes sense to think that clusters are the countries. Let's remember that our goal is to estimate the effect of education expenditure in the Gini of Latin American countries. Let's begin by plotting a simple scatterplot (Figure 5.17):

```
ggplot(welfare_no_na, aes(education_budget, gini)) +
  geom_point()
```

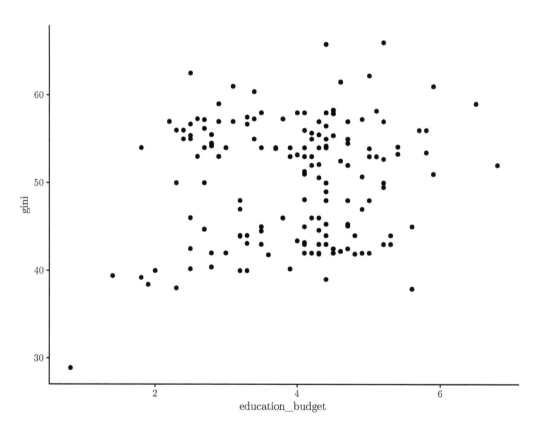

FIGURE 5.17: Relationship between education and Gini.

5.6 Testing OLS assumptions

Let's split this plot with country facets to see if, at first glance, clusters exist (Figure 5.18):

```
library(ggplot2)

ggplot(welfare_no_na, aes(education_budget, gini)) +
  geom_point() +
  facet_wrap(~country_id)
```

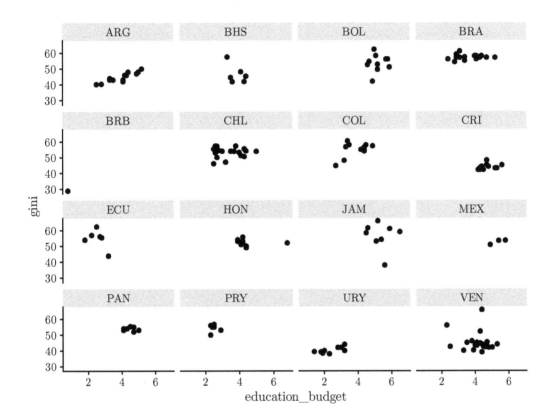

FIGURE 5.18: Relationship between education and Gini by country.

Visual inspection suggests that there is some clustering by country. That is, education expenditure by country usually stays within a range that slightly varies.

For doing the OLS estimation with clustered errors, we will use the `lm.cluster` command from the `miceadds` package. This command clusters the standard errors according to the indicated cluster variable. In short, what we are doing is to allow the presence of an error correlation within the clusters, in this case, countries (loosening the homoscedasticity assumption).

Robust standard errors by clusters can increase or decrease the standard errors. That is, the clustered standard errors can be larger or smaller than conventional standard errors. The direction in which the standard errors change depends on the sign of the intra-group error correlation. For this step, it is necessary to install the `miceadds` and the `multiwayvcoc` packages. With the "cluster" option, we indicate which variable will group the errors:

```
library(miceadds)
model_2_cluster <- miceadds::lm.cluster(
  data = welfare_no_na,
  formula = gini ~ 1 + education_budget + sector_dualism + foreign_inv +
    gdp + ethnic_diversity + regime_type + health_budget +
    socialsec_budget + legislative_bal,
  cluster = "country_id"
)

summary(model_2_cluster)
## R^2= 0.51
##
##
##                    Estimate Std. Error t value Pr(>|t|)
## (Intercept)         4.9e+01    3.56936 13.7775  3.5e-43
## education_budget    1.1e+00    0.62151  1.8321  6.7e-02
## sector_dualism     -1.5e-01    0.05207 -2.8472  4.4e-03
## foreign_inv         6.1e-01    0.11008  5.5403  3.0e-08
## gdp                 6.8e-05    0.00015  0.4576  6.5e-01
## ethnic_diversity    3.9e+00    1.69727  2.3025  2.1e-02
## regime_type        -1.6e+00    0.79724 -1.9525  5.1e-02
## health_budget      -9.5e-01    0.35512 -2.6694  7.6e-03
## socialsec_budget    1.6e-03    0.18080  0.0088  9.9e-01
## legislative_bal    -9.7e+00    4.38621 -2.2099  2.7e-02
```

When using clusters, the coefficient of our independent variable `education_budget` decreased from 1,56 to 1,19, but it maintained a high statistical significance (t-value of >2).

5.6.6 Normality in the error distribution

So far, we have reviewed and empirically evaluated – when possible – the five assumptions of the Gauss-Markov theorem that ensure that the OLS estimator is the Best Linear Unbiased Parameter (BLUE). However, these are not sufficient for making statistical inferences. For this, we need an additional assumption:

As we have previously learned, to test a hypothesis of individual significance of a coefficient estimated by OLS, we use the t statistics that allows us to contrast the empirical t value against a theoretical t value (called "critical value") given a certain level of significance (α). An alpha of 5% is commonly used (this is why we speak of statistical significance at a 95% level of confidence). However, in order to perform this hypothesis test, and thus make statistical inference, it is necessary to assume that the coefficient (β) follows a T-Student distribution. Only then we can perform the hypothesis test using the t statistic.

5.6 Testing OLS assumptions

The assumption that allows this is that of *normality in the error distribution*. Since the OLS estimator (β) is a linear combination of the errors ($Y = \beta_0 + \beta_1 x + u$), by assuming a normal distribution of the error (u), we can assume a normal distribution of the OLS estimator. Since the error and its variance are unknown, they are estimated using the regression residuals ($\hat{u}$), thus obtaining the standard error of the estimate. As estimations imply a loss in degrees of freedom (for each estimated parameter we lose one degree of freedom: n-k-1, n=sample size, k=number of estimated parameters - variables of the model-, 1=the intercept estimate, β_0), the distribution of the error and of the coefficients will no longer be normal, but T-Student ($\hat{\beta} t_{n-k-1}$). The following two commands allows us to check that the residuals of the estimated model through OLS follow a T-Student distribution (approximately normal).

- The `qqPlot` command comes by default in R, and it generates a normal probability plot that displays the distribution of data against a theoretical expected normal distribution. Therefore, what we need to examine in the graph is that the observations (which are the residuals) should remain in between the dotted lines (which delimit the normal distribution):

```
qqPlot(model_2$residuals, col.lines = "black")
## [1] 160 118
```

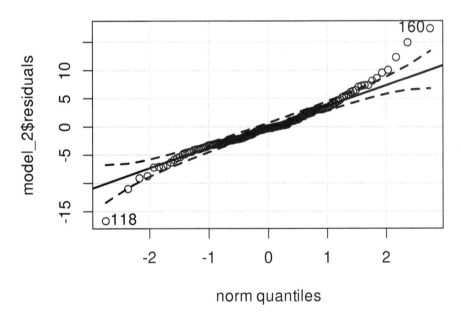

FIGURE 5.19: Normallity of the residuals. Note that the countries 160 and 118 are extreme values.

- The `ggdensity` command of the `ggpubr` package allows us to construct density plots. Thus, we can plot the residuals to visually evaluate if they follow an approximately normal distribution (Figure 5.20).

```
library(ggpubr)
ggdensity(model_2$residuals, main = "Density plot of the residuals")
```

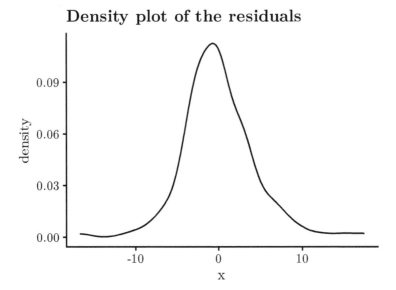

FIGURE 5.20: Normallity of the residuals test.

After evaluating the assumptions and finding the solutions (when necessary), we can have greater certainty in our estimation, and, as a result, in the relationship found between the variables. Nonetheless, a complete explanation of our discovery needs to go further into why and how the two variables relate to each other. Everything that we have learned will be helpful for the selection of study case selection Chapter 6. In Chapter 8, we will estimate models with binary dependent variables by Maximum Likelihood Estimation (MLE).

Exercise 5F.

- Before moving to the next chapter, make a scatter plot of the relationship between variable `gini` and variable `foreign_inv`. Add the code of the country to each observation.
- To model 1, add the `foreign_inv` variable as a control and interpret its coefficient.
- Make the corresponding tests to check that OLS assumptions are not violated.
- Using `htmlreg`, export the regression table to Word.

6

Case Selection Based on Regressions

Inés Fynn and Lihuen Nocetto[1]

Suggested readings

- Gerring, J. (2006). *Case Study Research: Principles and Practices.* Cambridge University Press, Cambridge.

- Lieberman, E. S. (2005). Nested Analysis as a Mixed-Method Strategy for Comparative Research. *American Political Science Review,* 99(3):435–452.

- Seawright, J. (2016). *Multi-Method Social Science: Combining Qualitative and Quantitative Tools.* Cambridge University Press, Cambridge.

Packages you need to install

- `tidyverse` (Wickham, 2019), `politicalds` (Urdinez and Cruz, 2020), `broom` (Robinson and Hayes, 2020).

This chapter will give you tools to, starting from what we have learned in the chapter of linear models, use regressions for selecting case studies. These techniques will be useful when you do mixed methods research. When we work with observational data (as opposed to experimental) linear regressions cannot, by themselves, answer causal inference questions.[2] That is, although they can unveil the existence of a relation between our independent and dependent variables, our investigation would be incomplete if we are not able to demonstrate, through other methods, how these variables are *causally* connected.

The method we use in our research should be guided by the question we want to answer. For example, if our interest is in understating what is the effect of a particular variable on inequality in Latin America and the Caribbean, we would use a *large-n* statistical analysis that would allow us to analyze as many countries as possible. In this way, in Chapter 5 we found that, on average, educational expenditure has a positive effect on inequality levels. Nonetheless, the discovery that greater expenditure generates greater levels of inequalities results somewhat intriguing and counterintuitive. A finding like this could have important repercussions in the elaboration of public policies and consequences for people's everyday lives.

---

[1] Institute of Political Science, Pontificia Universidad Católica de Chile. E-mails: ifynn@uc.cl and lnocetto@uc.cl.

[2] You might with good instrumental variables, but we do not cover these in this book.

Therefore, to move forward in our investigation it is desirable to answer, for example, *Why* does education positively affect inequality levels? That is, what is the *causal mechanism* that explains that in Latin America and the Caribbean a greater expenditure in education generates greater levels of inequality. To answer these types of questions, we usually resort to qualitative methods (such as in-depth case studies or process-tracing analysis[3]) that allow us to comprehend what are the types of processes that explain *why* and *how* the causal relation occurs. In this way, what we try to do is to *integrate* (Seawright, 2016) two research methods, in which one method poses the research question (derived from the statistical analysis), and the other seeks to answer it (through a case study). Another alternative to strengthen our investigation could be through *triangulation*, i.e., to approach the same research question but using different methods that, when taken as a whole, can generate a more complex and complete answer of our research question.

Despite your choice (integration or triangulation), the objective of a mixed research method design is to combine different methods to reach a more complex explanation of the phenomena we are studying. The objective of mixed methods is precisely to approach the same phenomena through different methodologies that allow capturing different angles or dimensions of one research problem. Although there are infinite ways of combining methods, some methods are more compatible with each other than others and, as a matter of fact, some combinations can cause more confusion than clarity (Lieberman, 2005).

In this section, we will learn a combination of methods that Lieberman (2005) has called *nested analysis*, which is the combination of statistical analysis of a large sample with in-depth study of one or more cases contained in that sample. In short, what we will do is to select cases (in this case, countries) from the estimation of our model.

After estimating the model, the first step for selecting cases is to estimate the residuals and the predicted values of the model for each of our observations. This is because, in order to select our study cases, we will compare what our model predicted against the observed values of each of these cases.

To obtain the residuals and the predicted values in R we will use the `augment` command from the `broom` package. This command creates a new dataset over the model that adds variables to the original dataset (for each observation): predicted values, standard errors, the residuals and standardized residuals, among other statistics. We will use model 2 we estimated in Chapter 5 with data from Huber et al. (2006) as an example. Residuals and predicted values:

```
library(tidyverse)
library(broom)
```

---

[3]A great example of process tracing applied to Latin American political science is the book by Pérez Bentancur, Piñeiro and Rosenblatt in the case of party activism in Uruguay (https://www.cambridge.org/core/books/how-party-activism-survives/93C5584DB63DF0A80B51F3EEB68BC8E9)

```
library(politicalds)
data("welfare")


welfare_no_na <- welfare %>%
  drop_na(gini, education_budget, foreign_inv, health_budget,
          socialsec_budget, population, sector_dualism,
          ethnic_diversity, gdp, regime_type, legislative_bal, repression)

model_2 <- lm(gini ~ 1 + education_budget + foreign_inv + health_budget +
                socialsec_budget + population + sector_dualism +
                ethnic_diversity + gdp + regime_type + legislative_bal +
                repression,
              data = welfare_no_na)

model_aug <- broom::augment(model_2, data = welfare_no_na)
model_aug
## # A tibble: 167 x 21
##    country_id  year population  gini sector_dualism   gdp foreign_inv
##    <chr>      <dbl>      <dbl> <dbl>          <dbl> <dbl>       <dbl>
## 1  ARG         1982       30.8  40.2           9.50 7711.       0.269
## 2  ARG         1983       30.9  40.4           8.36 7907.       0.178
## 3  ARG         1990       30.7  43.1           7.72 6823.       1.30
## # ... with 164 more rows, and 14 more variables
```

6.1 Which case study should I select for qualitative research?

Selected cases for an in-depth study should be chosen considering how these cases behave regarding the main variables of our research. Your best-case study depends on the objective for which the case was selected. In this way, case selection must be intentional and not random (Gerring, 2006). The following are different objectives for which cases are selected and implemented in R, based on our statistical model on the determinants of inequality in Latin America and the Caribbean.

6.1.1 Typical cases

One of the objectives of case selection is to examine in greater detail the mechanisms that connect the independent variable with the dependent variable. If this is our objective, then we want to select cases that are *typical* examples of the relations we found in the statistical analysis. Therefore, what we are looking for is a case with the lowest residual possible. That is, the case that our model predicted the best. These are also called *on the line cases* (cases that are over the regression line).

To identify this case we will plot, using the dataset created with the `augment()` function, the residuals. We will transform them into absolute value because, by default, there are always negative residuals (see Figure 6.1). Also, to identify cases, we will ask `ggplot()` to add the labels of the four (`top_n(-4. -resid_abs)`) countries (`mapping = aes(label = country_id)`) with the lowest residuals. We incorporate the horizontal line (`geom_hline(aes(yintercept = 0))`) to the graph to visualize where the residual are almost null (there you will find the cases that the model predicted perfectly: i.e. the *typical* case).

```
ggplot(data = model_aug, mapping = aes(x = .fitted, y = .resid)) +
  geom_point() +
  geom_hline(aes(yintercept = 0)) +
  geom_text(data = model_aug %>%
              mutate(.resid_abs = abs(.resid)) %>%
              top_n(-4, .resid_abs),
            mapping = aes(label = country_id))
```

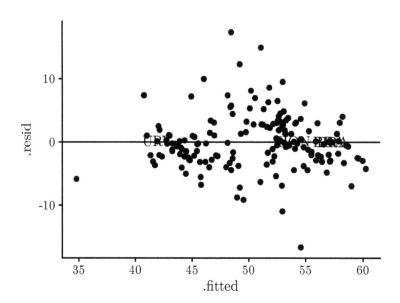

FIGURE 6.1: Top four typical cases.

According to the graph, Brazil (twice), Honduras and Uruguay are three typical cases of the estimated model on the determinants of inequality in Latin America and the Caribbean. That is, these are cases that explain their levels of inequality based on the variables of Model 2, whether inequality is high (Brazil, Honduras) or low (Uruguay).

6.1.2 Outliers

Outliers are cases that, given our model, present a non-expected behavior; they are outliers because they cannot be well explained through our model. In short, they

6.1 Which case study should I select for qualitative research?

are "theoretical anomalies" (Gerring, 2006, p. 106). In general, we select these types of cases to explore new hypotheses, which can eventually shed light into omitted variables in the statistical model (Figure 6.2). The selection of outliers works in the opposite way as typical cases selection: instead of selecting those with the lowest residual, we want to select the cases whose predicted value differs the most from the actual value (i.e. have the highest residuals).

```
ggplot(data = model_aug,
       mapping = aes(x = .fitted, y = .resid)) +
  geom_point() +
  geom_hline(aes(yintercept = 0)) +
  geom_text(data = model_aug %>%
              mutate(.resid_abs = abs(.resid)) %>%
              top_n(4, .resid_abs),
            mapping = aes(label = country_id))
```

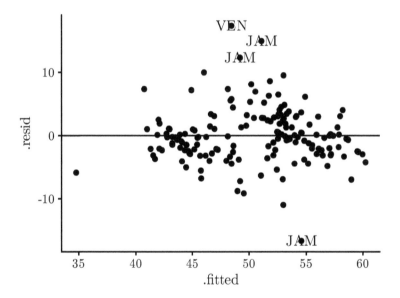

FIGURE 6.2: Top four outlier cases.

Jamaica (in several years) appears as a country badly explained by our model. It is a country that presents relatively low levels of inequality, yet the independent variables of the model do not account for these values. There is a particularly interesting year, 1993, in which its Gini score is 35.7, making it one of the most equitable of the sample. Surrounded by some of the most unequal countries in the world[4] we would need to add a new variable into our model in order to explain the Jamaican case.

---

[4]See https://www.businessinsider.com/most-unequal-countries-in-the-world-2011-10#35-philippines-gini-458-5

6.1.3 Influential Cases

Influential cases are those that present extreme values in either of the independent variables and have strong leverage in the regression. That is, these are cases that strongly influence the slope of regression we observe (remember that the slope is given by the regression coefficient β_i). These are cases that, as outliers, are also "unusual". While the selection of an outlier is used for exploring alternative hypothesis, the selection of an influential case helps to confirm our baseline hypothesis (Gerring, 2006). To identify influential cases, we can take two paths:

1) On the one hand, we can use the **dfbetas**, which are statistics that indicate how much the regression coefficient β_i changes in standard deviation units if the *i-th* observation was to be deleted. Therefore, we will have a *dfbeta* for each observation that indicates how much the β_i of the variable `education_budget` (education expenditure) would change if this observation was not included in the sample. Thus, the more the slope (β_i) varies with the absence of the observation, the more influential that observation will be.

Then, we want to select the cases that produce the greater changes in the standard deviation of β_i when they are eliminated from the sample. As we mentioned, influential cases are used to confirm our theories. At the same time, if the elimination of influential observations from the sample nullifies the relationships we found (if by removing the case β_i ceases to be significant), these cases are also useful for exploring new hypotheses or identifying variables that were omitted in the model.

```
model_aug %>%
  mutate(dfb_cseduc = as.tibble(dfbetas(model_2))$education_budget) %>%
  arrange(-dfb_cseduc) %>%
  slice(1:3) %>%
  dplyr::select(country_id, dfb_cseduc)
## # A tibble: 3 x 2
##   country_id dfb_cseduc
##   <chr>           <dbl>
## 1 BRB             0.483
## 2 JAM             0.298
## 3 VEN             0.241
```

2. Using **Cook's distance**, which is based on a similar logic as *dfbetas*. Cook's distance considers the values that every observation assumes for the independent and dependent variables to calculate how much the coefficients vary when each observation is absent from the sample. In short, this distance tells us how much each observation influences the regression as a whole: the greater the Cook's distance, the greater the contribution of the observation to the model's inferences. Cases with greater Cook's distance are central for maintaining analytical conclusions (especially with relatively small samples;

6.1 Which case study should I select for qualitative research?

with large samples it is less likely that cases with such an influence exist). Therefore, using Cook´s distance to select cases for an in-depth study can be relevant: if in the qualitative study of an influential case we cannot confirm our theory, it is unlikely that we can confirm it in other cases (Figure 6.3).

```
ggplot(data = model_aug,
       mapping = aes(x = .fitted, y = .cooksd)) +
  geom_point() +
  geom_text(data = model_aug %>% top_n(3, .cooksd),
            mapping = aes(label = country_id)) +
  labs(title = "Influential cases")
```

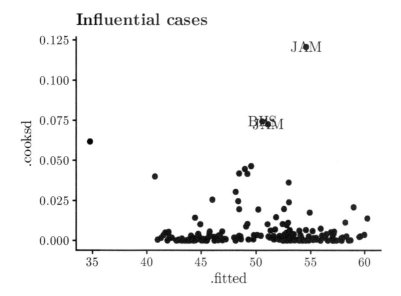

FIGURE 6.3: Top three influential cases.

Again, Jamaica stands out as a country to be studied in depth.[5]

6.1.4 Extreme cases

The selection of extreme cases involves identifying observations that are located far away from the mean of the distribution of the independent or dependent variable. The usefulness of these observations lies in the "oddity" of its values. This technique is recommended when there is not a strong theoretical assumption and, therefore, research is centered in theoretical building. An in-depth study of extreme cases is

---

[5]If you are already wondering what occurred in Jamaica, we recommend you this article: Handa, S., & King, D. (1997). Structural adjustment policies, income distribution and poverty: a review of the Jamaican experience. *World Development, 25*(6), 915-930.

rather exploratory: it is a way to evaluate and search for possible causes of y or possible effects of x. It is important to note that an extreme case may coincide with both a typical case and an outlier (Gerring, 2006).

A classic work of selection of extreme cases in the dependent variable is that of Theda Skocpol (1979) about social revolutions, where theory is developed based on three cases that present the most extreme value for a revolution (in fact, they are the only ones that present such value according to Skocpol).

6.1.4.1 Extreme cases in the independent variable: x

Let's observe how the independent variable behaves in Figure 6.4:

```
ggplot(welfare_no_na, aes(x = education_budget)) +
  geom_histogram(binwidth = 1) +
  labs(caption = "Source: Huber et al (2006)",
       x = "Education Expenditures (% of the GDP spent on Education)",
       y = "Frequence")
```

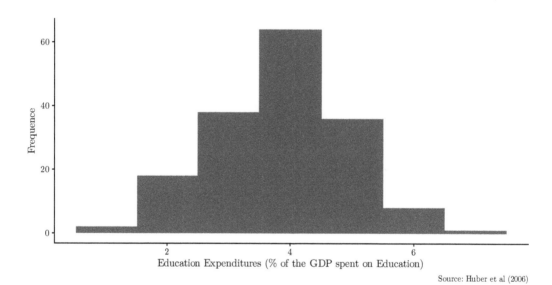

FIGURE 6.4: Histogram of the independent variable: education expenditures.

For selecting extreme cases on the independent variable, based on the estimated statistical model, we just need to calculate the difference -in absolute values- between the value of each observation and the sample mean of education expenditure. Then, the three cases that present the greatest difference from the sample mean are selected.

These steps are implemented in R in the following way:

```
mean(model_aug$education_budget, na.rm = T)
## [1] 4
```

6.1 Which case study should I select for qualitative research?

```
model_aug %>%
  mutate(dif_cseduc = abs(education_budget - mean(education_budget,
                                                  na.rm = T))) %>%
  top_n(3, dif_cseduc) %>%
  arrange(-dif_cseduc) %>%
  dplyr::select(country_id, year, education_budget, dif_cseduc)
## # A tibble: 3 x 4
##   country_id  year education_budget dif_cseduc
##   <chr>      <dbl>            <dbl>      <dbl>
## 1 BRB         1981              0.8       3.16
## 2 HON         2001              6.8       2.84
## 3 URY         1984              1.4       2.56
```

We graph in Figure 6.5 the results for a better visualization:

```
model_aug <- model_aug %>%
  mutate(dif_cseduc = education_budget - mean(education_budget, na.rm = T))

ggplot(data = model_aug,
       mapping = aes(x = .fitted, y = dif_cseduc)) +
  geom_point() +
  geom_text(data = model_aug %>% top_n(3, dif_cseduc),
            mapping = aes(label = country_id))
```

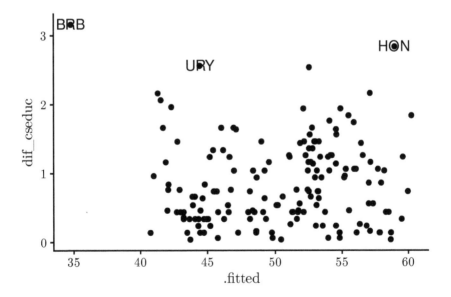

FIGURE 6.5: Top three extreme cases in education expenditures.

Barbados stands out for being an extreme case, since it is well below the sample mean. Honduras, on the contrary, is well above. It would be interesting to compare both. Considering that the third country is Uruguay, and that the three countries are relatively small economies, a question emerges that will surely make us improve the model: should we not control by the size of the economy, measured by its population? This doubt could lead us into a new model, where the statistical significance could change.

Exercise 6A. Select the extreme cases for the independent variable Foreign Direct Investment `foreign_inv`.

6.1.4.2 Extreme cases in the dependent variable y

The selection of extreme cases in the dependent variable is done in the same way as with extreme cases in x. The only difference is that now we calculate the difference -in absolute values- between the observed value of each observation and the sample mean in the dependent variable (Gini Index, in this example). Then, the three cases that present the greatest difference between the sample mean and its value of the dependent variable are selected.

```
mean(model_aug$gini, na.rm = T)
## [1] 50

model_aug %>%
  mutate(dif_gini = abs(gini - mean(gini, na.rm = T))) %>%
  top_n(3, dif_gini) %>%
  arrange(-dif_gini) %>%
  dplyr::select(country_id, gini, dif_gini)
## # A tibble: 3 x 3
##   country_id  gini  dif_gini
##   <chr>       <dbl> <dbl>
## 1 BRB         28.9  21.4
## 2 JAM         66    15.7
## 3 VEN         65.8  15.5
```

We can also graph it for a better visualization, as shown in Figure 6.6:

```
model_aug <- model_aug %>%
  mutate(dif_gini = abs(gini - mean(gini, na.rm = T)))

ggplot(data = model_aug,
       mapping = aes(x = .fitted, y = dif_gini)) +
  geom_point() +
  geom_text(data = model_aug %>% top_n(2, dif_gini),
            mapping = aes(label = country_id))
```

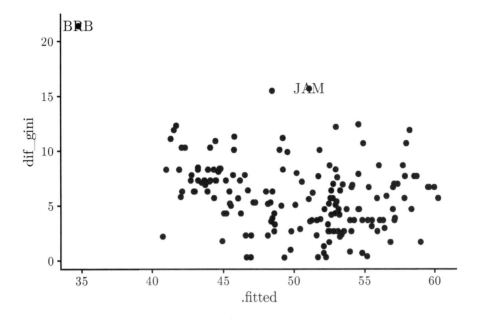

FIGURE 6.6: Top two extreme cases by Gini Index.

Again, Barbados and Jamaica appear as atypical cases in the dependent variable. Both have in common that they were Caribbean colonies of the British Empire. Perhaps, we could include this control for all the countries with this legacy and see how it adjusts to the new model? If great changes appear in the predicted values, we could explore with qualitative evidence the role that colonial institutions of the British Empire had over the inequality of these colonies.

6.1.5 Most similar cases

The selection of similar cases involves identifying two cases that are similar. If we are in the exploratory stage of our research project and we do not have strong theoretical assumptions (we have not yet identified a particular key independent variable), we look for a the pair of observations that are most similar in the independent variables but differ in the dependent variable. In this way, the objective will be to identify one or more factors that differ between the cases and that may explain the divergence of the result. This strategy is that of the direct method of agreement of Stuart Mill.

However, when we already have a theory about how an independent variable affects the dependent variable, the case selection consists in identifying two cases that are similar in all covariates but differ in the independent variable of interest. Here, the focus will be in confirming the argument and further exploring the causal mechanisms that connect the independent variable with the dependent one.

To select similar cases, it is recommended to use a *matching* technique (Gerring, 2006, p. 134). In simple words, this technique involves matching pairs of observations that are as similar as possible in all covariates yet differ in the independent variable of interest. To simplify the analysis, this independent variables tends to be dichotomous, emulating an experimental scenario where there is a treatment (1) and a placebo or control (0). In this way, the objective is to match pairs in which an observation belongs to the treatment group and the other one to the control group.

Since finding pairs that match in all covariates can be quite hard to achieve, a procedure of matching based on the *propensity scores* is often used. This procedure involves finding pairs of observations that have similar estimated probabilities of being in the treatment group (having a value of 1 in the independent variable of interest), conditioned by the control variables. To implement this case selection method in our investigation, we will create a treatment dummy variable (for the education expenditure variable), where 0 is an expenditure less or equal to the sample mean and 1 if the expenditure greater than the mean.

```
welfare_no_na <- welfare_no_na %>%
  mutate(treatment = if_else(education_budget > mean(education_budget),
                             1, 0))
```

Now that we have a treatment variable, we can estimate the propensity scores. That is, the probability of being in the treatment group (education expenditure greater than the sample mean) conditioned by the control variables of the model. Since our dependent variable is dichotomous, we estimate a Logit model (see Chapter 8).

```
m_propensityscore <- glm(treatment ~ sector_dualism + foreign_inv + gdp +
                           population + ethnic_diversity + regime_type +
                           regime_type * socialsec_budget + health_budget +
                           socialsec_budget + legislative_bal + repression,
                         data      = welfare_no_na,
                         family    = binomial(link = logit),
                         na.action = na.exclude)
```

As we did with the general model of the determinants of inequality, we will create a dataset with the `augment` command to save some statistics that will be useful for selecting the cases.

```
propensity_scores<- augment(m_propensityscore, data = welfare_no_na,
                            type.predict = "response") %>%
  dplyr::select(propensity_scores = .fitted, country_id, treatment, year,
                gini)
```

Now, we identify the cases with the lowest propensity scores for both the treatment group (high education expenditure) and the control group (low education expenditure) to decide the selection of cases. It should be noted that this can also be done for high propensity scores: the important thing is that they must have similar or "close" scores

6.1 Which case study should I select for qualitative research?

(equal probability of receiving treatment). Let's look at cases with low propensity scores, in the group of countries with expenditure higher than the sample mean:

```
propensity_scores %>%
  filter(treatment == 1) %>%
  arrange(propensity_scores) %>%
  dplyr::select(country_id, year, propensity_scores) %>%
  slice(1:2)
## # A tibble: 2 x 3
##   country_id  year propensity_scores
##   <chr>      <dbl>             <dbl>
## 1 BRA         1984            0.0815
## 2 MEX         2000            0.159
```

On the other hand, let's observe at the cases with a low propensity score among those countries with education expenditure below the sample average:

```
propensity_scores %>%
  filter(treatment== 0) %>%
  arrange(propensity_scores) %>%
  dplyr::select(country_id, year, propensity_scores) %>%
  slice(1:2)
## # A tibble: 2 x 3
##   country_id  year propensity_scores
##   <chr>      <dbl>             <dbl>
## 1 PRY         1994           0.00309
## 2 ARG         1982           0.00673
```

According to the results, both Brazil and Mexico could be selected to be compared with Paraguay or Argentina for carrying out in-depth most similar case studies. Taking into account its geographical proximity, we could compare Brazil to Argentina, and try to explain in which way education expenditure has impacted on income equity in both countries.

Exercise 6B. Select pairs of most similar cases taking Foreign Direct Investment `foreign_inv` as an independent variable (treatment).

6.1.6 Most different cases

The procedure for selecting most different cases involves an opposite logic to that of the most similar cases. Here we are looking for observations that are different in the control variables, but similar in the value assumed by the independent variable and the dependent variable. In short, what we are looking for different propensity scores but similarities in the independent and dependent variable.

It is worth noting that this type of case selection is useful when a "single causation" is assumed (Gerring, 2006, p. 143). That is, when the variation in the dependent variable is caused by a single variable (or when we are interested in explaining the effect of a single factor). If the interest is in inquiring about the combination of different causal factors, this procedure of case selection is not the most suitable. For selecting "most different" cases, we will also use the propensity scores, but now we are interested in selecting pairs with equal results in the dependent variable, as well as in the independent variable, and with very different propensity scores.

Let's see, then, which are these cases. First, we create a dummy variable for a Gini above or below the mean. Then, we identify the treated cases with low propensity scores (low probability of having an expenditure above the mean) having Gini values greater than the sample mean and values of expenditure on education also greater than the sample mean:

```
propensity_scores <- propensity_scores %>%
  mutate(gini = if_else(gini > mean(gini, na.rm = T), 1, 0))

propensity_scores %>%
  filter(gini == 1 & treatment==0) %>%
  arrange(propensity_scores) %>%
  dplyr::select(country_id, year, propensity_scores) %>%
  slice(1:2)
## # A tibble: 2 x 3
##   country_id  year propensity_scores
##   <chr>       <dbl>            <dbl>
## 1 PRY         1999          0.00953
## 2 PRY         1997          0.0221
```

Following that, we repeat the same process for the highest propensity scores (that is, where the probability of receiving treatment – having an expenditure on education greater than the mean- is very high). In other words, we identify the cases with the highest propensity score for Gini values greater than the sample mean and expenditure on education greater than the sample mean:

```
propensity_scores %>%
  filter(gini == 1 & treatment==0) %>%
  arrange(-propensity_scores) %>%
  dplyr::select(country_id, year, propensity_scores) %>%
  slice(1:2)
## # A tibble: 2 x 3
##   country_id  year propensity_scores
##   <chr>       <dbl>            <dbl>
```

```
## 1 HON        1994          0.983
## 2 HON        1996          0.969
```

Our results suggest that Paraguay could be selected to be compared with Honduras for carrying out an in-depth "most different" case study. Both have low education expenditure as percentage of their GDP, and both are highly unequal, yet they are very different in the controlling variables.

6.2 The importance of combining methods

To conclude, we consider important to insist on the relevance of combining methods to answer a research question. Although the appropriate methods will depend on your research question, the answer to a phenomenon requires both the identification of a relation between two (or more) variables and a detailed explanation about how these two variables link with each other and why the indicated effect occurs. To approach these two dimensions, it is necessary to combine different empirical strategies, in order to exploit the respective virtues and complement their weaknesses.

An estimation through OLS allows us to identify average relationships between two variables in a large number of observations, something that qualitative research cannot perform. Nonetheless, OLS cannot answer why or how these relations occur and, for that, a qualitative research design is necessary to deepen on the processes and agents that "explain" these relations. Of course, the process can also be performed the other way around: first, we could identify a relation between two variables starting from an in-depth case study, and then, test this relationship in other cases through a *large-n* quantitative study to evaluate the generalization of the discovery. In any case, the combination of methods is recommended to offer more complex and complete explanations about the phenomena we are interested in studying.

Exercise 6C.

- Estimate a model where the dependent variable is the Gini score (`gini`) and the independents Education Expenditure (`education_budget`), Health Expenditure (`health_budget`), Social Security Expenditure (`socialsec_budget`), GDP (`gdp`), and Foreign Direct Investment (`foreign_inv`).
- Select the typical, outliers and influential cases for this model. Which variables can be important for understanding the outliers?

    Now, let's suppose that your independent variable of interest is Foreign Direct Investment. Select extreme cases in x, extreme cases in y, most similar and most different cases.

7

Panel Data

Francisco Urdinez[1]

Suggested readings

- Beck, N. (2008). Time-series Cross-section Methods. In Box-Steffensmeier, J. M., Brady, H. E., and Collier, D., editors, *The Oxford Handbook of Political Methodology*, pages 475–493. Oxford University Press, Oxford.

- Beck, N. and Katz, J. N. (2011). Modeling Dynamics in Time-Series–Cross-Section Political Economy Data. *Annual Review of Political Science*, 14(1):331–352.

- Croissant, Y. and Millo, G. (2018). *Panel Data Econometrics with R*. John Wiley & Sons, Hoboken, NJ.

- Henningsen, A. and Henningsen, G. (2019). Analysis of Panel Data Using R. In Tsionas, M., editor, *Panel Data Econometrics: Theory*, pages 345–396. Academic Press, London.

Packages you need to install

- `tidyverse` (Wickham, 2019), `politicalds` (Urdinez and Cruz, 2020), `unvotes` (Robinson, 2017), `lubridate` (Spinu et al., 2020), `ggcorrplot` (Kassambara, 2019), `plm` (Croissant et al., 2020).

7.1 Introduction

In this chapter, we will further learn what we discussed in the OLS chapter. We will learn how to work with panel data, that is, how to add a temporal dimension into our linear regressions. We will show you how to diagnose if time has an effect in our OLS results and in which way, and we will (a) estimate models with fixed effects and random effects, (b) see how to diagnose unit roots, (c) how to create variables with lags or leads, and (d) how to calculate panel-corrected standard errors. Remember that this is an introductory chapter, so we recommend you consult the suggested bibliography for detailed questions.

---

[1] Institute of Political Science, Pontificia Universidad Católica de Chile. E-mail: furdinez@uc.cl. You can follow him on Twitter in @FranciscoUrdin.

We will work with a practical example of international relations: Historically, Latin American foreign policy has been strongly influenced by the United States, probably more than any other region of the world. However, in the last thirty years, this trend has weakened, and the influence is less evident now than during the Cold War. What is the reason behind this trend?

This question was addressed by Octavio Amorim Neto, a well-established Brazilian political scientist, in *From Dutra to Lula: The formation and determinants of Brazilian Foreign Policy* (De Dutra a Lula: a condução e os determinantes da política externa brasileira) (2012). This book has the advantage of answering this question using quantitative methodology, something unusual in a field dominated by historiographic works. Other articles then went on to further refine the arguments in the book, such as Neto and Malamud (2015) and Rodrigues et al. (2019). Let's open the dataset:

```
library(tidyverse)
```

```
library(politicalds)
data("us_brazil")
```

According to Amorim Neto, as Brazil transformed into a regional power, and its power grew it had a greater margin for moving away from the United States precepts. One way of approaching the abstract concept of "political proximity" in international relations is through the convergence of votes in the United Nations General Assembly. Countries vote, let's say, 20 times a year over different subjects. Thus, on those 20 votes, we can calculate how many times a country voted the same as another (for, against or abstention) and to express this similarity as a percentage. Figure 7.1 shows the percentage of votes in common between Brazil and the United States in the Assembly of 1945:

```
library(tidyverse)

ggplot(us_brazil) +
  geom_line(aes(x = year, y = vote)) +
  labs(x = "Year", y = "Vote convergence")
```

There is an outstanding tool called **unvotes**[2] in R for all of those who study the voting history of countries in the United Nations Assembly, which will be useful for us in this chapter.[3] Using **unvotes** we can plot the convergence of votes, and divide them by subjects.

```
library(unvotes)
library(lubridate)
```

---

[2] See https://cran.r-project.org/web/packages/unvotes/index.html
[3] This example was taken from a course by Mine Cetinkaya-Rundel; her website (http://www2.stat.duke.edu/~mc301/) has more useful resources that are worth looking at.

7.1 Introduction

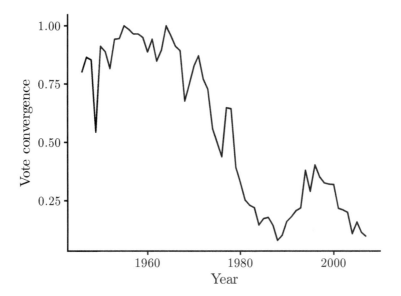

FIGURE 7.1: Vote convergence between Brazil and the US.

Additionally, we will make use of the `lubridate` package for the analysis. The `unvotes` package provides three datasets we can work with: `un_roll_calls`, `un_roll_call_issues`, and `us_latam`. Each of these datasets contains a variable called `rcid`, the roll call id, which can be used as a unique identifier to join them with each other. If you recall, in the advanced management chapter we taught you how to merge datasets using country codes.

- The `un_votes` dataset provides information on the voting history of the United Nations General Assembly. It contains one row for each country-vote pair.

- The `un_roll_calls` dataset contains information on each roll call vote of the United Nations General Assembly.

- The `un_roll_call_issues` dataset contains topic classifications of roll call votes of the United Nations General Assembly. Many votes had no topic, and some have more than one.

How has the voting record of the United States changed over time on a variety of issues compared to Brazil? (see Figure 7.2)

```
p_votes_us_br <- un_votes %>%
  filter(country %in% c("United States of America", "Brazil")) %>%
  inner_join(un_roll_calls, by = "rcid") %>%
  inner_join(un_roll_call_issues, by = "rcid") %>%
  mutate(issue = if_else(issue == "Nuclear weapons",
                         "Nuclear weapons", issue)) %>%
  group_by(country, year = year(date), issue) %>%
  summarize(votes = n(),
            percent_yes = mean(vote == "yes")) %>%
  filter(votes > 5)
```

```
ggplot(p_votes_us_br,
       mapping = aes(x = year, y = percent_yes, color = country)) +
  geom_point() +
  geom_smooth(method = "loess", se = F) +
  facet_wrap(~ issue, ncol = 2) +
  scale_color_grey() +
  labs(x = "Year", y = "% Yes", color = "Country")
```

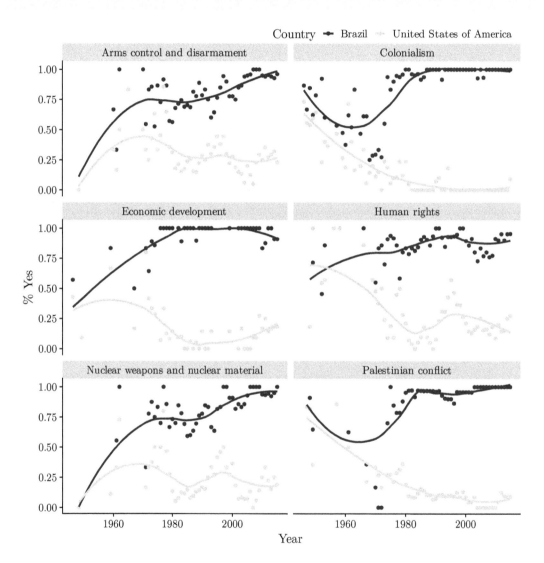

FIGURE 7.2: Percentage of positive votes in the UN General Assembly (1946 to 2015).

Let's consider the hypothesis of Amorim Neto (2012). His argument is that this distancing was caused by an increase in Brazil's power, which gave it greater autonomy. The discussion on how to measure power in international relations is another subject

7.1 Introduction

by itself. One of the most popular variables for measuring it was created by Singer et al. (1972), known as the CINC Index. This is a relative index: for each year, the index measures the share of power each country has of the total global power. How has Brazil developed since 1945? (Figure 7.3)

```
ggplot(us_brazil) +
  geom_line(aes(x = year, y = country_power)) +
  labs(x = "Year", y = "Country power index")
```

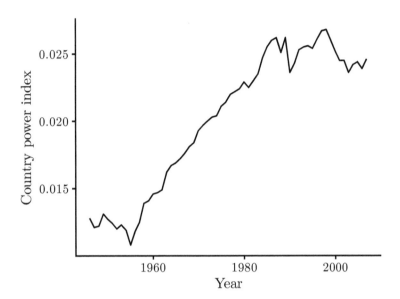

FIGURE 7.3: Brazilian international power.

Alright! A priori, Amorim Neto's hypothesis seems to have empirical support: As Brazil became more powerful, it had more margin to develop an autonomous foreign policy from the United States. If we observe the correlation between both variables, we notice that this one is high and negative (-0.89): higher values of vote convergence are correlated to lower values of power. We will use the ggpubr package for "Publication Ready Plots" (Figure 7.4):

```
library(ggpubr)

ggscatter(us_brazil, x = "country_power", y = "vote",
          add = "reg.line",  add.params = list(color = "black",
                                               fill = "lightgray"),
          conf.int = TRUE) +
  stat_cor(method = "pearson", label.x = 0.015)
```

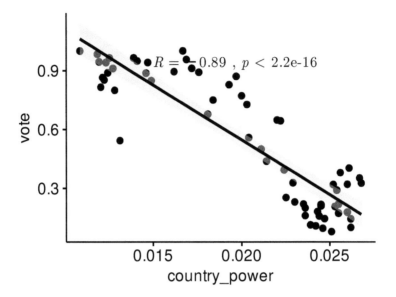

FIGURE 7.4: Correlation between UN votes convergence and share of international power.

In conclusion, this is the central argument of the book by Amorim Neto. Despite the large influence this analysis had over the studies that came after, there are some issues that partially invalidate the conclusion, which will be useful to exemplify in this chapter.

7.2 Describing your panel dataset

In political science, it is common to encounter the distinction between panel and Time Series Cross-Sectional (TSCS) data. Beck (2008) explains that a panel refers to a dataset with an N much larger than T, for example, a survey of 10,000 people during three consecutive years. On the other hand, a dataset in TSCS format often contains a larger T than N, and its units are fixed, that is, they are repeated in time. The distinction is clear in a text by Beck in the *Annual Review of Political Science*:

> Panel data are repeated cross-section data, but the units are sampled (usually they are survey respondents obtained in some random sampling scheme), and they are typically observed only a few times. TSCS units are fixed; there is no sampling scheme for the units, and any resampling experiments must keep the units fixed and only resample complete units (Freedman & Peters 1984). In panel data, the people observed are of no interest; all inferences

of interest concern the underlying population that was sampled, rather than being conditional on the observed sample. TSCS data are exactly the opposite; all inferences of interest are conditional on the observed units (2001, p. 273).

For practical purposes, the notation of both is the same, and the concepts tend to be used interchangeably:
$$y_{it} = x_{it}\beta + \epsilon_{it}$$
Where
$$x_{it}$$
is a vector of exogenous variables and observations are indexed by both unit (i) and time (t).

For reviewing the most common routines of panel models, we will use a similar dataset as that of Amorim Neto, but with eleven South American contries. We now have a base in TSCS format: a sample of eleven countries between 1970 and 2007, 38 years. If you master the introductory contents in this chapter, the following step will be to work through Croissant and Millo's (2018) book, who are the creators of the most used package for panel in R, plm.

```
data("us_latam")
```

```
us_latam %>%
  count(country)
## # A tibble: 10 x 2
##   country        n
##   <chr>      <int>
## 1 Argentina     38
## 2 Bolivia       38
## 3 Brazil        38
## # ... with 7 more rows
```

Let's observe the generalizability of the hypothesis (see Figure 7.5). The book only analyzes one case, but we want to know if an eleven-country panel can help us strengthen those findings and gain external validity. If we observe the voting behavior of these eleven Latin American countries in the United Nations, we will notice a similar pattern between them. It seems that voting convergence dropped between 1945 and 1990, then it rose during the 1990s, and then dropped again at the start of the 2000s. This step is always recommended to be done before moving on to the regressions. There are two ways of plotting the independent and dependent variables over time using ggplot2, as you previously learned.

7.2.1 Option A. Line graph

```
ggplot(us_latam, aes(x = year, y = vote,
                    color = country, linetype = country, group = country)) +
  geom_line() +
  labs(x = "Year", y = "% Yes", color = "", linetype = "")
```

FIGURE 7.5: Line trends for the evolution of UNGA vote convergence with the US by country.

7.2.2 Option B. Box plot

```
ggplot(us_latam, aes(x = factor(year), y = vote)) +
  geom_boxplot() +
  scale_x_discrete(breaks = seq(1970, 2007, by = 5)) +
  labs(x = "Year", y = "% Convergence with the US")
```

Also, as we observed for the Brazilian case, we can look at the proximity of votes between the eleven countries and the United States by separating the votes by subject using unvotes (Figure 7.7).

7.2 Describing your panel dataset

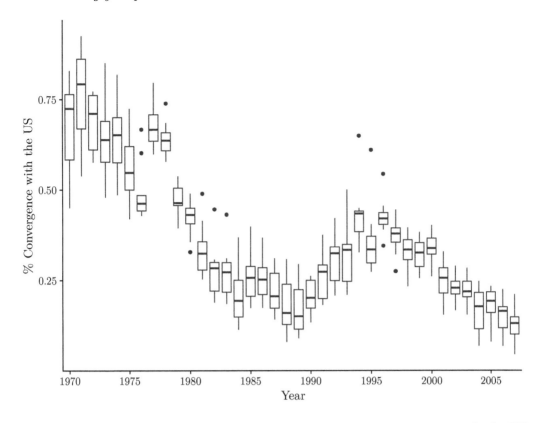

FIGURE 7.6: Boxplots for the evolution of UNGA vote convergence with the US by country.

```
p_votes_countries <- un_votes %>%
  filter(country %in% c("United States of America", "Brazil", "Bolivia",
                        "Argentina", "Chile", "Peru", "Ecuador", "Colombia",
                        "Venezuela", "Paraguay", "Uruguay")) %>%
  inner_join(un_roll_calls, by = "rcid") %>%
  inner_join(un_roll_call_issues, by = "rcid") %>%
  mutate(issue = if_else(issue == "Nuclear weapons",
                         "Nuclear weapons", issue)) %>%
  group_by(country, year = year(date), issue) %>%
  summarize(votes = n(),
            percent_yes = mean(vote == "yes")) %>%
  filter(votes > 5)

ggplot(p_votes_countries,
       mapping = aes(x = year, y = percent_yes,
                     linetype = country, color = country)) +
  geom_smooth(method = "loess", se = F) +
  facet_wrap(~issue, ncol = 2) +
  labs(x = "Year", y = "% Convergence with the US", color = "", linetype = "")
```

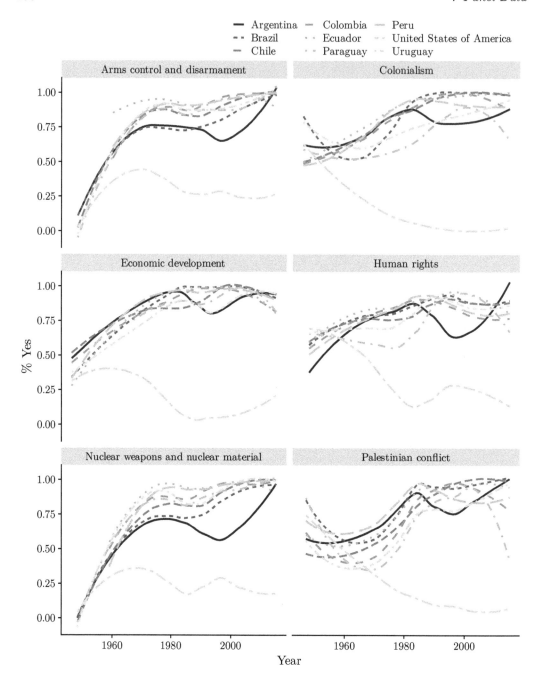

FIGURE 7.7: Percentage of positive votes in the UN General Assembly (1946 to 2015).

Once you have observed the behavior of your dependent variable, you will probably want to replicate the same exercise for your independent variable. In our case, the power of the country (Figure 7.8). When observing the behavior of the independent variable of our example (the power of the country) we will notice that, while Colombia is in line with what we observed for Brazil (growth in power over the years), other countries have weakened since 1970 (for example, Argentina), and the majority of them have remained virtually stable (for example, Chile, Uruguay and Peru). This heterogeneity of behaviors challenges the findings of Amorim Neto for Brazil.

```
us_latam %>%
  filter(country != "Brazil") %>%
  ggplot(aes(x = year, y = country_power)) +
  geom_line() +
  facet_wrap(~country, nrow = 3)
```

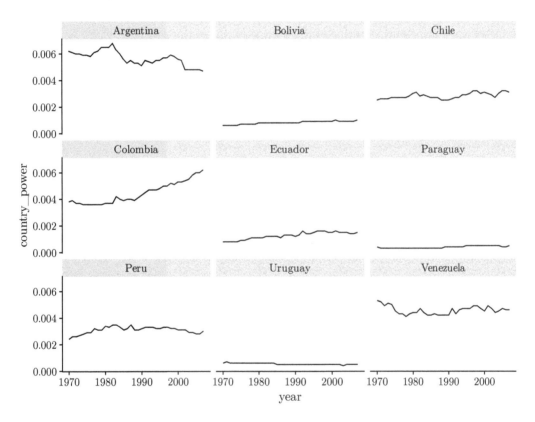

FIGURE 7.8: Country power index evolution in time for South American countries.

Exercise 7A. Using the `welfare` dataset of the OLS chapter (Latin America Welfare Dataset, 1960-2014, Huber et al., 2006), plot the behavior of the Gini Index (`gini`) in Latin America over time.

7.3 Modelling group-level variation

A panel dataset (or TSCS) will contain data for N units in T moments. In this chapter, N is countries, but in your dataset they could be political parties, counties, legislators or any other observational unit of interest. In this chapter, T is years, but in your dataset they could be months, semesters or decades. We obviously need regularity in the periodization. If we have years, it is important that the panel does not mix two different temporal units (for example, data by year and decade).

If our dataset does not contain missing values, we will say that our panel is balanced, since we have the same quantity of T for each N. If, on the contrary, there are missing values, our panel will not be balanced.[4] By having data that varies over time and/or among the individuals, our model will have more information than in cross-section models, thus obtaining more efficient estimators. Panel data will allow you to control by existing heterogeneities among the units that do not vary over time, and to reduce the bias of omitted variables, as well as to test the hypothesis of the behavior of observations over time.

Some of your variables will be fixed over time, that is, will not vary in T, just in N. For example, a country's geographical surface or distance to another country will not change under normal conditions (it could vary if there was a war and the national boundaries changed, for example). In the case of individuals, some attributes such as nationality or gender tend to remain the same over time. In political science, we will commonly have more Ns than Ts in our panel, although there might be exceptions. We will be interested in two types of effects on the dependent variable: effects on Y that vary in t but not in i, and effects that vary in i but not in t. Effects that vary in i *and* in t are considered in the error term $\epsilon_{i,t}$. If we ignore these effects, we will obtain biased coefficients. Let's observe it graphically!

- This is a model in which we ignore the existence of eleven countries in the sample, and we treat each observation as independent.

```
pooled <- lm(vote ~ country_power, data = us_latam)

summary(pooled)
## 
## Call:
## lm(formula = vote ~ country_power, data = us_latam)
## 
## Residuals:
##     Min      1Q  Median      3Q     Max
## -0.3194 -0.1465 -0.0538  0.0957  0.5657
## 
```

---

[4]In the advance data management chapter you will learn how to impute missing values.

7.3 Modelling group-level variation

```
## Coefficients:
##                Estimate Std. Error t value Pr(>|t|)
## (Intercept)       0.381      0.012   31.67   <2e-16 ***
## country_power    -3.564      1.473   -2.42    0.016 *
## ---
## Signif. codes:  0 '***' 0.001 '**' 0.01 '*' 0.05 '.' 0.1 ' ' 1
##
## Residual standard error: 0.19 on 378 degrees of freedom
## Multiple R-squared:  0.0152, Adjusted R-squared:  0.0126
## F-statistic: 5.85 on 1 and 378 DF,  p-value: 0.016
```

- This is a model in which we incorporate an intercept to each country, assuming that our variables vary between observations

```
manual_fe <- lm(vote ~ country_power + factor(country), data = us_latam)

summary(manual_fe)
##
## Call:
## lm(formula = vote ~ country_power + factor(country), data = us_latam)
##
## Residuals:
##      Min       1Q   Median       3Q      Max
## -0.3129  -0.1332  -0.0384   0.1002   0.5908
##
## Coefficients:
##                    Estimate Std. Error t value Pr(>|t|)
## (Intercept)          0.8145     0.0768   10.60  < 2e-16 ***
## country_power      -78.8200    12.5378   -6.29  9.2e-10 ***
## [ reached getOption("max.print") -- omitted 9 rows ]
## ---
## Signif. codes:  0 '***' 0.001 '**' 0.01 '*' 0.05 '.' 0.1 ' ' 1
##
## Residual standard error: 0.18 on 369 degrees of freedom
## Multiple R-squared:  0.125, Adjusted R-squared:  0.101
## F-statistic: 5.25 on 10 and 369 DF,  p-value: 3.18e-07
```

If we extract the predicted values from this model we can compare the difference it generates in the `country_power` coefficient by providing this information to the model:

```
us_latam <- us_latam %>%
  mutate(hat_fe = fitted(manual_fe))
```

As you can observe in the following figure (Figure 7.9), the coefficient of the pooled model for `country_power` is of -3.56, while for the model that accounts for heterogeneity of units is of -78.8. This correction is achieved by incorporating an intercept for each observation. The figure shows the slope for each country once we added the dummies, and the slope is less steep when we do a pooled regression (dashed line). The specificity that includes dummies for each observation is known as fixed effects model, and it is a workhorse for modeling panel data.

```
ggplot(data = us_latam, aes(x = country_power, y = hat_fe,
                            label = country, group = country)) +
  geom_point() +
  # add country-specific lines
  geom_smooth(method = "lm", se = F, color = "black") +
  # add pooled line
  geom_smooth(mapping = aes(x = country_power, y = hat_fe), inherit.aes = F,
              method = "lm", se = T, color = "black", linetype = "dashed") +
  # label lines
  geom_text(
    data = us_latam %>%
      group_by(country) %>%
      top_n(1, country_power) %>%
      slice(1),
    mapping = aes(label = country), vjust = 1
  )
```

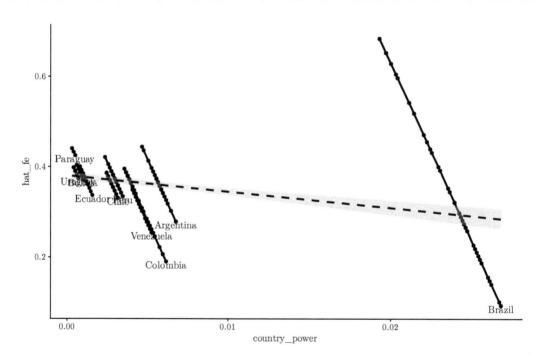

FIGURE 7.9: By adding country dummies the slope of the linear relation changes substantially.

7.4 Fixed vs. random effects

In political science, both specifications for modeling variation between groups in panel data are models of fixed or random effects. It is not yet clear to applied researchers when one or the other should be used. A good discussion of this subject is delivered by Clark and Linzer (2015). Fixed effects models add a dichotomous (dummy) variable for each unit. Random effects models assume that the variation between units follows a probability distribution, typically normal, with parameters estimated from the data.

The `index` function is necessary to inform which is our N and which is our T, in this case `c("code", "year")`. `plm`, by default, generates fixed effects models, so we do not need to use the `model` option. If we compare the `fe` model that we will generate with the `manual_fe` model, you will notice that the `country_power` coefficient is the same.

```
library(plm)
```

```
fe <- plm(vote ~ country_power, data = us_latam, index = c("code", "year"))
```

```
summary(fe)
## Oneway (individual) effect Within Model
##
## Call:
## plm(formula = vote ~ country_power, data = us_latam, index = c("code",
##     "year"))
##
## Balanced Panel: n = 10, T = 38, N = 380
##
## Residuals:
##    Min. 1st Qu.  Median 3rd Qu.    Max.
## -0.3129 -0.1332 -0.0384  0.1002  0.5908
##
## Coefficients:
##                Estimate Std. Error t-value Pr(>|t|)
## country_power    -78.8       12.5   -6.29  9.2e-10 ***
## ---
## Signif. codes:  0 '***' 0.001 '**' 0.01 '*' 0.05 '.' 0.1 ' ' 1
##
## Total Sum of Squares:    13.5
## Residual Sum of Squares: 12.2
## R-Squared:      0.0967
## Adj. R-Squared: 0.0723
## F-statistic: 39.5215 on 1 and 369 DF, p-value: 9.17e-10
```

If you want to obtain the fixed effects of the intercepts to compare them with the manual model, you need to use the `fixef()` function.

```
fixef(fe, type = "dfirst")
##      2      3      4      5      6      8      9     10     11
## -0.364  1.391 -0.232 -0.136 -0.352 -0.350 -0.204 -0.143 -0.385
```

In contrast, for the random effects we need to specify `model = "random"`.

```
re <- plm(vote ~ country_power, data = us_latam,
          index = c("code", "year"), model = "random")
```

```
summary(re)
## Oneway (individual) effect Random Effect Model
##    (Swamy-Arora's transformation)
##
## Call:
## plm(formula = vote ~ country_power, data = us_latam, model = "random",
##     index = c("code", "year"))
##
## Balanced Panel: n = 10, T = 38, N = 380
##
## Effects:
##                  var std.dev share
## idiosyncratic 0.03311 0.18195  0.99
## individual    0.00017 0.01304  0.01
## theta: 0.0853
##
## Residuals:
##    Min.  1st Qu.  Median 3rd Qu.    Max.
## -0.3149 -0.1449 -0.0551  0.0974  0.5653
##
## Coefficients:
##                Estimate Std. Error z-value Pr(>|z|)
## (Intercept)      0.3814     0.0131   29.10   <2e-16 ***
## country_power   -3.7487     1.6057   -2.33     0.02 *
## ---
## Signif. codes:  0 '***' 0.001 '**' 0.01 '*' 0.05 '.' 0.1 ' ' 1
##
## Total Sum of Squares:    13.9
## Residual Sum of Squares: 13.7
## R-Squared:      0.0142
## Adj. R-Squared: 0.0116
## Chisq: 5.45058 on 1 DF, p-value: 0.0196
```

7.4.1 Hausman test

We assume that the fixed effects model is consistent for the true parameters, and the random-effects model is a fully efficient specification of the individual effects under the assumption that they are random and follow a normal distribution. It is assumed that the fixed effects model compute the always-consistent estimator, while the random effects model will compute the estimator that is consistent *and efficient* under H_0. The command to perform this test is `phtest`, which is part of the `plm` package.

```
phtest(fe, re)
##
##  Hausman Test
##
## data:  vote ~ country_power
## chisq = 36, df = 1, p-value = 2e-09
## alternative hypothesis: one model is inconsistent
```

Under the current specification, our initial hypothesis that the individual-level effects are adequately modeled by a random-effects model is clearly rejected, with a p-value below the threshold of 0.05. In this case, you must remain with the fixed effects model.

Exercise 7B. Use the `welfare` dataset of the OLS chapter (Latin America Welfare Dataset, 1960-2014, by Evelyne Huber and John D. Stephens) to estimate a fixed-effect and a random-effects model in which your dependent variable is the Gini Index (`gini`) Then, perform a Hausman especification test.

7.5 Testing for unit roots

Many political-data time series exhibit trending behavior or non-stationarity behavior in the mean. Yet, non-stationarity is not always checked. This is not nuance, as non-stationarity can lead to spurious regressions, and there are considerable chances to obtain falsely significant results. Two common trend removal or de-trending procedures are first differencing and time-trend regression. Unit root tests can be used to determine if trending data should be corrected to render the data stationary. A dynamic panel data model is one which contains (at least) one lagged dependent variable and a first difference model is one in which both dependent and independent variables are expressed as Δ, i.e. as $X_t - (X_{t-1})$. In a sense, a lagged dependent variable introduces history into the model. After you include it, the dependent variable is influenced not only by the current value of the independent variable (X_t), but also by values of the independent variable in the past, (X_{t-1}, X_{t-2}, etc).

Let's explore our data and test for unit roots using our fixed effects specification.

```
fe <- plm(vote ~ country_power, data = us_latam, index = c("code", "year"))

summary(fe)
## Oneway (individual) effect Within Model
##
## Call:
## plm(formula = vote ~ country_power, data = us_latam, index = c("code",
##     "year"))
##
## Balanced Panel: n = 10, T = 38, N = 380
##
## Residuals:
##     Min.  1st Qu.   Median  3rd Qu.     Max.
## -0.3129  -0.1332  -0.0384   0.1002   0.5908
##
## Coefficients:
##                Estimate Std. Error t-value Pr(>|t|)
## country_power    -78.8       12.5   -6.29  9.2e-10 ***
## ---
## Signif. codes:  0 '***' 0.001 '**' 0.01 '*' 0.05 '.' 0.1 ' ' 1
##
## Total Sum of Squares:    13.5
## Residual Sum of Squares: 12.2
## R-Squared:      0.0967
## Adj. R-Squared: 0.0723
## F-statistic: 39.5215 on 1 and 369 DF, p-value: 9.17e-10
```

If we explore how highly correlated our dependent and independent variables are with our time variable, we might get a sense of how it works. The figure we create with `ggcorrplot` shows that `country_power` has a very strong positive correlation with the `year`, while the `vote` variable has a very negative correlation with the `year` variable (see Figure 7.10). This is a warning sing, as it might be that `country_power` might not explain `vote` after all, which would turn our initial findings spurious.

```
library(ggcorrplot)

corr_selected <- us_brazil %>%
  select(year, country_power, vote) %>%
  # calculate correlation matrix and round to 1 decimal place:
  cor(use = "pairwise") %>%
  round(1)

ggcorrplot(corr_selected, type = "lower", lab = T, show.legend = F)
```

7.5 Testing for unit roots

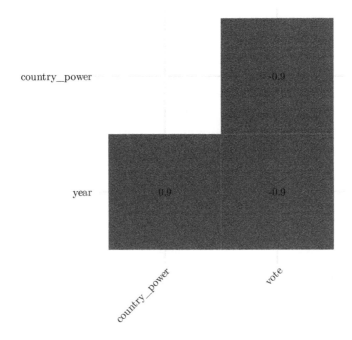

FIGURE 7.10: Correlation with time variable.

For diagnosing unit-roots we will use the `purtest` function. There are multiple tests, each one with different specifications. Thus, reviewing all their differences exceeds the purpose of this chapter. We will use two, which are suited for our goal of detecting unit roots in our dependent and independent variables.

Testing unit roots in our dependent variable:

```
# option Levin et al. (2002)
purtest(vote ~ 1, data = us_latam,
        index = c("country", "year"), pmax = 10, test = "levinlin",
        lags = "AIC", exo = "intercept")
##
##  Levin-Lin-Chu Unit-Root Test (ex. var.: Individual Intercepts)
##
## data:  vote ~ 1
## z = -2, p-value = 0.03
## alternative hypothesis: stationarity

# option Im et al. (2003)
purtest(vote ~ 1, data = us_latam,
        index = c("country", "year"), pmax = 10, test = "ips",
        lags = "AIC", exo = "intercept")
##
##  Im-Pesaran-Shin Unit-Root Test (ex. var.: Individual
##  Intercepts)
```

```
##
## data:  vote ~ 1
## Wtbar = -1, p-value = 0.1
## alternative hypothesis: stationarity
```

In this case, just one of the models, the one from `ips`, indicates the existence of unit roots. Although the evidence is not absolute, it is worth correcting for possible unit roots in `vote`. Now, let's observe our dependent variable.

Testing unit roots in our independent variable, using two alternative methods (Levin et al., 2002; Im et al., 2003):

```
# option Levin et al. (2002)
purtest(country_power ~ 1, data = us_latam,
        index = c("country", "year"), pmax = 10, test = "levinlin",
        lags = c("AIC"), exo = c("intercept"))
##
##  Levin-Lin-Chu Unit-Root Test (ex. var.: Individual Intercepts)
##
## data:  country_power ~ 1
## z = -0.6, p-value = 0.3
## alternative hypothesis: stationarity

# option Im et al. (2003)
purtest(country_power ~ 1, data = us_latam,
        index = c("country", "year"), pmax = 10, test = "ips",
        lags = "AIC", exo = "intercept")
##
##  Im-Pesaran-Shin Unit-Root Test (ex. var.: Individual
##  Intercepts)
##
## data:  country_power ~ 1
## Wtbar = -0.4, p-value = 0.4
## alternative hypothesis: stationarity
```

In this case, both models are clear in indicating unit roots in the variable of countries' power. We will try to solve the problem by specifying two models, one dynamic and the other of first differences. In this chapter we do not have the extension for covering which one is preferable in each situation, so we will assume that both are equally valid. To specify these models, we first need to learn how to create lagged variables (t-1, t-2, etc.), leads variables (t+1, t+2, etc.) and first differences (Δ).

7.5.1 Creation of lags and leads

The following chunk creates variables that we will most likely need. First, we sort (`arrange`) our panel according to our N and T. This is an important step so that the

7.5 Testing for unit roots

dataset does not mix up our observations. Then, we group them according to our N, in this case `country`. Finally, we create the variables. After the function (for example, `lag`) we specify that it is a lag in relation to t-1. For a greater lag, we would change the 1 for another value. We have created one of each type so that you know how to do it, but we will not use the `vote_lead1` variable.

```
us_latam <- us_latam %>%
  arrange(country, year) %>%
  group_by(country) %>%
  mutate(vote_lag1 = dplyr::lag(vote, 1),
         vote_lead1 = dplyr::lead(vote, 1),
         vote_diff1 = c(NA, diff(vote))) %>%
  ungroup()
```

Now we do the same for the `country_power` variable. We will create a lag and the difference (Δ).

```
us_latam <- us_latam %>%
  arrange(country, year) %>%
  group_by(country) %>%
  mutate(country_power_lag1 = dplyr::lag(country_power, 1),
         country_power_diff1 = c(NA, diff(country_power))) %>%
  ungroup()
```

- Dynamic model

Our dynamic model will include the lag of the dependent variable. Thus, `vote_lag1` is a predictor of vote in the present time. You will notice that by including this variable, Amorim Neto's hypothesis that the greater the power of the country, the lower the convergence of votes in the UNGA, is left with no empirical support. That is why it is essential that when you use panel data you always run this kind of test!

```
fe_lag <- plm(vote ~ vote_lag1 + country_power, data = us_latam,
              index = c("code", "year"))
summary(fe_lag)
## Oneway (individual) effect Within Model
##
## Call:
## plm(formula = vote ~ vote_lag1 + country_power, data = us_latam,
##     index = c("code", "year"))
##
## Balanced Panel: n = 10, T = 37, N = 370
##
## Residuals:
##      Min.   1st Qu.    Median   3rd Qu.      Max.
## -0.20668  -0.04013  -0.00833   0.02616   0.28549
##
```

```
## Coefficients:
##                 Estimate Std. Error t-value Pr(>|t|)
## vote_lag1         0.8912    0.0218   40.90   <2e-16 ***
## country_power    -3.0418    5.7186   -0.53    0.6
## ---
## Signif. codes:  0 '***' 0.001 '**' 0.01 '*' 0.05 '.' 0.1 ' ' 1
##
## Total Sum of Squares:    12.3
## Residual Sum of Squares: 1.98
## R-Squared:       0.839
## Adj. R-Squared:  0.834
## F-statistic: 934.399 on 2 and 358 DF, p-value: <2e-16
```

The same occurs when testing the models of first differences, so, the hypothesis of the chapter is left with no support.

- First-difference model

```
fe_diff <- plm(vote_diff1 ~ country_power_diff1, data = us_latam,
               index = c("code", "year"))
```

```
summary(fe_diff)
## Oneway (individual) effect Within Model
##
## Call:
## plm(formula = vote_diff1 ~ country_power_diff1, data = us_latam,
##     index = c("code", "year"))
##
## Balanced Panel: n = 10, T = 37, N = 370
##
## Residuals:
##      Min.    1st Qu.    Median    3rd Qu.      Max.
## -1.59e-15 -2.93e-18 -1.66e-19  2.00e-18  6.46e-17
##
## Coefficients:
##                       Estimate Std. Error   t-value Pr(>|t|)
## country_power_diff1   1.00e+00   5.77e-17  1.73e+16   <2e-16 ***
## ---
## Signif. codes:  0 '***' 0.001 '**' 0.01 '*' 0.05 '.' 0.1 ' ' 1
##
## Total Sum of Squares:    2.13
## Residual Sum of Squares: 2.54e-30
## R-Squared:       1
## Adj. R-Squared:  1
## F-statistic: 3.00273e+32 on 1 and 359 DF, p-value: <2e-16
```

7.6 Robust and panel-corrected standard errors

Finally, let's compare the model that does not address unit roots (`fe`) with the ones that correct it (`fe_lag`) through the Wooldridge Test for AR(1) errors in FE panel models:

```
pwartest(fe)
##
##   Wooldridge's test for serial correlation in FE panels
##
## data:  fe
## F = 4437, df1 = 1, df2 = 368, p-value <2e-16
## alternative hypothesis: serial correlation
```

```
pwartest(fe_lag)
##
##   Wooldridge's test for serial correlation in FE panels
##
## data:  fe_lag
## F = 0.7, df1 = 1, df2 = 358, p-value = 0.4
## alternative hypothesis: serial correlation
```

We obviously have covered these contents without going into much detail or explanation. One could write an entire book about this chapter. To further learn about this topic, it would be great if you could read the suggested readings at the start of this chapter, especially Beck and Katz (2011)[5].

Exercise 7C. Use the Latin America Welfare Dataset (`welfare`) to create lagged variables in t-1 y t-10 of the Gini Index (`gini`). Incorporate both variables into your model and diagnose unit roots.

7.6 Robust and panel-corrected standard errors

7.6.1 Robust standard errors

You have probably used Stata at some point for doing panel analysis, or maybe one of your co-authors wants to replicate your results in Stata. If you want to report robust standard errors equivalent to those of the "robust" option of Stata, you need

---
[5]Beck and Katz (2011).

to calculate them with `coeftest` and to define `type"sss"`, which corresponds to the same small-sample correction for panel data as Stata does.

```
library(lmtest)
```

If you wanted to do it for original fixed effects (`fe`) you can do it with the following line of code:

```
coeftest(fe, vcov. = function(x){vcovHC(x, type = "sss")})
##
## t test of coefficients:
##
##                Estimate Std. Error t value Pr(>|t|)
## country_power    -78.8       19.4   -4.06   6e-05 ***
## ---
## Signif. codes:  0 '***' 0.001 '**' 0.01 '*' 0.05 '.' 0.1 ' ' 1
```

And to compare them with the original standard errors:

```
summary(fe)
## Oneway (individual) effect Within Model
##
## Call:
## plm(formula = vote ~ country_power, data = us_latam, index = c("code",
##     "year"))
##
## Balanced Panel: n = 10, T = 38, N = 380
##
## Residuals:
##    Min. 1st Qu.  Median 3rd Qu.    Max.
## -0.3129 -0.1332 -0.0384  0.1002  0.5908
##
## Coefficients:
##                Estimate Std. Error t-value Pr(>|t|)
## country_power    -78.8       12.5   -6.29  9.2e-10 ***
## ---
## Signif. codes:  0 '***' 0.001 '**' 0.01 '*' 0.05 '.' 0.1 ' ' 1
##
## Total Sum of Squares:    13.5
## Residual Sum of Squares: 12.2
## R-Squared:      0.0967
## Adj. R-Squared: 0.0723
## F-statistic: 39.5215 on 1 and 369 DF, p-value: 9.17e-10
```

It is also possible that you want to calculate panel-corrected standard errors, equivalent to the `xtpcse` command of Stata. These standard errors have been popularized in political science since the classic paper by Beck and Katz (1995). In 2011 a package

7.6 Robust and panel-corrected standard errors

was published for calculating them in R (Bailey and Katz, 2011). In the paper, the authors explain that these standard errors are useful for those that work with panels of "states or nations":

> "Time-series–cross-section (TSCS) data are characterized by having repeated observations over time on some set of units, such as states or nations. TSCS data typically display both contemporaneous correlation across units and unit level heteroskedasity making inference from standard errors produced by ordinary least squares incorrect. Panel-corrected standard errors (PCSE) account for these these deviations from spherical errors and allow for better inference from linear models estimated from TSCS data." (2011, 1)

For using Panel-corrected standard errors (PCSE) you need to use the `vcov = "vcovBK"` option within the `coeftest()` function. You will need to set the option `cluster = "time"`:

```
coeftest(fe, vcov = vcovBK, type = "HC1", cluster = "time")
##
## t test of coefficients:
##
##               Estimate Std. Error t value Pr(>|t|)
## country_power   -78.8      11.3    -6.96  1.6e-11 ***
## ---
## Signif. codes:  0 '***' 0.001 '**' 0.01 '*' 0.05 '.' 0.1 ' ' 1
```

Voilá!

Up to this point, we have covered the basic functions that you need to reckon for analyzing your panel data. To further your knowledge please refer to the recommended bibliography at the beginning of the chapter. We hope you have found this chapter useful.

Exercise 7D. In the model you estimated in the previous exercise, calculate the panel corrected standard errors.

8
Logistic Models

Francisco Urdinez[1]

Suggested readings

- Agresti, A. (2007). *An Introduction to Categorical Data Analysis*. Wiley-Interscience, Hoboken, NJ, 2nd edition.
  - Chapters 3, 4 and 5 – "Generalized Linear Models"; "Logistic Regression"; "Building and Applying Logistic Regression Models."
- Glasgow, G. and Alvarez, R. M. (2008). Time-series Cross-section Methods. In Box-Steffensmeier, J. M., Brady, H. E., and Collier, D., editors, *The Oxford Handbook of Political Methodology,* pages 513–529. Oxford University Press, Oxford.
- Long, J. S. (1997). *Regression models for categorical and limited dependent variables*. SAGE, Thousand Oaks, CA.
  - Chapters 3 and 4 - "Binary Outcomes"; "Hyphotesis Testing and Goodness of Fit."

Packages you need to install

- `tidyverse` (Wickham, 2019), `politicalds` (Urdinez and Cruz, 2020), `ggcorrplot` (Kassambara, 2019), `margins` (Leeper, 2018), `prediction` (Leeper, 2019), `texreg` (Leifeld, 2020), `jtools` (Long, 1997), `skimr` (Waring et al., 2020), `pscl` (Jackman et al., 2020), `DescTools` (Signorell, 2020), `broom` (Robinson and Hayes, 2020), `plotROC` (Sachs, 2018), `separationplot` (Greenhill et al., 2020).

8.1 Introduction

In Chapter 6 you learned how to run linear regressions when you have continuous dependent variables in a simple manner and covering the most useful packages available in R. In this chapter you will learn how to estimate regression models when you have dichotomous dependent variables (also called binary, or *dummy* variables). These

---
[1] Institute of Political Science, Pontificia Universidad Católica de Chile. E-mail: furdinez@uc.cl. Twitter: @FranciscoUrdin.

variables assume one of two values, commonly 0 and 1. As in the previous chapters, we will not cover substantial aspects of the theory behind each model, nor will we break down the formulas in detail. For this, we suggest the references above that will help you accompany what we will discuss.

8.2 Use of logistic models

Models for dichotomous dependent variables are used to estimate the probability of the occurrence of an event. In our dataset, we code as '1' the cases in which the event occurs, and '0' when it does not occur. For example, if my variable was "countries with Free Trade Agreements with the United States", Chile and Mexico would be coded with a '1', Argentina and Brazil with a '0'. These models estimate probabilities, that is, what is the probability that a '1' is observed given certain characteristics of the observations in our sample. The whole discussion of how to estimate the probability of a sample is very interesting, and you can read it in chapters 3, 4 and 5 of Agresti (2007). It is also important to understand the distinction between probability and likelihood. A probability is estimated from a "population" of which we know its "parameters", while a likelihood estimates the values of the parameters for which the observed result best fits them (see Figure 8.1).

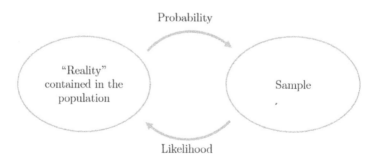

FIGURE 8.1: The two-way road of probability and likelihood.

When you have a dichotomous dependent variable that you want to model through logit, it is assumed that it has a Bernoulli distribution with a probability of $Y = 1$ that you do not know. Thus, we estimate our unknown probability by means of Maximum Likelihood Estimation, which will give us a certain linear combination of independent variables of our choice (see Figure 8.2). A very good exercise to understand how to estimate a parameter with a binomial distribution through Maximum Likelihood is offered by RPubs[2].

---

[2]See https://rpubs.com/felixmay/MLE

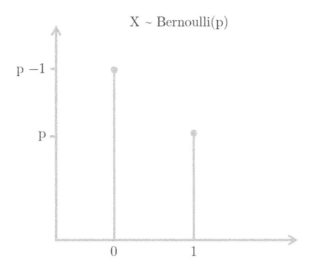

FIGURE 8.2: Bernoulli distribution.

Logistic models have grown tremendously in popularity in Political Science. Google Scholar reports that from the search for "logit" + "political science" a total of 17,100 references were published between 2008 and 2018. The previous decade (1998-2007) reports 8,900 articles, and the previous one (1988-1997), 2,100. Political Science has extended the use of Logit models, above Probit models, largely because the former allows the calculation of *odds ratios*. Almost all econometric manuals discuss the differences and similarities between Probit and Logit, although for the practical purposes of estimating a model, both are very similar and report almost identical coefficients. Therefore, since they are methods that lead to very similar results, we will only use Logit in this chapter.

The Logit and Probit models differ in their link functions. Logit bears its name because the function is given by the natural logarithm of the odds ("log odds" → logit!).

$$ln(odds) = ln(\frac{p}{1-p})$$

By clearing the terms, we can calculate the inverse of its function, so that we will have

$$logit^{-1}(\alpha) = \frac{1}{1+e^{-\alpha}} = \frac{e^\alpha}{1+e^\alpha}$$

Where α is the linear combination of independent variables and their coefficients. The inverse of the Logit will give us the probability of the dependent variable being equal to 1 given a certain combination of values for our independent variables (Figure 8.3).

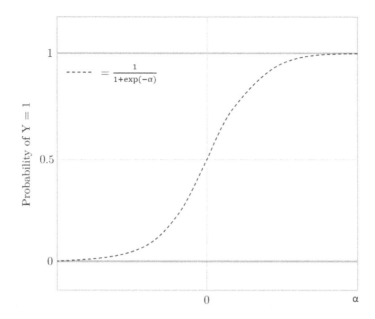

FIGURE 8.3: Inverse logit function.

If you delve into the recommended literature, you will notice that the function is undefined at 0 and 1, that is, that the probability is infinitely close to the limit without ever touching it. If we modeled the probabilities by means of Ordinary Least Squares, we would obtain probabilities greater than 100% and smaller than 0%, which is conceptually impossible. The sigmoid function of Logit prevents this from happening. What we will be estimating is the probability that, given a certain combination of attributes (independent variables), a certain observation assumes that the dependent variable is equal to 1. This probability can be compared to the real value that the dependent variable has assumed for each observation. For this reason, many textbooks refer to the logit models as a "classification" model, since what we estimate is the probability of each observation being classified as $Y = 1$.

8.3 How are probabilities estimated?

As we mentioned before, probabilistic models have gained enormous preeminence in Political Science in recent years, and that is why you are probably looking for a guidebook to know what to do and what not to do when you have a dichotomous dependent variable. We will illustrate a step by step in R using as an example the dataset of the book *Democracies and Dictatorships in Latin America: Emergence, Survival, and Fall* by Scott Mainwaring and Aníbal Perez-Liñan (2013). Throughout the book, the authors analyze which variables help explain why democratic breakdowns occurred in Latin America throughout the 20th century and the beginning of the

8.3 How are probabilities estimated?

21st. In Chapter 4, the authors explore what factors explain the survival of political regimes. While they try several complex models, some logistic and others survival models (we will cover them in this book in the next chapter); We will illustrate a very simple example so that you can accompany us from your computer step by step. To load the chapter dataset, use our `politicalds` package.

Then, we will load the dataset called, `dem_breakdown`.

```
library(politicalds)
data("dem_breakdown")
```

Now, the dataset has loaded in our R session:

```
ls()
## [1] "dem_breakdown"
```

Assuming that the dependent variable has the value '1' if the country suffers the breakdown of its democratic political regime and '0' if not, we can ask a couple questions. What effect does having a national constitution that grants great constitutional powers to the executive branch on the probability of a democratic breakdown? if those powers are small, is the probability of a democratic breakdown greater or lesser? As the authors argue, these powers can be measured by means of an index created by Shugart and Carey (1992) (in Chapter 15 we will illustrate how to create indexes in R).

First of all, you want to explore your data. We want to see how many countries have suffered a democratic breakdown in the sample. Do you remember what was explained in Chapter 2? We start by loading the `tidyverse` and then we use `count()`. The country with the most democratic breakdowns is Peru with 6, followed by Argentina and Panama, with 5 each.

```
library(tidyverse)
```

```
dem_breakdown %>% filter(breakdown == 1) %>% count(country_name)
## # A tibble: 18 x 2
##   country_name     n
##   <chr>        <int>
## 1 Argentina        5
## 2 Bolivia          2
## 3 Brazil           1
## # ... with 15 more rows
```

The first model we are going to test has the dependent variable of democratic breakdown ("breakdown") being predicted by the index of Shugart and Carey (1992) of executive powers (`pres_power_index`). The `glm` function requires that we define our data and the model, because with the same function for Generalized Linear Models you can use Probit, Poisson and other less common functions[3] in Political Science.

```
model_1 <- glm(breakdown ~ pres_power_index,
               data   = dem_breakdown,
               family = binomial("logit")) # we use the logistic link
               function
```

As we saw in previous chapters, the `summary` function allows us to quickly see the results of a regression model:

```
summary(model_1)
##
## Call:
## glm(formula = breakdown ~ pres_power_index, family = binomial("logit"),
##     data = dem_breakdown)
##
## Deviance Residuals:
##     Min       1Q   Median       3Q      Max
## -0.727  -0.295  -0.269  -0.223    2.792
##
## Coefficients:
##                   Estimate Std. Error z value Pr(>|z|)
## (Intercept)        -0.2393     0.9638   -0.25   0.8039
## pres_power_index   -0.1914     0.0648   -2.96   0.0031 **
## ---
## Signif. codes:  0 '***' 0.001 '**' 0.01 '*' 0.05 '.' 0.1 ' ' 1
##
## (Dispersion parameter for binomial family taken to be 1)
##
##     Null deviance: 217.84  on 643  degrees of freedom
## Residual deviance: 209.56  on 642  degrees of freedom
##   (1572 observations deleted due to missingness)
## AIC: 213.6
##
## Number of Fisher Scoring iterations: 6
```

The coefficient for the variable `pres_power_index` is negatively associated with the probability of occurrence of a regime break (-0.191), and it is statistically significant (p=0.00312). Now, unlike the OLS models of the previous chapter –where we could use the regression coefficients to directly interpret the effect of the independent variable on the dependent variable—-in the case of logistic regressions this is not as straightforward. We must transform the coefficients into probabilities or odds. Given that the logit link function is the logarithm of the quotient of probabilities, we have that:

$$ln(\frac{p}{1-p}) = \beta_0 + \beta_1 x_1$$

---

[3] See https://www.rdocumentation.org/packages/stats%20/versions/3.5.1/topics/family

8.3 How are probabilities estimated?

Clearing ln, we obtain:

$$\left(\frac{p}{1-p}\right) = e^{\beta_0 + \beta_1 x_1}$$

And clearing the terms, once again, we obtain:

$$\hat{p} = \frac{e^{\beta_0 + \beta_1 x_1}}{1 + e^{\beta_0 + \beta_1 x_1}}$$

What we want, then, is to transform the coefficients as R reports them into a probability of the dependent variable assuming the value '1'. We know that the independent variable `pres_power_index` is an index that at a higher value means greater concentration of power for the executive *vis a vis* the legislature, therefore the regression coefficient indicates that the lower the concentration of power of the executive, the higher the probability of a regime break. This will be intuitive to anyone who is familiar with the literature. The book sample covers 20 Latin American countries between 1900 and 2010, and the index ranges from a minimum of 5 (Haiti, in several years) to a maximum of 25 (Brazil in 1945) (see Figure 8.4).

```
dem_breakdown %>%
  select(country_name, year, pres_power_index) %>%
  arrange(pres_power_index) %>%
  slice(1)
## # A tibble: 1 x 3
##   country_name  year pres_power_index
##   <chr>        <dbl>            <dbl>
## 1 Haiti         1987                5
```

Tip: If we put a (-) before the variable within the `arrange()` function, it orders from highest to lowest.

```
dem_breakdown %>%
  select(country_name, year, pres_power_index) %>%
  arrange(-pres_power_index) %>%
  slice(1)
## # A tibble: 1 x 3
##   country_name  year pres_power_index
##   <chr>        <dbl>            <dbl>
## 1 Brazil        1945               25
```

We can see the distribution of the variable plotting a histogram. Most observations have values of 16 in the index and the variable assumes a normal distribution. We will make a histogram with `ggplot2`, which we covered in depth in Chapter 3 (Figure 8.4):

```
ggplot(dem_breakdown, aes(x = pres_power_index)) +
  geom_histogram(binwidth = 2) +
  labs(x = "Presidential Power Index", y = "Frequency")
```

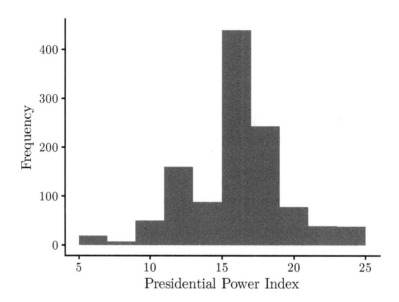

FIGURE 8.4: Histogram of the Shugart and Carey index.

How can we calculate to what extent the probability of a democratic breakdown is affected if the level of concentration of the executive power 'jumps' from a score of 5 (the minimum observed) to one of 25 (the maximum observed), given that we do not control for anything else in the regression? To do this calculation we can replace the values from our last formula, in which we isolated on the left side of the formula a $\hat{p}$. First we must calculate what the probability of suffering a regime break is at a Shugart and Carey index (SC index) level of 5 and at a level 25, respectively. Then we calculate the difference of both probabilities. Hence, we have for an index of 5 that

$$\hat{p} = \frac{e^{(0+(-0.019*5))}}{1+e^{(0+(-0.019*5))}}$$

You will notice that the value corresponding to the intercept is equal to 0 because that coefficient is not statistically significant. We know that for a SC index of 5, after making the calculation in the formula above, the probability of a breakdown is equal to 0.47 or 47%.

8.3 How are probabilities estimated?

Tip. We recommend that you stop reading for a second and do the calculation of the function in pencil and paper.

If we repeat the process for a `pres_power_index` value of 25, the probability drops to 38%. With the probabilities we can calculate odds, which are simply $\frac{p}{1-p}$. Thus, the odd for a value 5 of the SC index is 0.90, while for an SC index of 25 it is 0.62.

$$odd_a = \frac{0.47}{1 - 0.47}$$

$$odd_b = \frac{0.38}{1 - 0.38}$$

The usefulness of the odds is that they allow you to calculate odds ratios. What is the point of calculating an odds ratio? Let's see. When we calculate the probability of a change in the Shugart and Carey index from 23 to 24, the magnitude of this probability will be different than a change in the probability given that the index goes from, let's suppose, 12 to 13. This is because the effects of the independent variable on the probability that the dependent variable is =1 are not linear (remember the" S "function of Figure 8.3). However, the odds ratios have the useful property of being able to reflect changes regardless of the curvature of the function, that is, they are *constant* changes. Thus, we can express the effect of the variable without having to specify a certain value for it. Therefore, in many articles you will see results expressed as *odds ratios* as well as in probability changes. Both complement well, indeed.

Let's see how the calculation of odd ratios would look like when following the example we just done using Mainwaring and Pérez-Liñan's dataset. We said that the odd is given by $\frac{p}{1-p}$. Hence, a ratio of odds would be expressed as $\frac{\frac{p_1}{1-p_1}}{\frac{p_2}{1-p_2}}$.

Let's suppose that in 1992 a country had an SC index of 15, and that in 1993 that index went to 16. How much does the probability of a democratic breakdown change assuming everything else remained constant?

$$Pr(democratic breakdown)_{country,1992} = \frac{e^{(0+(-0.019*15))}}{1 + e^{((0+(-0.019*15))}} = 0.42$$

$$Pr(democratic breakdown)_{country,1993} = \frac{e^{(0+(-0.019*16))}}{1 + e^{(0+(-0.019*16))}} = 0.43$$

The probability differs little, falling by 2.4%, which seems to be a small effect. The odds ratio is calculated as the ratio of both odds, so we get

$$\frac{0.42}{0.43} = 0.97$$

In this way, an odds ratio greater than 1 expresses a positive change, while if smaller than 1 (between 0 and 1) represents a negative change in the estimated odds. If we did the same exercise for other values of the Shugart and Carey index, for example, a change from 3 to 4 or from 23 to 24, the odds ratio would also give 0.97.

Exercise 8A. Take a minute to do an exercise before continuing. 1. Open the dataset `lapop` from the book's package: `data("lapop")`. This is a 2018 survey of Latin American public opinion regarding political issues. 2. The variable `satisfied_dem` assumes 1 if the person believes that democracy is, despite their problems, the best existing form of government. Calculate how much the probability that this answer is 1 changes depending on her confidence in the media (`trust_media`, measures on a scale of 1 to 7). 3. What is the reason for one more year of education opportunities? In which country is the effect greater, in Brazil or in Mexico?

R offers packages to make this analysis easy. We can easily plot the odds ratios using `ggplot2` combined with `margins` and `prediction`. We can calculate predicted probabilities, and we can also carry out easy tests to know the explanatory capacity of our models. Using the same dataset, we will give you an example of a typical routine, which you can replicate on your computer using your own data.

The steps to follow are: (a) estimate your models, (b) create formatted tables to include in your articles, (c) create figures to visualize the magnitude of the coefficients by means of odds ratios, (d) plot predicted probabilities for variables of interest, and explore interactions, and (e) calculate explanatory capacity of the models (percentage correctly predicted, AIC, BIC, ROC curves, Brier scores or separation plots, which we will explain below).

8.4 Model estimation

To exemplify this step, we will estimate two more models apart from model 1. Model 2 will have as independent variables the presidential power index of Shugart and

8.4 Model estimation

Carey and the `regime_age` of the political regime, measured in years. Let's look at the distribution of the variable `regime_age` (Figure 8.5):

```
ggplot(dem_breakdown, aes(x = regime_age)) +
  geom_histogram(binwidth = 5) +
  labs(x = "Age of the political regime", y = "Frequency")
```

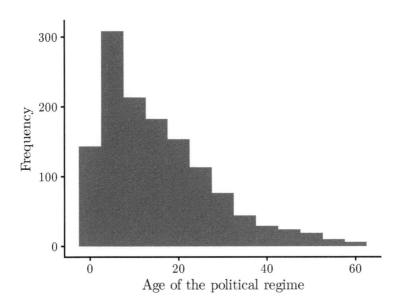

FIGURE 8.5: Histogram of the age of the political regime.

Let's look at the model with this new variable:

```
model_2 <- glm(breakdown ~ pres_power_index + regime_age,
               data   = dem_breakdown,
               family = binomial("logit"))
```

After running the regression, with `summary()` we get the coefficients:

```
summary(model_2)
```

```
## 
## Call:
## "(...truncated)"
## 
## Deviance Residuals:
##     Min      1Q  Median      3Q     Max
## -0.734  -0.302  -0.266  -0.218   2.828
## 
## Coefficients:
```

```
##                 Estimate Std. Error z value Pr(>|z|)
## (Intercept)     -0.22220    0.96764   -0.23   0.8184
## pres_power_index -0.18897   0.06581   -2.87   0.0041 **
## regime_age      -0.00408    0.01961   -0.21   0.8350
## ---
## Signif. codes:  0 '***' 0.001 '**' 0.01 '*' 0.05 '.' 0.1 ' ' 1
##
## (Dispersion parameter for binomial family taken to be 1)
##
##     Null deviance: 217.84  on 643  degrees of freedom
## Residual deviance: 209.52  on 641  degrees of freedom
##   (1572 observations deleted due to missingness)
## AIC: 215.5
##
## Number of Fisher Scoring iterations: 6
```

Model 3 adds a third variable, dem_index, that corresponds to the Freedom House score[4] of quality of democracy (see Figure 8.6):. The higher the score, the performance of a country's democracy would be expected to improve.

Note: These are example models; we are not concerned about proving causality. See Chapter 10 for more information.

```
ggplot(dem_breakdown, aes(x = dem_index)) +
  geom_histogram(binwidth = 1) +
  labs(x = "Freedom House Index", y = "Frequency")
```

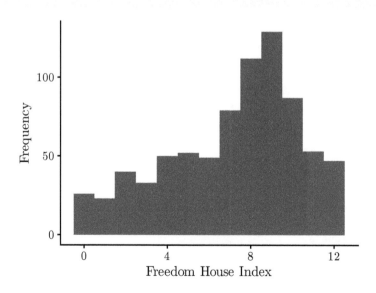

FIGURE 8.6: Histogram of the FH democracy index.

---

[4]See https://freedomhouse.org/report/methodology-freedom-world-2017

8.4 Model estimation

You will surely want to add a correlation matrix to your analysis to diagnose how the independent variables are related before including them in the model. To do this, we can use the `ggcorrplot()` package that allows you to create correlation matrices with ease. We see that the correlation between `regime_age`, `dem_index` and `pres_power_index` is very low, yet slightly positive (Figure 8.7).

```
library(ggcorrplot)

corr_selected <- dem_breakdown %>%
  dplyr::select(regime_age, dem_index, pres_power_index) %>%
  # calculate correlation matrix and round to 1 decimal place:
  cor(use = "pairwise") %>%
  round(1)

ggcorrplot(corr_selected, type = "lower", lab = T, show.legend = F)
```

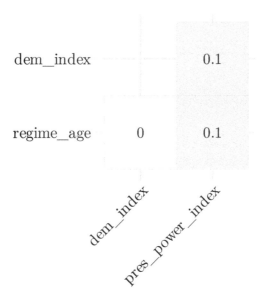

FIGURE 8.7: Correlation matrix of selected variables.

Model 3 will then look as follows:

```
model_3 <- glm(breakdown ~ pres_power_index + regime_age + dem_index,
               data   = dem_breakdown,
               family = binomial("logit"))


summary(model_3)
##
## Call:
## glm(formula = breakdown ~ pres_power_index + regime_age + dem_index,
```

```
##     family = binomial("logit"), data = dem_breakdown)
## 
## Deviance Residuals:
##     Min       1Q   Median       3Q      Max
## -1.7864  -0.0008  -0.0001   0.0000   1.8940
## 
## Coefficients:
##                  Estimate Std. Error z value Pr(>|z|)
## (Intercept)        15.360      6.584    2.33    0.020 *
## pres_power_index   -0.217      0.166   -1.31    0.189
## regime_age          0.166      0.112    1.48    0.138
## dem_index          -3.748      1.505   -2.49    0.013 *
## ---
## Signif. codes:  0 '***' 0.001 '**' 0.01 '*' 0.05 '.' 0.1 ' ' 1
## 
## (Dispersion parameter for binomial family taken to be 1)
## 
##     Null deviance: 71.271  on 421  degrees of freedom
## Residual deviance: 12.113  on 418  degrees of freedom
##   (1794 observations deleted due to missingness)
## AIC: 20.11
## 
## Number of Fisher Scoring iterations: 12
```

Now the variable `pres_power_index` is no longer statistically significant, however, both `dem_index` and the intercept are significant.

8.5 Creating tables

Here we will see how to create editable tables for academic articles using the `texreg` package of Philip Leifeld[5]. Basically, the function contains three packages, one to export tables to html (`htmlreg()`), one to export tables to LaTeX, and a third to see them in RStudio (called `screenreg()`). Through `htmlreg` we can export our tables formatted in html and from there incorporate them in our articles directly.

Once we estimated the models will go in the table, we group them into a list using the `list` function. This saves time, because instead of having to write the name of the models, we will simply refer to the set `mp_models`:

```
library(texreg)
mp_models <- list(model_1, model_2, model_3)
```

---

[5]See https://www.rdocumentation.org/collaborators/name/Philip%20Leifeld

8.5 Creating tables

To export the table to html we must define the option `file` and a name for the html file. In this example we will name it `table_1`. This html file can be dragged to your Word to insert the tables directly into the article. In the example, We will use two function options to further improve the table. The first is `custom.model.names =` that allows to name the models and the second is `custom.coef.names =` that allows to name the coefficients. So, the full command would be:

```
htmlreg(mp_models,
        custom.model.names = c("Model 1", "Model 2", "Model 3"),
        custom.coef.names = c("Intercept", "Shugart & Carey Index",
                              "Regime age", "Freedom House Index"),
        file = "table_1.html") # name of the new html file
```

This lines creates an html file in your project folder. If you right click on the file, you can open it in Word or any text processor.

Before exporting it, we can also see a screen version of the table using `screenreg()`. In this case, let's see how it looks without the `custom.coef.names =` option. You can use this command in your work projects, and once you decide on the best table, you export it with `htmlreg()`.

```
screenreg(mp_models,
          custom.model.names = c("Model 1", "Model 2", "Model 3"),
          custom.coef.names = c("Intercept", " Shugart & Carey Index",
                                "Regime age", "Freedom House Index")
          )
##
## ===========================================================
##                          Model 1      Model 2      Model 3
## -----------------------------------------------------------
## Intercept                 -0.24        -0.22       15.36 *
##                           (0.96)       (0.97)      (6.58)
##  Shugart & Carey Index    -0.19 **     -0.19 **    -0.22
##                           (0.06)       (0.07)      (0.17)
## Regime age                             -0.00        0.17
##                                        (0.02)      (0.11)
## Freedom House Index                                -3.75 *
##                                                    (1.51)
## -----------------------------------------------------------
## AIC                       213.56       215.52      20.11
## BIC                       222.50       228.92      36.29
## Log Likelihood           -104.78      -104.76      -6.06
## Deviance                  209.56       209.52      12.11
## Num. obs.                 644          644         422
## ===========================================================
## *** p < 0.001; ** p < 0.01; * p < 0.05
```

Notice that `pres_power_index` is no longer statistically significant when we control for `dem_index` in Model 3. In this model, `dem_index` becomes the only statistically significant variable. Also, notice that the number of observations falls significantly by including the variable `dem_index`, which makes it difficult to compare the models. The `skimr:skim()` function will be very useful to diagnose *missing values*, as we can see in Figure 8.8. All three variables have a very high number of missing levels, for `dem_index` of almost 2/3 of the total observations (1436 out of 2216).[6]

```
> skimr::skim(dem_breakdown)
-- Data Summary ---------------------
                        Values
Name                    dem_breakdown
Number of rows          2216
Number of columns       10

Column type frequency:
  character             2
  numeric               8

Group variables         None

-- Variable type: character ---------------------------------------
  skim_variable    n_missing complete_rate  min  max empty n_unique whitespace
1 country_name             0            1    4   18     0       20          0
2 president_name           0            1   10   64     0      540          0

-- Variable type: numeric -----------------------------------------
  skim_variable    n_missing complete_rate   mean    sd     p0     p25     p50     p75    p100 hist
1 year                     0          1     1955.  32.0   1900    1927    1955    1983    2010
2 breakdown             1268          0.428 0.0485 0.215     0       0       0       0       1
3 regime_age             896          0.596 15.4  12.4       1       5      12      22      61
4 dem_index             1436          0.352 7.16   3.15      0       5       8       9      12
5 growth_10y               1          1.00  0.0229 0.0307 -0.144 0.00573  0.0176  0.0352  0.141
6 x_oil_min             1508          0.319 0.0486 0.0727    0    0.00318 0.0110  0.0709  0.437
7 gini                  1551          0.300 51.2   6.37   30.0    46.5    51.9    56.5    67.8
8 pres_power_index      1054          0.524 16.3   3.30      5      15      16      18      25
```

FIGURE 8.8: Skim of our dataset.

In the highly recommended book *The Chicago Guide to Writing about Multivariate Analysis*, Jane Miller (2013) puts a lot of emphasis on the importance of understanding the difference between *statistical* significance and *substantive* significance when interpreting regressions: not because a coefficient is statistically significant the magnitude of the effect will be as expected, nor does it mean that the finding is scientifically relevant. What are some good options to grasp the magnitude of our effects? Odds ratios and visual plots of probabilities are good options.

To use odds ratios we want to show you how you can change the coefficients you get directly from the logistic regression, which are *log odds*, and replace them with *odds ratios*. To do this, you can take advantage of the arguments `override.coef`, `override.se` and `override.pvalues` of `screenreg()`. Odds ratios, as we saw before, are simply the exponentiated coefficients. Calculating their standard errors and p-values, meanwhile, is a bit more complex: we need to access the variance-covariance matrix of the model. Luckily this has already been done by Andrew Heiss[7], whom we thank for sharing the function. We slightly adapt these functions, which

---

[6]This might be problematic, yet sometimes there is nothing we can do about missing data. Yet, we cover a few possible solutions in Chapter 11.

[7]See https://www.andrewheiss.com/blog/2016/04/25/convert-logistic-regression-standard-errors-to-odds-ratios-with-r/

8.5 Creating tables

you can easily enter into your analysis as long as you have loaded the book package, `politicalds`.

```
screenreg(model_3,
          custom.model.names = "Model 3 - Odds Ratios",
          override.coef     = exp(coef(model_3)),
          # the following function, odds_*, are in the package of the book
          override.se       = odds_se(model_3),
          override.pvalues = odds_pvalues(model_3),
          # also, we will omit the coefficient of the intercept
          omit.coef = "Inter")
## 
## =======================================
##                   Model 3 - Odds Ratios
## ---------------------------------------
## pres_power_index    0.80 ***
##                    (0.13)
## regime_age          1.18 ***
##                    (0.13)
## dem_index           0.02
##                    (0.04)
## ---------------------------------------
## AIC                20.11
## BIC                36.29
## Log Likelihood     -6.06
## Deviance           12.11
## Num. obs.          422
## =======================================
## *** p < 0.001; ** p < 0.01; * p < 0.05
```

When obtaining a table like the one we just created, we run into two challenges that we will address in the next subsections: The first is to know if the magnitude of the effects is substantive from a scientific point of view. For example, what would we say if the variable `dem_index` is statistically significant yet we find that if a country goes from the worst score (0) to the best score (12) in `dem_index`, the probability of a democratic breakdown falls just by 0.03%? We would probably say that, despite having a statistically significant effect, our variable has no substantive significance.

The second challenge we will face is to compare our models to decide which one has the best fit, that is, which one has the best explanatory capacity. Unlike OLS, we do not have R^2, so other measures are used. Let's see!

Exercise 8B. Using the Latinobarometer dataset, choose three variables that you think can predict `pro_dem` and interpret the model with `summary`. If you dare, create tables with `texreg`. The available variables are: `regime_age` (age of the respondent), `ideol` (ideology, where 1 is extreme left and 10 extreme right), educ (years of education of the respondent) and socioecon_status (1, very good - 5, very bad).

8.6 Visual representation of results

We can visually represent the table we made with texreg using the prediction and margins packs. If you have used Stata in the past, this command will be familiar.

To visualize magnitudes, a figure is much more intuitive than a table. Figure 8.9 shows the predicted probability of a democratic breakdown according to the presidential powers index.

```
library(margins)
library(prediction)

predict_model_2 <- prediction::prediction(
  model_2,
  at = list(pres_power_index = unique(model.frame(model_2)$pres_power_index))
)
```

```
summary(predict_model_2)
```

```
## # A tibble: 15 x 7
##   `at(pres_power_index)` Prediction     SE     z       p    lower upper
##                    <dbl>      <dbl>  <dbl> <dbl>   <dbl>    <dbl> <dbl>
## 1                      5      0.227  0.117  1.95  0.0512 -0.00119 0.456
## 2                      8      0.143 0.0587  2.43  0.0149  0.0279  0.258
## 3                      9      0.121 0.0449  2.71 0.00681  0.0335  0.209
## # ... with 12 more rows
```

The summary offers us a tibble where we have the predicted probability of our dependent variable for each observed value of pres_power_index and its statistical significance. This easily becomes a figure, and for this there are two alternatives that we recommend (Figures 8.9 and 8.10).

```
# option 1
figure_op_1 <- ggplot(summary(predict_model_2),
              aes(x = `at(pres_power_index)`, y = Prediction,
                  ymin = lower, ymax = upper,
                  group = 1)) +
  geom_line() +
  geom_errorbar(width = 0.2) +
  theme(axis.text.x = element_text(angle = 90)) +
```

8.6 Visual representation of results

```
        labs(x = "Shugart index",
             y = "Pred. probability of Democratic Breakdown")

figure_op_1


# option 2
cdat <- cplot(model_2, "pres_power_index", what = "prediction",
              main = "Pr(Breakdown)", draw = F)


ggplot(cdat, aes(x = xvals)) +
  geom_line(aes(y = yvals)) +
  geom_line(aes(y = upper), linetype = 2)+
    geom_line(aes(y = lower), linetype = 2) +
  geom_hline(yintercept = 0) +
  labs(x = "Shugart index",
       y = "Pred. probability of Democratic Breakdown")
```

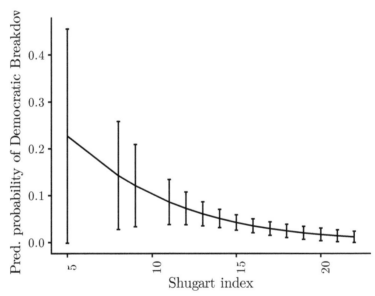

FIGURE 8.9: Option 1. Model 2 based on Mainwaring and Pérez Liñan (2013), pred. probability of a democratic breakdown.

Both options lead to the same result, the format of the figure can be customized to your preference with the syntax of ggplot2. what would you say about this effect in terms of scientific significance? If we were to interpret the magnitude of the SC index on the probability of a regime break, we would probably say that its fall produces a considerable effect in countries ranging between values 6 and 10 of the index, but for values larger than 10 the increase in the index almost does not affect the probability

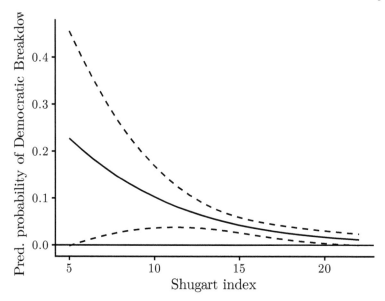

FIGURE 8.10: Option 2. Model 2 based on Mainwaring and Perez Liñan (2013), pred. probability of a democratic breakdown.

of a breakdown. Indeed, for countries with a presidential strength index greater than 10, the probability of democratic breakdown is still very low.

What is the average effect of the increase in a unit of the independent variable on the probability of occurrence of the dependent variable? The *Average Marginal Effect (AME)* are used to visualize these effects, and are achieved using the `margins()` function of the homonymous package and `ggplot2` (see Figure 8.11).

```
marginal_ef <- margins(model_2) %>%
  summary() %>%
  as_tibble()
```

```
ggplot(marginal_ef, aes(x = factor, y = AME, ymax = upper, ymin = lower)) +
  geom_point() +
  geom_linerange() +
  geom_hline(yintercept = 0) +
  scale_x_discrete(labels = c("Age of regime"," Shugart index")) +
  labs(x = "")
## $x
## [1] ""
##
## attr(,"class")
## [1] "labels"
```

8.6 Visual representation of results

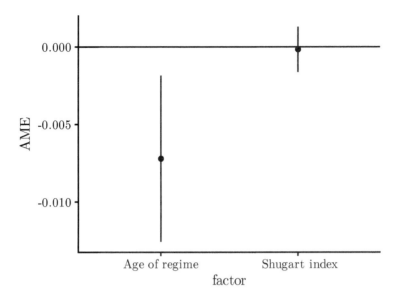

FIGURE 8.11: Average Marginal Effect of Model 2 based on Mainwaring and Perez Liñan (2013).

We may also be interested, not in the average effect, but in the complete marginal effect of a variable. Given the non-linearity of these models, the marginal effect of a variable on the probability of occurrence of the dependent variable is neither constant nor is it significant throughout the entire variable. For this we use the `cplot()` function of the `margins` package and then customize it with `ggplot2` options (see Figure 8.12).

```
marginal_pres_power_index <- cplot(
  model_2, "pres_power_index",
  what = "effect", main = "ME(Breakdown)", draw = F
)
```

```
ggplot(marginal_pres_power_index, aes(x = xvals)) +
  geom_line(aes(y = yvals)) +
  geom_line(aes(y = upper), linetype = 2)+
  geom_line(aes(y = lower), linetype = 2) +
  geom_hline(yintercept = 0) +
  labs(x = "Shugart index", y = "Marginal effect")
```

We can also graph the coefficients as *odds ratios*: remember that odds ratios smaller than 1 are negative effects and greater than 1 are positive. The coefficient is expressed as its average value and its 95% confidence interval. If the coefficient is statistically significant, your confidence interval will not pass the line at 1. If, on the contrary, they are not significant, the effect will cross the line. Compare the three models. To plot odds ratios we use the `jtools` package again, this time with its `plot_summs()`

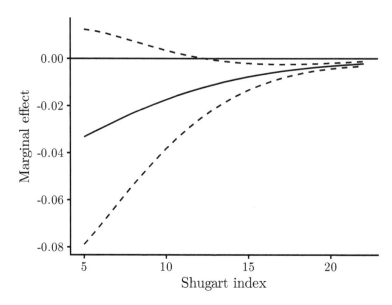

FIGURE 8.12: Marginal effect of presidential power index in Model 2 on Mainwaring and Perez Liñan (2013).

function. Since the option `exp = T` is true, we are exponentiating the coefficients. If we did `exp = F`, we would see the log-odds of the table (Figure 8.13).

```
library(jtools)

plot_summs(model_1, model_2, model_3,
           exp = T, scale = F, inner_ci_level = .9,
           coefs = c("Shugart index" = "pres_power_index",
                     "Age of regime" = "regime_age",
                     "Freedom House index" = "dem_index"),
           model.names = c("Model 1", "Model 2", "Model 3")) +
  labs(x = "Exponentiated coefficients", y = "")
```

```
library(jtools)

plot_summs(model_1, model_2, model_3, exp = T, scale = F,
           inner_ci_level = .9,
           coefs = c("Shugart index" = "pres_power_index",
                     "Age of regime" = "regime_age",
                     "Freedom House index" = "dem_index"),
           model.names = c("Model 1", "Model 2", "Model 3"),
           colors = c("black", "darkgray", "gray")) +
  labs(x = "Exponentiated coefficients", y = "")
```

It is increasingly common to find these types of figures instead of tables, and personally we think they are preferable. A precursor in the discipline of the use of figures and

8.6 Visual representation of results

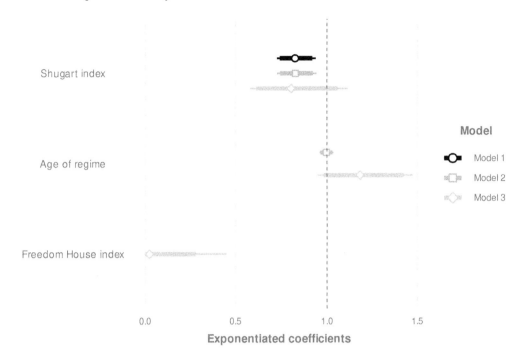

FIGURE 8.13: Odds ratios of Model 3, based on Mainwaring and Pérez Liñán (2013).

infographics was Edward Tufte[8], who was mentioned in Chapter 3. Political science, however, had not paid much attention to the presentation of results through figures until about two decades ago. Today, the trend[9] in the discipline is avoid using tables when they are not essential. Packages like `margins` and `jtools` have facilitated this task.

Let's look at predicted probabilities for the Freedom House score: We note that the relationship between the probability of occurrence of a democratic breakdown and the institutional quality score of Freedom House is negative. When a country has a Freedom House score of 2.5, the probability of a breakdown is 100%. When the score is 5, the probability is almost nil. Thus, a small change in this index significantly affects the probability of a regime break (Figure 8.14).

```
cdat <- cplot(model_3, "dem_index", what = "prediction", draw = F)


ggplot(cdat, aes(x = xvals)) +
  geom_line(aes(y = yvals)) +
  geom_line(aes(y = upper), linetype = 2)+
  geom_line(aes(y = lower), linetype = 2) +
```

---

[8]See http://pages.mtu.edu/~hcking/Tufte_hKing.pdf
[9]See https://www.princeton.edu/~jkastell/Tables2Graphs/graphs%20.pdf

```
geom_hline(yintercept = 0) +
labs(x = "Freedom House index",
    y = "Pred. probability of Democratic Breakdown")
```

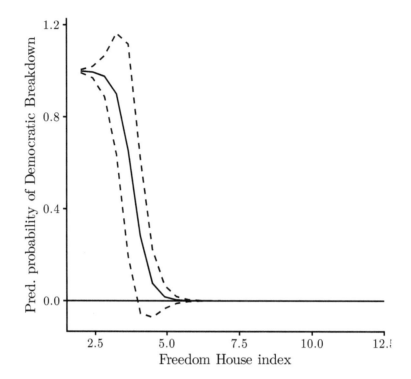

FIGURE 8.14: Predicted probabilities of FH values in Model 3, based on Mainwaring and Pérez Liñán (2013).

Exercise 8C. With the regressions you have run using the Latinobarometer data, create three plots to visualize the effects, whether predicted probabilities, marginal effects or coefficients expressed as odds ratios. Do your findings have substantive significance?

8.6.1 Visualization of interactions between variables

This topic may be useful if you already have some knowledge about what an interactive variable is and what it is used for. If this is not the case, and you are not interested in knowing, you can move on to the next section on adjusting the models without problems.

8.6 Visual representation of results

What would an interactive hypothesis look like in logistic models? Suppose that after months of exploring political variables that predict the occurrence of a regime breakdown in Latin America (regime_age, pres_power_index and dem_index), we decided to explore whether economic variables can have explanatory power. After extensively studying the existing literature, you formulate the hypothesis that the economic inequality in society (which we will measure with the Gini index) can affect the probability of a democratic breakdown. For that we will be still using the dataset dem_breakdown, but now with the economic variables (growth_10y, x_oil_min and gini). Thus, lets do another model, this time incorporating the gini index:

```
model_4 <- glm(breakdown ~ gini, data = dem_breakdown,
               family = binomial("logit"))
summary(model_4)
```

```
##
## Call:
## "(...truncated)"
##
## Deviance Residuals:
##     Min      1Q  Median      3Q     Max
## -0.334  -0.253  -0.226  -0.204   2.898
##
## Coefficients:
##              Estimate Std. Error z value Pr(>|z|)
## (Intercept)  -1.2591     2.4405   -0.52     0.61
## gini         -0.0460     0.0488   -0.94     0.35
##
## (Dispersion parameter for binomial family taken to be 1)
##
##     Null deviance: 110.28  on 442  degrees of freedom
## Residual deviance: 109.40  on 441  degrees of freedom
##   (1773 observations deleted due to missingness)
## AIC: 113.4
##
## Number of Fisher Scoring iterations: 6
```

But that's not all, because you also have an interactive hypothesis, which argues that the effect of economic inequality in society on the probability of a democratic breakdown will depend on how much the "cake" has grown in order to be distributed (that is, on growth of the recent economy). Thus, we will interact gini with growth_10y that measures the economic growth in the ten years prior to the democratic breakdown. We will also add some controls, which do not deserve too much discussion.

As you can see, the growth_10y*gini interaction is statistically significant. How do we interpret this coefficient?

```
model_5 <- glm(breakdown ~ growth_10y * gini + x_oil_min + pres_power_index,
               data = dem_breakdown,
               family = binomial("logit"))

summary(model_5)
## 
## Call:
## glm(formula = breakdown ~ growth_10y * gini + x_oil_min + pres_power_index,
##     family = binomial("logit"), data = dem_breakdown)
## 
## Deviance Residuals:
##     Min      1Q  Median      3Q     Max
## -0.581  -0.226  -0.178  -0.143   3.009
## 
## Coefficients:
##                    Estimate Std. Error z value Pr(>|z|)
## (Intercept)          6.5797     4.8787    1.35   0.177
## growth_10y        -256.1668   128.7769   -1.99   0.047 *
## gini                -0.1371     0.0788   -1.74   0.082 .
## x_oil_min           -1.6673     4.7359   -0.35   0.725
## pres_power_index    -0.2393     0.1320   -1.81   0.070 .
## growth_10y:gini      5.4901     2.6101    2.10   0.035 *
## ---
## Signif. codes:  0 '***' 0.001 '**' 0.01 '*' 0.05 '.' 0.1 ' ' 1
## 
## (Dispersion parameter for binomial family taken to be 1)
## 
##     Null deviance: 94.80  on 425  degrees of freedom
## Residual deviance: 87.54  on 420  degrees of freedom
##   (1790 observations deleted due to missingness)
## AIC: 99.54
## 
## Number of Fisher Scoring iterations: 7
```

The need to include interactive variables to test conditional hypotheses in binary models has been the subject of great academic debate in recent years. Unlike linear models estimated from OLS, where there is consensus on this need (partly from Brambor et al., 2006). In OLS, due to the construction of the estimator itself, and the assumption of linearity in the parameters, the effects of the independent variables on the dependent variable are always constant effects and independent of the value taken by the other covariates of the model.

Logit models, on the other hand, are interactive by nature: the effect of an independent variable on the probability of occurrence of the dependent variable, depends on the values of the other covariates of the model, as exemplified with Model 1 at the beginning of the chapter. There are two views about this in political science: According to Berry et al. (2010), whether or not to include an interactive term in a logistic regression will depend largely on the hypothesis or theory that predicts the effect of the interaction. In

8.6 Visual representation of results

particular, it will depend on if the hypothesis predicts an effect on the latent dependent variable, or if it is on the probability of occurrence of the event. If the hypothesis refers to the first scenario, then there is no further discussion in the literature, since the situation is analogous to the case of a continuous dependent variable estimated with an OLS regression. Therefore, it is necessary to incorporate a multiplication term with its respective coefficient to evaluate the interaction.

Rainey (2016), meanwhile, argues that even when the interaction hypothesis about the probability of occurrence of the dependent variable is based on the notion of compression, it is necessary to incorporate terms of interaction in the models. Rainey's recommendation (2016) is to always incorporate terms of interaction because if they do not include the logistic regression model, it cannot represent reasonable situations that would be inconsistent with the interactive theory.

While this debate remains open, in this chapter we recommend that you always include interactions if you have theoretical reasons to do so. For practical purposes the `margins` package that we will use to visualize this type of interactive effects requires this term.

How do we interpret this coefficient? Results show that the inequality of society increases the probability of a democratic breakdown only when the economy has grown in the last ten years. We also confirm that in economic recessions the probability of a democratic breakdown increases sharply but only if society has a Gini index of less than 50 points, that is, in more equitable societies. To plot this findings we use the `persp()` function of the `margins` package from our model that has the interaction (see Figure 8.15).

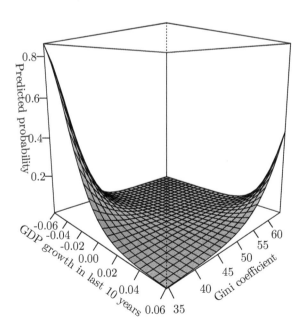

FIGURE 8.15: Interaction between Gini coefficient and GDP growth in three dimensions.

```
persp(model_5, "growth_10y", "gini", what = "prediction",
      xlab = "GDP growth in last 10 years",
      ylab = "Gini coefficient",
      zlab = "Predicted probability",
      family = "Latin Modern Roman")
```

An alternative way to see the results of the interaction is in two dimensions. For this we use the `image()` function. Both are equivalent, it will depend on our aesthetic preference (Figure 8.16).

```
image(model_5, xvar = "growth_10y", yvar = "gini",
      xlab = "GDP growth in the last 10 years", ylab = "Gini coefficient")
```

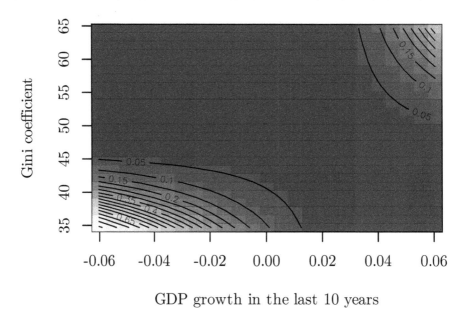

FIGURE 8.16: Interaction between Gini coefficient and GDP growth in two dimensions.

8.7 Measures to evaluate the fit of the models

Once you have analyzed the substantive significance of the models by means of tables and plotting the predicted probabilities and the marginal effects, you might be interested in the fit of the models. Just as in OLS, when R^2 and Mean Root Square Error are used, there is a series of statistics designed to know which of the logistic models has the best fit.

8.7 Measures to evaluate the fit of the models

8.7.0.1 Pseudo-R^2

Many times, you will see that logistic regressions are accompanied by Pseudo-R^2. It is not our favorite measure for logit models, but it is commonly used. To understand how Pseudo-R^2 is interpreted (McFadden's is normally used) it is important to understand how it differs from an R^2 in linear regressions (you can use this link[10] to accompany its interpretation in the OLS chapter). The formula, in this case is

$$psR^2 = 1 - \frac{ln\hat{L}(Complete model)}{ln\hat{L}(Model with intercept only)}$$

Where $\hat{L}$ is the likelihood estimated by the model. Basically, what the formula is doing is comparing the model with all our covariates to the model that barely has the intercept, to see how much the explanatory capacity of it improves if we include independent variables. Since L ranges between 0 and 1, its log is less than or equal to 0. Thus, the smaller the ratio, the greater the difference between the chosen model and the model with just the intercept. The function is taken from the `pscl` package. Models 1 and 2 have almost the same explanatory capacity, although one more variable has been incorporated in the second model. Instead, by adding `dem_index` to Model 3, our Pseudo-R^2 jumps up at 0.96.

```
library(pscl)
pR2(model_1)[["McFadden"]]
## fitting null model for pseudo-r2
## [1] 0.038
pR2(model_2)[["McFadden"]]
## fitting null model for pseudo-r2
## [1] 0.038
pR2(model_3)[["McFadden"]]
## fitting null model for pseudo-r2
## [1] 0.83
```

An adjusted $pseudo - R^2$ could also be implemented. Adjusted means that this is a version that penalizes the amount of covariates. As in the adjusted R^2, we add to the formula the term c, which is the amount of covariates, and hence we have

$$Pseudo - R^2 = 1 - \frac{ln\hat{L}(Complete model) - c}{ln\hat{L}(Model with intercept only)}$$

Although the R^2 always increases when adding covariates to the model, if we use adjusted measures, we will be taking this aspect into account.

```
library(DescTools)
PseudoR2(model_1, "McFadden")
## McFadden
##     0.43
PseudoR2(model_2, "McFadden")
```

---

[10]See http://setosa.io/ev/ordinary-least-squares-regression/

```
## McFadden
##     0.43
PseudoR2(model_3, "McFadden")
## McFadden
##     0.97
```

8.7.0.2 AIC

Almost always, tables with logistic regression models report AIC (Akaike Information Criterion). These allow the comparison of different models run for the same sample and decide which one has a better explanatory power. The AIC, like the $Pseudo-R^2$, uses information of $ln(\hat{L})$. What AIC does is measure the "distance" that exists between the true parameters and the estimators of the model, by means of a mathematical distance called Kullback-Leibler divergence. The smaller this distance, the better the model. It is very useful when comparing different models to argue which one should be taken as a reference, and it is calculated as

$$AIC = 2p - 2ln(\hat{L})$$

p is the number of regressors including the intercept, and $\hat{L}$ is the likelihood estimated by the model. Of the three models, clearly the one with the best fits falls under this criterion is the third.

```
AIC(model_1)
## [1] 214
AIC(model_2)
## [1] 216
AIC(model_3)
## [1] 20
```

8.7.0.3 BIC

The BIC (Bayesian Information Criterion), like AIC, is a criterion for comparing models according to their fit. For practical purposes, and not to enter into the differences between AIC and BIC, it is important to know that BIC penalizes the complexity of the model more rigorously than AIC, since its formula is

$$BIC = ln(n)p - 2ln(\hat{L})$$

where it adds to the formula n, which is the number of observations in the sample. The results are similar to those we obtained through AIC, that is, the third model shows the best fit.

```
BIC(model_1)
## [1] 222
```

```
BIC(model_2)
## [1] 229
BIC(model_3)
## [1] 36
```

For practical purposes, it is recommended to use AIC or BIC but it is not necessary to report both in a table with several models.

8.7.0.4 Percentage of correct predictions

Perhaps, a better name for it would be "Percentage of correct classifications". This is one of our favorite diagnose for goodness of fit in logit models, since it is the most intuitive of all. To understand the percentage of correct predictions in a model, it is important to be clear that a model produces four possible combinations for each observation:

	"1"	"0"
The model predicts "1"	True Positive	False Positive
The model predicts "0"	False Negative	True Negative

FIGURE 8.17: Classification table from which the percentage of correct predictions is calculated.

Any observation will be classified as "correct" if it corresponds to the upper left box (true positive) or the lower right (true negative). The percentage of observations that belong to these two boxes determines the percentage of correct predictions in the model. As a standard criterion, if the estimated probability for an observation is greater than or equal to 50%, it is estimated to be a positive probability, and if it is less than 50%, it will be a negative probability.

In order to calculate the correct predictions, we need to do several steps, but if you use the commands we offer you below, you will be able to do it with your own dataset when you work by yourself. First of all, we need to use the `broom` package, which transforms the models we have saved into datasets in `tidy` format as we saw in Chapter 5. We will calculate the predicted values of model 3 as an example that can be repeated for the other models. The first thing we need is the predicted values, that is, the probability assigned to each observation that Y=1. In the tidy we got using `broom` this variable is called `.fitted`.

```
library(broom)

pred_model_3 <- augment(model_3, type.predict = "response")
```

```
pred_model_3
## # A tibble: 422 x 12
##   .rownames breakdown pres_power_index regime_age dem_index .fitted
##   <chr>         <dbl>            <dbl>      <dbl>     <dbl>   <dbl>
## 1 75                0               19          1         8 8.47e-9
## 2 76                0               19          2         8 1.00e-8
## 3 77                1               19          3         3 6.19e-1
## # ... with 419 more rows, and 6 more variables
```

The `dplyr` function allows us to transform our new base obtained through `broom`. One of the basic functions of `dplyr` is `select()`, which allows us to choose variables by their names. We need to identify the predicted values and the dependent variable (in this case `breakdown`) in order to compare the probability assigned by the model with the real value of the variable. The `mutate` function of `dplyr` will allow us to create a binary variable in order to know if the model has correctly predicted each observation. The cut-off point is 0.5, that is, if the estimated probability is equal to or greater than 0.5, the model is considered to have predicted the occurrence of the event, and if it is less than that value, it is considered to have predicted the non-occurrence of the event. Working with arbitrary cut-off values has disadvantages, which as we will see below are resolved by ROCs. Finally, we create a variable that we call `pcp` (*p*ercentage of *c*orrectly *p*redicted) that shows the proportion of true positives and true negatives estimated by model 3. The result shows that of the 422 observations of model 3, 99.5% have been correctly predicted.

```
pred_model_3 %>%
  dplyr::select(.rownames, breakdown, .fitted) %>%
  mutate(predict_binary = if_else(.fitted >= 0.5, 1, 0),
                   predict_binary = if_else(breakdown ==
                                              predict_binary, 1, 0)) %>%
  summarize(pcp = mean(predict_binary))
## # A tibble: 1 x 1
##     pcp
##   <dbl>
## 1 0.995
```

8.7.0.5 Brier Score

This is another adjustment measure, less frequent in Political Science yet very useful. The closer the Brier score is to 0, the better the fit of the model. In general, one does not use all the measures that we are reviewing (AIC, BIC, Brier, etc.), but chooses two or three that are to one's liking. We think that situations in which you want to "punish" wrong predictions this is an ideal alternative since its formula is given by:

$$BS = \frac{1}{N}\sum(\hat{p} - x)^2$$

8.7 Measures to evaluate the fit of the models

where N is the number of observations, $\hat{p}$ is the predicted probability for each observation, and x is the actual value of the observation in our dataset. The score is the average for all observations in the sample. Which of the three models has the lowest score?

```
BrierScore(model_1)
## [1] 0.038
BrierScore(model_2)
## [1] 0.038
BrierScore(model_3)
## [1] 0.0048
```

8.7.0.6 ROC plot

Another one of our favorite measures for its versatility is the ROC plot. Compared to the percentage of correctly predicted observations, ROC curves have the advantage of not defining an arbitrary cutoff from which it can be decided whether the observation has been correctly or incorrectly classified. Its disadvantage is that it is an extra figure that you should include in your document, which you may want to add to an appendix. To interpret these figures, what matters most is the area below the diagonal curve that crosses the figure. The larger the area under the curve, the better the fit of the model. If you want to read more about it, the area forms a score called AUC score (which comes from *Area Under the Curve*). Let's build it with the geom_roc() function of the plotROC package. What we will do first is to create a dataset with the results of the three regression models. We will call the base pred_models.

```
library(plotROC)
library(broom)

pred_models <- bind_rows(augment(model_1, response.type = "pred") %>%
                           mutate(model = "Model 1"),
                         augment(model_2, response.type = "pred") %>%
                           mutate(model = "Model 2"),
                         augment(model_3, response.type = "pred") %>%
                           mutate(model = "Model 3"))
```

Once the object was created with the information of the three models, we proceed to create the figure using gglpot, that you learned in Chapter 3. On the vertical axis we have the *sensitivity* of the model, while on the horizontal axis we have (1-*specificity*) of the model. The sensitivity is the ratio between the true positives (that is, those observations predicted as "1", which were really "1" in the dataset), and the sum of the true positives plus the false negatives (those predicted as "0", which really were "1"). The specificity is the ratio between the true negatives (those observations predicted as "0" that were "0" in the dataset) and the sum of the false positives (those observations predicted as "1" that really were "0") added to the true negatives (Figure 8.18).

```
roc <- ggplot(pred_models, aes(d = breakdown, m = .fitted,
                               linetype = model)) +
  geom_roc(n.cuts = 0) +
  geom_abline(slope = 1) +
  labs(x = "1 - Specificity", y = "Sensitivity", linetype = "") +
  theme(legend.key.width = unit(2, "cm")) # make legend readable

roc
```

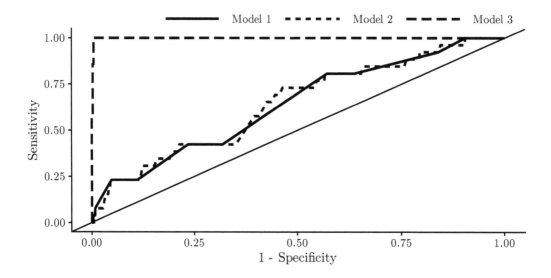

FIGURE 8.18: ROC plot of our models.

In order to get the AUC, we use the function `calc_auc()` on top of the object that we called `roc`.

```
calc_auc(roc)
## # A tibble: 3 x 3
##   PANEL group   AUC
##   <fct> <int> <dbl>
## 1 1         1 0.639
## 2 1         2 0.644
## 3 1         3 0.999
```

Exercise 8D. 1. Use the dataset `lapop` from the book's package: `data(lapop)`. 2. The variable `satisfied_dem` assumes 1 if the person believes that democracy is, despite their problems, the best form of existing government. Depending on her confidence in the media (`trust_media`, measures on

8.7 Measures to evaluate the fit of the models

a scale of 1 to 7), calculate how much the probability that this answer is 1 changes. 3. Estimate a model that predicts, the best you can, the dependent variable. 4. Plot the ROC curve of the model.

8.7.0.7 Separation plots

The separation plot was proposed by three political scientists in 2011 (Greenhill et al., 2011) and refers to the idea that we mentioned at the beginning of the chapter. This idea is that logistic models are classification models. The separation plot mixes the usefulness of both the correctly predicted percentages and the ROCs. These are figures that allow us to see the information in a classification table, but where we can also identify which observations are the worst predicted.

How do we interpret the figure? If the model predicted our dependent variable perfectly, all the observations in light gray would be on the left side, and all the black observations on the right side, since on the horizontal axis of the figure we have the predicted probability for each observation (see Figure 8.19).

```
library(separationplot)
separationplot(
  pred    = predict.glm(model_1, type = "response"),
  actual  = as.vector(model_1$y),
  type    = "line", BW = T, show.expected = T,
  newplot = F, line = F,
  heading = "Model 1"
)

separationplot(
  pred    = predict.glm(model_3, type = "response"),
  actual  = as.vector(model_3$y),
  type    = "line", BW = T, show.expected = T,
  newplot = F, line = F,
  heading = "Model 3"
)
```

The advantage of the separation plot is that each observation is included in the figure in the form of a vertical line that will be light gray if the observation assumes `breakdown = 0` and black if `breakdown = 1`. Notice how we use the argument `type = "line"` in the options, so that each observation is included as a line. A criticism that can be made to this attractive visual tool is that the greater the number of observations in our sample, the more difficult it becomes to interpret the figure. If the observations were too many it would be advisable to use the `type = "band"` argument, which facilitates reading for models with very large samples.

The little black triangle that we see below the figure is the point from which the black observations should begin if they were perfectly classified from the light gray ones.

Model 1

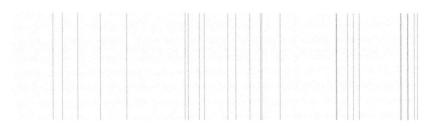

Model 3

FIGURE 8.19: Separation plots of models 1 and 3.

If you look at the separation plot of model 1, you will see that there are black lines on the left side of the figure. These are observations for which `breakdown = 1` and yet they have been estimated to have a very low probability of a democratic breakdown. Model 1, just with one independent variable (`pres_power_index`) predicts the reality of democratic breakdowns in Latin America quite badly. You will notice that model 3 has rated the observations much better than model 1. In fact, model 3, with only 3 independent variables, has been able to classify the dependent variable almost perfectly.

Exercise 8E.

- Add two independent variables to the model 3 we reviewed in the chapter and interpret the coefficients as *odds ratios*.
- Graph these coefficients using `ggplot2`.
- Diagnose the model fit with a ROC and a separation plot.

9
Survival Models

Francisco Urdinez[1]

Suggested readings

- Allison, P. D. (2014). *Event History and Survival Analysis*. SAGE, Thousand Oaks, CA, 2nd edition.
- Box-Steffensmeier, J. M. and Jones, B. S. (2004). *Event History Modeling: A Guide for Social Scientists*. Cambridge University Press, Cambridge.
- Broström, G. (2012). *Event History Analysis with R*. CRC Press, Boca Raton, FL.
- Golub, J. (2008). Discrete Choice Methods. In Box-Steffensmeier, J. M., Brady, H. E., and Collier, D., editors, *The Oxford Handbook of Political Methodology*, pages 530–546. Oxford University Press, Oxford.

Packages you need to install

- `tidyverse` (Wickham, 2019), `politicalds` (Urdinez and Cruz, 2020), `skimr` (Waring et al., 2020), `countrycode` (Arel-Bundock, 2020), `ggalt` (Rudis et al., 2017), `survival` (Therneau, 2020), `survminer` (Kassambara et al., 2020), `texreg` (Leifeld, 2020).

9.1 Introduction

There is a series of recurring questions regarding political data analysis we have not addressed yet. In many occasions we are interested in knowing why certain events last what they last, or why they take longer than others to occur. Why is peace so long-lasting in certain countries, while others are in constant war? What was the possibility of social unrest in Venezuela in 2018? Why do some legislators stay in office for many consecutive periods, while others do not even get reelected once? How long does it take for a union to strike during an economic crisis?

---

[1] Institute of Political Science, Pontificia Universidad Católica de Chile. E-mail: furdinez@uc.cl. Twitter: @FranciscoUrdin.

All these questions regard the duration of an event. The moment an event occurs is part of the answer we are looking for, so we need a model that allows us to reach that answer. Janet Box-Steffensmeier, one of the main names in Political Science for this method, refers to them as "event history models" (2004), although a good part of the literature calls them survival models or duration models. Though in Political Science these models are not used as much as one might think (in the end, almost every question we ask ourselves can be reformulated into questions about the duration of an event), medical science has explored these methods in depth, and many of the references found in R about these models' accessory packages are located in biostatistical and medic subjects. Hence, "survival models" is the name frequently used for referencing these models, since in medicine they were first used to model which variables affected the survival of sick patients.

We can work with two types of datasets to address survival problems. On the one hand, we can have a dataset in panel format, in which, for a given moment, our dependent variable encodes if the event has occurred (=1) or not (=0). Thus, for example, we can have a sample of twenty countries for fifty years (1965-2015) in which our variable of interest is whether the country has implemented a constitutional reform. The independent variable will assume the value "1" for the year 1994 in Argentina[2], but it will be "0" for the rest of the years in this country. On the other hand, we can have a cross-sectional dataset in which every observation appears encoded just once. In this case, we need, a variable that will tell us if in the period of interest the event occurred or not for every observation (for example, Argentina should be encoded as 1), and an extra variable that encodes the "survival" time of each observation, that is, how much time passed until that event finally happened. For the case of Argentina, this variable will encode 29 (years), which is the time it took for a constitutional reform to be implemented since 1965. The selection of the starting year, as you can suspect, is a decision of the researcher but it has an enormous effect on our results. Quite often the starting date ends up defined by the availability of data, which deviates from the ideal we would like to model.

Let's suppose that we ask ourselves the same question David Altman (2019) did: "Why some countries take less time than others in implementing instances of direct democracy?" For that, we have a dataset in a panel format that starts in the year 1900 and ends in 2016 for 202 countries (some observations, such as the Soviet Union, are transformed into other observations from the given year in which they ceased to exist). While taking a quick look at his data, one notes something that will probably happen to you in your dataset. By the last year, in 2016, only a small percentage of countries had implemented these type of mechanisms (27% to be precise), yet the dataset is censored, meaning that we do not know what has happened in the countries that had not implemented direct democracy mechanisms by that year. How do we estimate *when* those observations which have not "died" yet will do? It is a recurrent question in Political Science that we will answer thanks to these types of models. We will estimate the time it will take to "die" for every censored country in our sample (with the data we give to the model, which will always be insufficient).

---

[2]See https://en.wikipedia.org/wiki/1994_amendment_of_the_Constitution_of_Argentina

9.1 Introduction

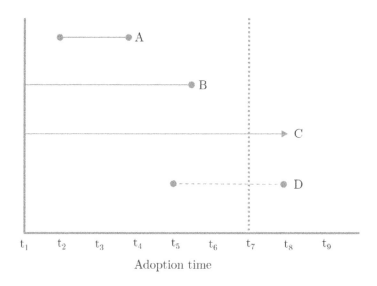

FIGURE 9.1: Example of observations present in a survival dataset.

In our datasets you can have four types of observations (see figure 9.1): (a) those that are already in the sample for the moment we have data, although we will not always know for how long they have "existed". They are, in the figure, the observations B and C. In Altman's datasets, for example, Mexico already existed as a political entity in 1900, when his dataset starts (we know that the First Federal Republic existed as a political entity since October of 1824, so that Mexico would be encoded as existing since that date). We also know that in 2012, for the first time, Mexico implemented an initiative of direct democracy, which defines as positive the occurrence of the event we are interested in measuring. Thus, Mexico would be like observation B in the figure; (b) Some observations will be present since the start of the sample and will exist until the last moment without registering the event of interest. That is the case of observation C in the figure. In Altman's sample, an example would be Argentina, which is registered in the dataset since 1900 (had already been "born") and had not registered instances of direct democracy until the last year of the sample (it did not "die"), which transforms it into a censored observation.

For practical purposes, it does not make a difference what happened onward from the year in which our base ends. For example, in the figure, our dataset covers until t_7, and we know that in t_8 observation C had not died yet, and observation D did in t_8. In our dataset, both C and D will be censored observations in t_7; (c) Some observations may enter "late" in the sample, as is the case with observations A and D. For example, Slovenia entered in Altman's sample in 1991, which is the year when it became independent from Yugoslavia and was "born" as a country; (d) Some observations, regardless of when they enter the sample, will "die" during the analyzed period. For example, A and B die during the period we have measured between t_1 and t_7. For observation D, we do not register its death. There is a case not considered in the example, that of observations that are born and die successively throughout the study period. For them, we have to decide whether we treat them as independent observations, or we model the possibility of multiple deaths per observation. If so, the probability of dying a second time will be conditioned by the probability of having died (and when!) for the first time. This is a somewhat more complex type of case, and it will not be covered in this chapter.

9.2 How do we interpret hazard rates?

Survival models are interpreted as the possibility that, at a given moment, the event of interest occurs given that it has not occurred yet. This possibility is named "hazard rate". We begin by knowing that we have a variable, that will be called T, which represents a random positive value, and that has a probability distribution (corresponding to the probability of the event to occur in each possible moment) that we will name $f(t)$. This probability can be expressed in a cumulative way as a cumulative density $F(t)$. As expressed in the following formula, $F(t)$ is given by the probability that, in the time of survival, T is less than or equal to a specific time t :

$$F(t) = \int_0^t f(u)d(u) = Pr(T) \leq t)$$

The survival function $\hat{S}(t)$, which is a key concept in these models, is related to $F(t)$, since:

$$\hat{S}(t) = 1 - F(t) = Pr(T \geq t)$$

In other words, the survival function is the inverse probability of $F(t)$, since it refers to the probability that the survival time T is greater than or equal to a time t of interest. For the specific example of Altman, one might ask what is the probability that a country will not implement a mechanism of direct democracy (which would be equivalent to "surviving" such implementation) being that it has already survived them for 30 years. As more and more countries in the sample implement initiatives of direct democracy, the probability of survival decreases.

The coefficients of the survival models are often interpreted as "hazard rates", which is the quotient of the probability of the event happening and the survival function:

$$h(t) = \frac{f(t)}{S(t)}$$

Then, the hazard rate indicates the rate at which the observations "die" in our sample in a given t, considering that the observations have survived until t. Later, we will see how in Altman's example we can interpret the coefficients of our regressions as hazard rates. All in all, the hazard rate $h(t)$ is the hazard that the event occurs in a given time interval, which is determined by:

$$f(t) = \lim_{\triangle x \to 0} \frac{P(t + \triangle t > T \geq t)}{\triangle t}$$

9.3 Cox's model of proportional hazards

There are two types of survival models, the so-called parametric models and the semi-parametric models. The former corresponds to those that make assumptions about the characteristics of the population to which the sample is representative of. In this case, the assumptions are about the "baseline hazard", that is, about the hazard of the event occurring when all our independent variables are equal to zero. The most common survival model for this category is Weibull's model. On the other hand, semi-parametric models do not make any type of assumptions about the base function, as this is estimated from the data. The most famous example of this specification is Cox's model.

The *Oxford Handbook of Political Methodology* dedicates an entire chapter (Golub, 2008) to discussing survival models, in which the author takes a strong position in favor of semi-parametric models. On the one hand, since no assumptions are made about the baseline hazard, its estimates are more precise. In a parametric estimate, choosing a wrong baseline hazard implies that all our analysis will be biased. Deciding a priori the baseline hazard function in a Weibull model should be oriented by theoretical reasons regarding the effect of our independent variable in the survival probability of the observation (see Figure 9.2). Nonetheless, our theory does not often define those assumptions. Choosing a specification by Cox's saves us from making a costly decision.

Another advantage of the semi-parametric models over the parametric models has to do with the proportionality of hazards. Both models, parametric and semi-parametric, assume that the hazard rate between any two observations in the sample are constant throughout the whole survival period (i.e., constant relative hazard). That is, it is assumed that the hazard function for each individual follows the same function over time. This is a strong assumption in political science research, since observations change over time and are different from one another. Think about Altman's thesis, for example. One can theorize that the probability of a direct democracy initiative in a given year in a given country will be affected by the strenght of its democratic institutions, which can proxied with a certain type of standard variable such as the 21 scale-points of Polity IV[3] or the most recent measurement of V-Dem[4]. Then, we

---

[3] See https://www.systemicpeace.org/polity/polity4.htm
[4] See https://www.v-dem.net

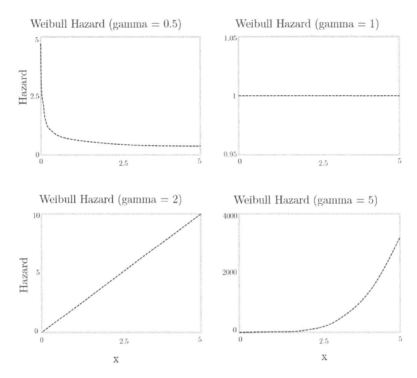

FIGURE 9.2: Different baseline hazards in Weibull's model.

could theorize that, the greater a democracy's institutional strength, the greater the probability of implementing mechanisms of direct democracy.

However, the values of these variables not only differ across countries, but vary over time in a certain country. For example, take Colombia, where the V-Dem variable "vdem_score" suffered advances and setbacks between 1900 and 2016 (see Figure 9.3). Every time the values of this variable changes, the hazard rates necessarily change for direct democracy to occur in Colombia, violating the proportional hazard assumption.

The advantage of Cox's model over its parametric counterparts is the existence of tests to identify if a variable in our model breaks the proportional hazard assumption. If they do we can correct it by generating interactions between these variables and temporal variables. In this way, we allow two types of coefficients to be present in our model: coefficients that stay constant over time and coefficients that change over time. For example, our theory could predict that an abrupt expansion in the quality of democratic institutions in a country would increase the hazard rate of implementing direct democracy and that this effect would fade within four or five years. When you specify your model, it is important to think about which variables can be assumed to remain constant in their hazards, and which not.

The recommendation given in the *Oxford Handbook* for a good implementation of survival models is as follows: (a) First, given the advantages of semi-parametric models over parametric, it is recommended the use of Cox over Weibull or another parametric model; (b) Once we have defined our dependent variable (the event), and

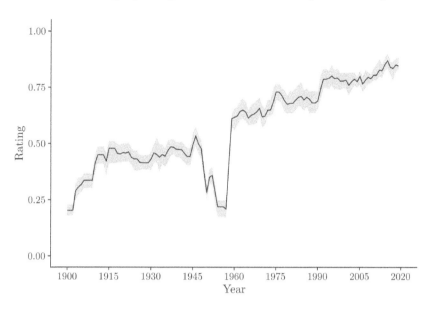

FIGURE 9.3: Colombia's values for polyarchy according to V-Dem.

the time of "birth" and "death" of each observation, we can specify our model; (c) The coefficients must be interpreted as hazard rates, which requires the exponentiation of the base coefficients obtained in R; (d) Once we have the model that we think is the right one based on our theoretical intuitions, it is necessary to test that none of the coefficients violates the proportional hazard assumption. To do so, we run a Grambsch and Therneau test, or the analysis of Schoenfeld residuals; (e) Once the problematic coefficients are identified, we will interact them with the natural logarithm of the variable that measures the duration of the event. By doing this, there will be coefficients whose effects fade or grow over time. When the problematic coefficients have been corrected, we can proceed to interpret our model and the survival function of the model.

9.4 Estimating Cox Models in R

Let's go back to the question that Altman made in Chapter 3 of *Citizenship and Contemporary Direct Democracy* (2019), entitled "Catching on: waves of adoption of citizen-initiated mechanisms of direct democracy since World War I". The question is: why do some countries take less time than others in implementing instances of direct democracy?. Let's start by loading `tidyverse` and our dataset (the latter, from our package `politicalds`):

```
library(tidyverse)
library(survival)


library(politicalds)
data("direct_democracy")
```

Now the dataset has been loaded into our R session:

```
ls()
## [1] "direct_democracy"
```

The variables we have in the dataset are the following:

- Dependent variable, which registers the occurrence of the event, in this case, the adoption of a direct democracy mechanism - `direct_dem`
- Year - `year`
- Country name - `country_name`
- The country goes through a quick process of democratization - `rapid_positive_dem`
- The country goes through a quick process of democracy decay - `rapid_negative_dem`
- Memory of previous instances of direct democracy - `memory`
- Democracy score of the country - `vdem_score`
- Effect of the diffusion of capabilities - `capab_diffusion`
- Effect of the diffusion of occurrences - `ocur_difussion`
- Country's natural logarithm of the total population - `log_pop`
- Dummy for British ex - colonies - `colony_gbr`
- Dummy for ex members of the USSR - `colony_ussr`

Throughout the example we will use the following packages: `skimr`, `countrycode`, `survival`, `rms`, `survminer`, `ggalt`, `tidyverse` and `texreg`. We will load them individually for you to see what they are for.

If we use `skimr::skim()`, as we did in previous chapters, we can observe that it is a dataset in a balanced panel format (see Figure 9.4). That is, we have a variable "country_name" which is repeated along a "year" variable.

Countries "enter" to the dataset when they start existing as independent countries. For example, let's look at Albania, which was born as a country in 1912 after the Balkan Wars:

```
direct_democracy %>%
  filter(country_name == "Albania")
## # A tibble: 105 x 23
##   country_name  year direct_dem rapid_positive_~ rapid_negative_~ memory
##   <chr>        <dbl>      <dbl>            <dbl>            <dbl>  <dbl>
## 1 Albania       1912          0                0                0      0
```

9.4 Estimating Cox Models in R

```
> skimr::skim(direct_democracy)
-- Data Summary ----------------------
                           Values
Name                       direct_democracy
Number of rows             13885
Number of columns          23

Column type frequency:
  character                1
  numeric                  22

Group variables            None

-- Variable type: character ------------------------------------------------
  skim_variable n_missing complete_rate min max empty n_unique whitespace
1 country_name          0           1     4  32     0      201          0

-- Variable type: numeric --------------------------------------------------
   skim_variable      n_missing complete_rate  mean     sd      p0     p25     p50     p75    p100 hist
   year                       0          1    1973.   31.1   1900    1952    1979    1999    2016
   direct_dem                 3       1.00    0.103   0.304     0       0       0       0       1
   rapid_positive_dem         0          1    0.0395  0.195     0       0       0       0       1
   rapid_negative_dem         0          1    0.0224  0.148     0       0       0       0       1
   memory                     3       1.00    0.343   0.424     0       0     0.1    0.88       1
   vdem_score               643      0.954    0.430   0.296  0.00987  0.177  0.334   0.717      1
   capab_diffusion          384      0.972    0.345   0.784     0    0.0381  0.125   0.411    18.6
   ocur_difussion           396      0.971    0.0709  0.131     0   0.00947 0.0297  0.0728    2.32
   c_pos_capabilities       384      0.972    0.0102  0.167     0       0       0       0      16.5
   c_pos_occurrences        396      0.971   0.00231  0.0203    0       0       0       0     0.818
   c_pos_memory               3       1.00    0.0167  0.123     0       0       0       0       1
   log_pop                  522      0.962    15.4    2.04    8.74   14.4    15.6    16.7     21.1
   colony_gbr                 0          1    0.273   0.445     0       0       1       1       1
   c_fra                      0          1    0.126   0.332     0       0       0       0       1
   colony_ussr                0          1    0.0387  0.193     0       0       0       0       1
   c_spa                      0          1    0.172   0.377     0       0       0       0       1
   c_usa                      0          1    0.0265  0.154     0       0       0       0       1
   c_ned                      0          1    0.0335  0.180     0       0       0       0       1
   c_prt                      0          1    0.0331  0.179     0       0       0       0       1
   c_bel                      0          1    0.0120  0.109     0       0       0       0       1
   c_ahe                      0          1    0.0214  0.145     0       0       0       0       1
   c_ote                      0          1    0.0768  0.266     0       0       0       0       1
```

FIGURE 9.4: Skim of our dataset.

```
## 2 Albania       1913           0                  0                      0       0
## 3 Albania       1914           0                  0                      0       0
## # ... with 102 more rows, and 17 more variables
```

For the correct functioning of the models in R, countries should exit the analysis (and the dataset!) when they "die". In this case, death occurs when countries adopt mechanisms of direct democracy. Albania, following the example, should stop existing in 1998, and should not last until 2016 as it does in the dataset. Thus, we need to create a second version of our dataset where this has already been corrected.

Note: If your dataset is in this format since the beginning, you can skip this step.

Once the dataset is ready, it should look as follows (see Figure 9.5). NA denotes missing values, 0 if the country in question has not implemented instances of direct democracy, 1 if it has, and when it does, it should exit our sample.

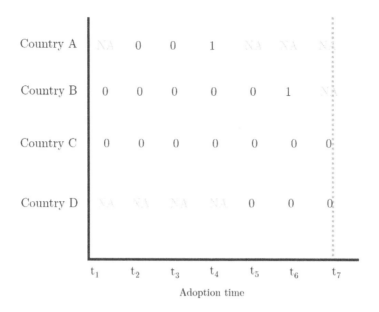

FIGURE 9.5: How your dataset should look.

```
direct_democracy_b <- direct_democracy %>%
  group_by(country_name) %>%
  # that the cumulative sum of the dummy is maximum 1
  filter(cumsum(direct_dem) <= 1) %>%
  ungroup()
```

What we are doing is filtering the data so that when the event of interest occurs (in this case is the `direct_dem` variable) the following periods of time are replaced with NAs. If we compare the case of Albania for the original dataset and the new dataset, we will see the key difference. In the first dataset we have observations for 1999 and 2000, after the event occurred. In the second dataset, Albania "dies" in 1998:

```
direct_democracy %>%
  filter(country_name == "Albania" & direct_dem == 1)
## # A tibble: 19 x 23
##   country_name  year direct_dem rapid_positive_~ rapid_negative_~ memory
##   <chr>         <dbl>      <dbl>            <dbl>            <dbl>  <dbl>
## 1 Albania       1998           1                0                0      1
## 2 Albania       1999           1                0                0      1
## 3 Albania       2000           1                0                0      1
## # ... with 16 more rows, and 17 more variables

direct_democracy_b %>%
  filter(country_name == "Albania" & direct_dem == 1)
## # A tibble: 1 x 23
##   country_name  year direct_dem rapid_positive_~ rapid_negative_~ memory
```

```
##   <chr>       <dbl>       <dbl>            <dbl>              <dbl>  <dbl>
## 1 Albania      1998           1                0                  0      1
## # ... with 17 more variables
```

To summarize, we transformed our dataset into an unbalanced panel, where countries enter the dataset when they start existing as such, and leave, either when they adopt direct democracy mechanisms, or when the dataset ends its temporal extension (in 2016). In this way, our dataset gets closer to Figure 9.1, which is used to exemplify the different types of observations.

Exercise 9A. Take a minute to do an exercise before we continue. In the previous chapter we used the dataset from the book "Democracies and Dictatorships in Latin America: Emergence, Survival, and Fall" (Mainwaring and Pérez-Liñán, 2013). Use that same dataset to see if it is ready to be used with survival models.

What if we tried something similar to Figure 9.1 using the data from David Altman's book? These types of figures are called Gantt charts, and they can be recreated with ggplot2, although is fair to say that it is necessary to follow some steps that can be a bit difficult. We hope that with this example you can recreate the plot with your own data, since it is a figure that will be beneficial for your readers.

First, we need to create a dataset that, for every country, registers the year of entry and the year of exit from the dataset. We are also interested in knowing why the country exits: does it adopt a direct democracy mechanism, or the dataset just ends? We will create a subset called `gantt_chart_df` where we will keep only three variables of the dataset, which are the country name `country_name`, the year `year`, and the dependent variable `direct_dem`. We will also remove from the dataset those observations that for the first year have already "died". For example, Switzerland had already implemented direct democracy mechanisms before 1900, so from the first year of the dataset to the last the dependent variable will be 1:

```
gantt_plot_df <- direct_democracy_b %>%
  # the variables we are interested in
  dplyr::select(country_name, year, direct_dem) %>%
  group_by(country_name) %>%
  filter(year == min(year) | year == max(year)) %>%
  # we need to remove the observations for countries that "are born" with
  #direct democracy:
  filter(!(year == min(year) & direct_dem == 1)) %>%
  summarize(year_enters = min(year),
```

```
                 year_exits  = max(year),
                 exits_bc_dd = max(direct_dem)) %>%
  ungroup()
gantt_plot_df
## # A tibble: 193 x 4
##   country_name year_enters year_exits exits_bc_dd
##   <chr>              <dbl>      <dbl>       <dbl>
## 1 Afghanistan         1919       2016           0
## 2 Albania             1912       1998           1
## 3 Algeria             1962       2016           0
## # ... with 190 more rows
```

The countries that exit because of direct democracy ("die") are:

```
gantt_plot_df %>% filter(exits_bc_dd == 1)
## # A tibble: 48 x 4
##   country_name year_enters year_exits exits_bc_dd
##   <chr>              <dbl>      <dbl>       <dbl>
## 1 Albania             1912       1998           1
## 2 Belarus             1991       1995           1
## 3 Belize              1981       2008           1
## # ... with 45 more rows
```

We can identify in a new variable the geopolitical region of each country thanks to the `countrycode::countrycode()` function (we explain this in greater detail in Chapter 11). This package is of great utility for those who do comparative politics or international relations because it makes very easy the task of giving codes to countries. This package assigns each country to its region of belonging automatically:

```
library(countrycode)

gantt_plot_df_region <- gantt_plot_df %>%
  mutate(region = countrycode(country_name,
                              origin = "country.name", dest = "region"))

gantt_plot_df_region
## # A tibble: 193 x 5
##   country_name year_enters year_exits exits_bc_dd region
##   <chr>              <dbl>      <dbl>       <dbl> <chr>
## 1 Afghanistan         1919       2016           0 South Asia
## 2 Albania             1912       1998           1 Europe & Central Asia
## 3 Algeria             1962       2016           0 Middle East & North Af~
## # ... with 190 more rows
```

With our dataset already assembled we can plot it thanks to `ggalt::geom_dumbbell()`:

```
library(ggalt)
gantt_plot <- ggplot(data    = gantt_plot_df_region,
                     mapping = aes(x     = year_enters,
                                   xend  = year_exits,
                                   y     = fct_rev(country_name),
                                   color = factor(exits_bc_dd))) +
  geom_dumbbell(size_x = 2, size_xend = 2)

gantt_plot
```

In the vertical axis we have the countries in alphabetical order, and in the x-axis there are two variables that feed the plot. On one side, the year of birth (`year_enters`), and on the other, its year of death (`year_exits`). Also, there is a third informative variable in the color of the line, which denotes if the country implemented or not an instance of direct democracy (`exits_bc_dd`). The countries in a darker tone are the ones which implemented such an instance between 1900 and 2016. Although the figure is of an enormous visual utility, we must admit that there are too many countries for it to be included in the body of an article. Indeed if you are reading from our printed book this figure is almost impossible to grasp. A possible solution is to plot by regions. You might remember the function `filter()` from Chapter 2. For example, let's filter countries in Latin America & Caribbean:

```
gantt_plot_la <- ggplot(
  data = gantt_plot_df_region %>%
    filter(region == "Latin America & Caribbean"),
  mapping = aes(x     = year_enters, xend = year_exits,
                y     = fct_rev(country_name),
                color = fct_recode(factor(exits_bc_dd)))
) +
  geom_dumbbell(size_x = 2, size_xend = 2)

gantt_plot_la
```

We could add the years as labels to improve the reading of the figure, alongside some cosmetic adjustments (using all we have learned in Chapter 3):

```
gantt_plot_la <- gantt_plot_la +
  geom_text(aes(label = year_enters), vjust = -0.4) +
  geom_text(aes(x = year_exits, label = year_exits), vjust = -0.4) +
  labs(x = "year", y = "",
       color = "Do they adopt direct democracy?") +
  theme(axis.text.x = element_blank())

gantt_plot_la
```

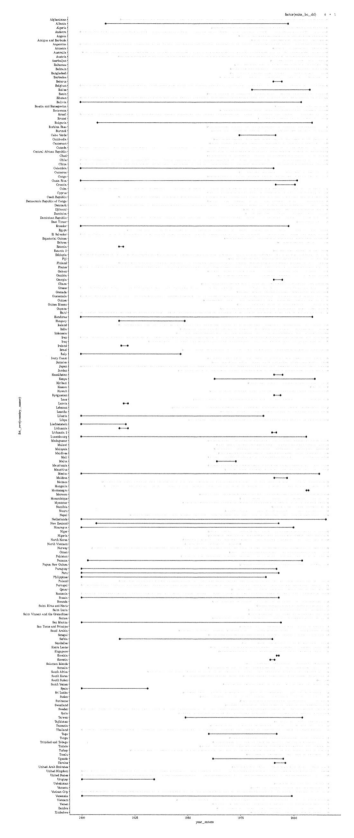

FIGURE 9.6: Gantt plot for all observations.

9.4 Estimating Cox Models in R

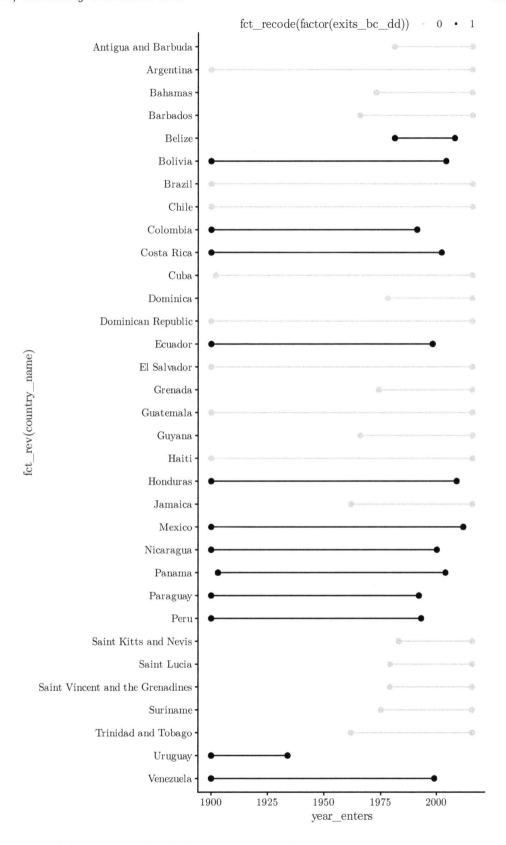

FIGURE 9.7: Gantt plot for Latin America, simple version.

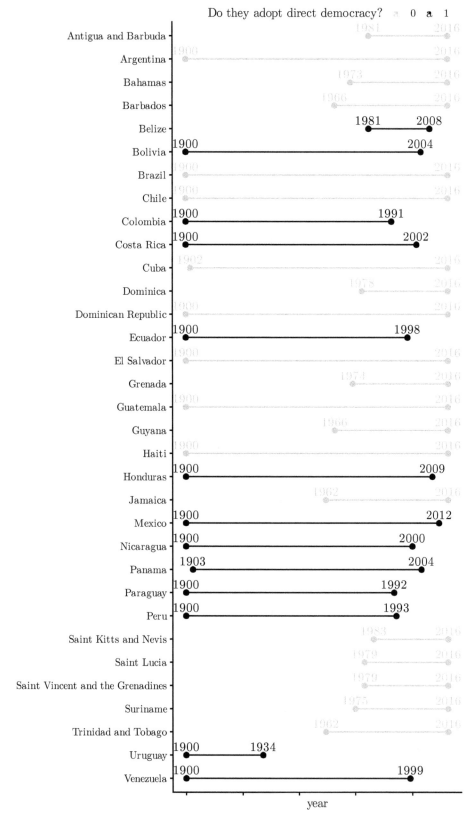

FIGURE 9.8: Gantt plot for Latin America, with starting and ending years.

9.4 Estimating Cox Models in R

Exercise 9B. With the dataset from the book "Democracies and Dictatorships in Latin America: Emergence, Survival, and Fall" (Mainwaring & Pérez-Liñán 2013) create a Gantt chart like the previous figure, plotting the democratic breaks in Mexico.

In addition to Gantt charts, it is common that people who work with survival models show graphs of survival curves comparing two groups of interest. For example, David Altman asks himself if there was a difference in the 20th century between countries that had rapid democratization processes and those which took decades to do so in how quickly direct democracy was implemented. These types of figures do not have inferential value, but great descriptive value. We will estimate a non-parametric survival curve, using the Kaplan-Meier method.

Following this, we will make another modification to our dataset. You will probably have to make it with your own data, so pay attention. Survival models do not work with traditional panel data, like our dataset. We need to convert them into "time at risk" format. What does this mean? It means that we need two new variables, one that states how long the observation carries the risk of dying at the beginning of each t, `risk_time_at_start`, and another variable that does the same at the end of each t, `risk_time_at_end`. From now on, these two variables will be used, in the scripts of our survival analysis.

```
direct_democracy_c <- direct_democracy_b %>%
  group_by(country_name) %>%
  # we will eliminate the first year for each country;
  # as it is not at risk of dying yet!
  filter(year != min(year)) %>%
  mutate(risk_time_at_end   = c(1:n()),
         risk_time_at_start = c(0:(n() - 1))) %>%
  ungroup() %>%
  dplyr::select(country_name, year, risk_time_at_start, risk_time_at_end,
                everything())
```

```
direct_democracy_c
## # A tibble: 11,995 x 25
##    country_name  year risk_time_at_st~ risk_time_at_end direct_dem
##    <chr>        <dbl>            <int>            <int>      <dbl>
## 1 Afghanistan   1920                0                1          0
## 2 Afghanistan   1921                1                2          0
## 3 Afghanistan   1922                2                3          0
## # ... with 11,992 more rows, and 20 more variables
```

Let's run the non-parametric survival curve, using the Kaplan-Meier method.

```
km <- survfit(Surv(time = risk_time_at_start, time2 = risk_time_at_end,
             event = direct_dem) ~ rapid_positive_dem,
        type      = "kaplan-meier",
        conf.type = "log",
        data      = direct_democracy_c)
```

Now we can create our plot with `survminer::ggsurvplot()` (see Figure 9.9).

```
library(survminer)
ggsurvplot(km, data = direct_democracy_c,
        conf.int = T, risk.table = T,
        break.x.by = 20,
        legend.labs = c("Fast democratization = 0",
                        "Fast democratization = 1"),
        legend.title = "")
```

Altman holds that countries that suffered democratic "shocks" were faster in implementing direct democracy mechanisms. The figure confirms his intuition, because we observe that the probability of survival of a country (read it as the probability of a country to persist without implementing mechanisms of direct democracy) decreases to a half on the first four years following a democratic shock. On the other hand, countries that carried out slow democratic processes did not suffer this effect.

Exercise 9C. Using the same data as the previous exercise: How does the Kaplan-Meier curve compare between countries that received high political support from United States and those who received low political support? For that, use the variable `us_t`, which is an index from 0 to 1, where 1 denotes constant support from the United States in a country, and 0 denotes that no support was offered from the United States. To compare the two groups in the Kaplan-Meier curve, create a *dummy* that assumes 1 if the support is greater than a 0.75 and 0 otherwise.

9.5 Tools to interpret and present hazard ratios

We will use Altman's dataset to estimate some models and run Grambsch and Therneau's to test if the coefficients violate the proportional hazard assumption. We are not replicating the models from his chapter because they are a little more complex (they include interactions), but we are just using his baseline models as a reference.

9.5 Tools to interpret and present hazard ratios

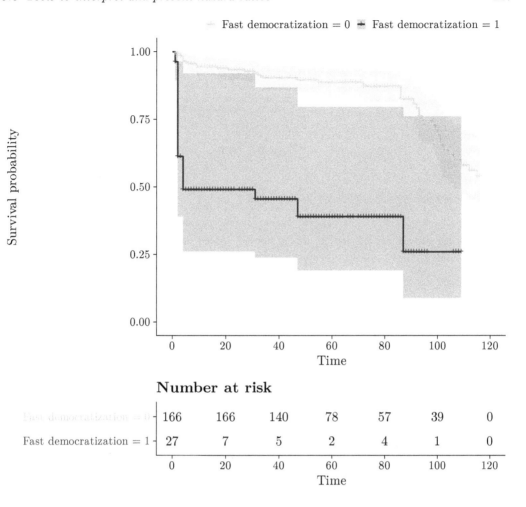

FIGURE 9.9: Kaplan-Meier Survival estimates.

Note: It is important to make a statement about replicability: If an author used Stata (for example, the `stcox` command) there will be some slight differences in the results obtained using R.

First model:

```
cox_m1 <- coxph(
  Surv(risk_time_at_start, risk_time_at_end, direct_dem) ~
    rapid_positive_dem + rapid_negative_dem + memory + vdem_score,
  data   = direct_democracy_c,
  method = "breslow"
)
```

If you wish, you can choose to express your model as a figure (see Figure 9.10). Given that we will estimate five models, we will use a table instead, yet it is good for you to know ggforest():

```
ggforest(cox_m1)
```

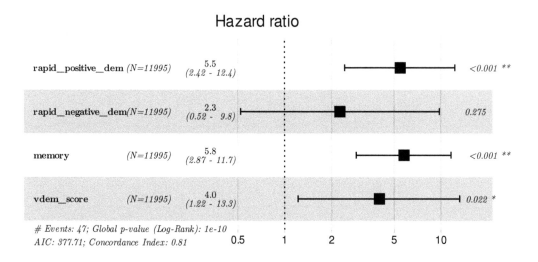

FIGURE 9.10: Graphical representation of hazards rates of Model 1.

Using the function cox.zph() we run the proportional hazards test:

```
test_cox_m1 <- cox.zph(cox_m1)
```

Let's examine the global value of the test. Its p-value is greater than the 0.05 cutoff.

If you wish, you can choose to plot the results of the test using ggcoxzph(). We will not do it in the following models, yet these figures can go to an appendix (Figure 9.11).

```
ggcoxzph(test_cox_m1, point.col = "black")
```

Let's estimate a second model:

```
cox_m2 <- coxph(
  Surv(risk_time_at_start, risk_time_at_end, direct_dem) ~
    rapid_positive_dem + rapid_negative_dem + memory + vdem_score +
    capab_diffusion,
  data    = direct_democracy_c,
  method  = "breslow"
)
```

9.5 Tools to interpret and present hazard ratios

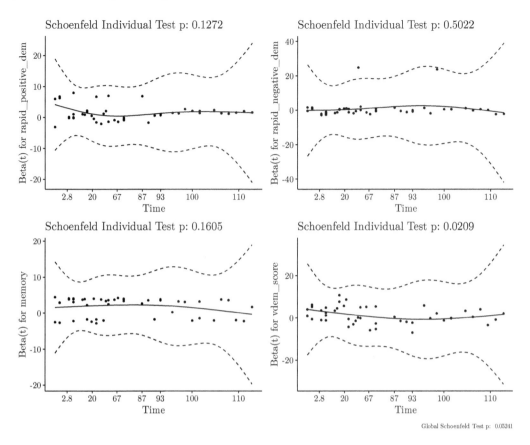

FIGURE 9.11: Proportional hazards (Schoenfeld) tests.

And its proportional hazard test:

```
cox.zph(cox_m2)
##                     chisq df      p
## rapid_positive_dem  1.693  1 0.1932
## rapid_negative_dem  0.718  1 0.3968
## memory              1.623  1 0.2026
## vdem_score          6.594  1 0.0102
## capab_diffusion     8.135  1 0.0043
## GLOBAL             14.825  5 0.0111
```

In this case, the global test is barely superior to the p 0.05 cutoff, so the proportional hazard assumption is not yet violated. However, there is a significant variable in its Chi-squared: `capab_diffusion`, with a p of 0.02.

Now, a third model:

```
cox_m3 <- coxph(
  Surv(risk_time_at_start, risk_time_at_end, direct_dem)
```

```
    rapid_positive_dem + rapid_negative_dem + memory + vdem_score +
    capab_diffusion + ocur_difussion,
  data    = direct_democracy_c,
  method = "breslow"
)
```

And its proportional hazard test:

```
cox.zph(cox_m3)
##                      chisq df     p
## rapid_positive_dem   2.092  1 0.148
## rapid_negative_dem   0.473  1 0.491
## memory               1.697  1 0.193
## vdem_score           4.515  1 0.034
## capab_diffusion      4.264  1 0.039
## ocur_difussion       2.089  1 0.148
## GLOBAL              12.785  6 0.047
```

Here, the test reveals a similar scenario to the first model, i.e. without problems.

The fourth model:

```
cox_m4 <- coxph(
  Surv(risk_time_at_start, risk_time_at_end, direct_dem) ~
    rapid_positive_dem + rapid_negative_dem + memory + vdem_score +
    capab_diffusion + ocur_difussion + log_pop,
  data    = direct_democracy_c,
  method = "breslow"
)
```

And its proportional hazard test:

```
cox.zph(cox_m4)
##                      chisq df     p
## rapid_positive_dem   2.084  1 0.149
## rapid_negative_dem   0.445  1 0.505
## memory               1.677  1 0.195
## vdem_score           4.533  1 0.033
## capab_diffusion      4.224  1 0.040
## ocur_difussion       2.059  1 0.151
## log_pop              2.817  1 0.093
## GLOBAL              17.867  7 0.013
```

Note that the global p-value is very close to the cutoff, but it is still above 0.05. Variables `vdem_score` and `log_pop` are violating the proportional hazard assumption, since its p-values are less than 0.05.

9.5 Tools to interpret and present hazard ratios

The fifth model:

```
cox_m5 <- coxph(
  Surv(risk_time_at_start, risk_time_at_end, direct_dem) ~
    rapid_positive_dem + rapid_negative_dem + memory + vdem_score +
    capab_diffusion + ocur_difussion + log_pop + colony_gbr,
  data   = direct_democracy_c,
  method = "breslow"
)
```

Its proportional hazard test:

```
cox.zph(cox_m5)
##                    chisq df     p
## rapid_positive_dem 1.883  1 0.170
## rapid_negative_dem 0.380  1 0.538
## memory             1.274  1 0.259
## vdem_score         4.640  1 0.031
## capab_diffusion    3.960  1 0.047
## ocur_difussion     1.798  1 0.180
## log_pop            2.227  1 0.136
## colony_gbr         0.119  1 0.730
## GLOBAL            16.833  8 0.032
```

The fifth model presents the same scenario as the fourth model. We have two variables violating the proportional hazard assumption. The global test has a p-value of 0.16, so we should not worry about solving this violation. However, in the case that your work has a global p-value less than 0.05, we show you how to address the problem just as its recommended by the *Oxford Handbook* (the same book we based the theoretical discussion on at the start of the chapter). One way of solving it is by interacting the problematic variables with the natural logarithm of the temporal variable we created before.

If the global p-value was less than 0.05, the corrected fifth model would look like this:

```
cox_m5_int <- coxph(
  Surv(risk_time_at_start, risk_time_at_end, direct_dem) ~
    rapid_positive_dem + rapid_negative_dem + memory + vdem_score +
    capab_diffusion + ocur_difussion + log_pop + colony_gbr +
    vdem_score:log(risk_time_at_end) + log_pop:log(risk_time_at_end),
  data   = direct_democracy_c,
  method = "breslow"
)
```

You can see that the test no longer shows problems with the proportional hazard assumption.

```
cox.zph(cox_m5_int)
##                                        chisq df     p
## rapid_positive_dem                  1.13e+00  1 0.287
## rapid_negative_dem                  3.37e-01  1 0.562
## memory                              5.39e-01  1 0.463
## vdem_score                          5.88e-05  1 0.994
## capab_diffusion                     4.16e+00  1 0.041
## ocur_difussion                      1.15e+00  1 0.284
## log_pop                             6.80e-01  1 0.410
## colony_gbr                          5.27e-02  1 0.818
## vdem_score:log(risk_time_at_end)    4.75e-02  1 0.828
## log_pop:log(risk_time_at_end)       9.49e-01  1 0.330
## GLOBAL                              1.35e+01 10 0.195
```

Let's look at all the models together with **texreg**:

```
library(texreg)
```

For obtaining the *hazard ratios* in R we will need to exponentiate the coefficients, and then calculate the standard errors and p-values from a transformation of the variance-covariance matrix of the model. In Chapter 8 we learned how to make this transformation for logistic models when we wanted *odd ratios* (using the arguments `override.coef =`, `override.se =` and `override.pvalues =` of the `texreg` functions). This step is the same for survival models. The only difference for our current case is as follows: now we have various models for which we want *hazard ratios*, thus, we will use the iteration function `map()` so that the transformations of coefficients, standard errors and p-values are applied in each model (see Table 9.1):

```
list_models <- list(cox_m1, cox_m2, cox_m3, cox_m4, cox_m5, cox_m5_int)


texreg(
  l = list_models,
  custom.model.names = c("Model 1", "Model 2", "Model 3",
                         "Model 4", "Model 5", "Model 5b"),
  # adding custom names for the coefficients
  custom.coef.names = c(
    "Quick Democratization", "Quick Democracy Setback", "Memory",
    "Democracy", "Cappacity Diffusion", "Occurency Diffusion",
    "Population(ln)", "Was a British Colony",
    "Democracy x time at risk(ln)",
    "Population(ln) x time at risk(ln)"),
  override.coef    = map(list_models, ~ exp(coef(.x))),
  override.se      = map(list_models, ~ odds_se(.x)),
  override.pvalues = map(list_models, ~ odds_pvalues(.x)),
  caption = "Regression table for our survival models. Hazard ratios shown."
)
```

9.5 Tools to interpret and present hazard ratios

TABLE 9.1: Regression table for our survival models. Hazard ratios shown.

	Model 1	Model 2	Model 3	Model 4	Model 5	Model 5b
Quick Democratization	5.48*	5.43*	6.14*	6.07*	5.48*	3.93*
	(2.28)	(2.27)	(2.60)	(2.59)	(2.36)	(1.81)
Quick Democracy Setback	2.26	2.37	2.35	2.36	2.22	2.30
	(1.69)	(1.78)	(1.77)	(1.78)	(1.67)	(1.74)
Memory	5.79**	5.35**	5.55**	5.51**	5.04**	5.17**
	(2.07)	(1.92)	(2.00)	(1.99)	(1.84)	(1.90)
Democracy	4.03	3.07	3.81	3.91	5.20	379.22
	(2.45)	(1.93)	(2.41)	(2.60)	(3.59)	(636.11)
Cappacity Diffusion		1.60***	2.95**	2.94**	2.77**	2.70**
		(0.34)	(0.93)	(0.93)	(0.88)	(0.88)
Occurency Diffusion			0.03	0.03	0.02	0.04
			(0.05)	(0.06)	(0.04)	(0.08)
Population(ln)				1.01***	1.01***	1.57***
				(0.09)	(0.10)	(0.37)
Was a British Colony				0.45*	0.49*
				(0.21)	(0.24)
Democracy x time at risk(ln)						0.26*
						(0.12)
Population(ln) x time at risk(ln)						0.88***
						(0.06)
AIC	377.71	374.17	370.81	372.70	371.44	365.19
R²	0.00	0.00	0.01	0.01	0.01	0.01
Max. R²	0.04	0.04	0.04	0.04	0.04	0.04
Num. events	47	47	47	47	47	47
Num. obs.	11441	11269	11260	11229	11229	11229
Missings	554	726	735	766	766	766
PH test	0.05	0.01	0.05	0.01	0.03	0.19

***$p < 0.001$; **$p < 0.01$; *$p < 0.05$

Although we are not replicating the specifications from Altman, a first glance our results confirm his hypothesis that countries who suffered democratic "shocks" were faster in implementing direct democracy mechanisms. We have finished the example, so now we invite you to go thorugh the list of exercises before proceeding to the next chapter!

Exercise 9D.

- Using `survminer`, graph a Kaplan-Meier curve for the variable `colony_gbr`.
- Variable `colony_ussr` indicates countries that were part of the USSR. Graph a Kaplan-Meier curve for this variable. Add this variable into a sixth model, make the Grambsch and Therneau test and remake the model table with `texreg`.

- Do you have your own survival dataset? Please repeat all the exercise using your data and share your questions in our GitHub. If you do not have a dataset, use the dataset from the previous exercises and to try to identify the variable that most increases the risks of democratic breakdown in Latin America.

10
Causal inference

Andrew Heiss[1]

Suggested readings

- Elwert, F. (2013). Graphical Causal Models. In S. L. Morgan (Ed.), *Handbook of Causal Analysis for Social Research* (pp. 245–273). Springer.
- Hernán, M. A., & Robbins, J. M. (2020). *Causal Inference: What If*. CRC Press.
- Morgan, S. L., & Winship, C. (2007). *Counterfactuals and Causal Inference: Methods and Principles for Social Research*. Cambridge University Press.
- Pearl, J., Glymour, M., & Jewell, N. P. (2016). *Causal Inference in Statistics: A Primer*. Wiley.
- Pearl, J., & Mackenzie, D. (2018). *The Book of Why: The New Science of Cause and Effect*. Basic Books.
- Rohrer, J. M. (2018). Thinking Clearly About Correlations and Causation: Graphical Causal Models for Observational Data. *Advances in Methods and Practices in Psychological Science, 1*(1), 27–42.

Packages you need to install

- `tidyverse` (Wickham, 2019), `politicalds` (Urdinez and Cruz, 2020), `ggdag` (Barrett, 2020), `dagitty` (Textor and van der Zander, 2016), `MatchIt` (Ho et al., 2018), `broom` (Robinson and Hayes, 2020), `texreg` (Leifeld, 2020).

10.1 Introduction

One of the most repeated phrases in any introductory statistics class is the warning that "correlation is not causation." In political science research, though, we are often con-

---

[1] Andrew Young School of Policy Studies, Georgia State University. E-mail: aheiss@gsu.edu. Twitter: @andrewheiss.

cerned about the causes of social and political phenomena. Does government spending on education decrease social inequality? Does increased executive power cause regime collapse? Does increased ethnofractionalization cause genocide? Do international development projects reduce poverty or increase health? These are important questions, but using the statistical methods covered in this book, we can only talk about these relationships using associational language. In Chapter 5 we found that education expenditures are associated with increased inequality, and in Chapter 8 we found that increased concentration of executive power is associated with a lower probability of the regime collapsing. Though we were very careful to not use causal language when interpreting these regression coefficients, we are ultimately concerned with causation, especially if we have the ability to influence policy. If a development intervention *causes* improvements in health, it would be valuable to roll it out on a large scale.

It is easy enough to run a regression that includes two seemingly unrelated variables and find that they are actually significantly correlated. For instance, in the United States, per capita consumption of mozzarella cheese correlates strongly ($r = 0.959$) with the number of civil engineering doctorates awarded (see Figure 10.1). Mathematically there is no difference between the strong relationship between cheese consumption and doctoral degrees in civil engineering and the strong relationship between education spending and social inequality. Both relationships are defined by a single number: a regression coefficient. However, the relationship between cheese and degrees does not mean that increased consumption of cheese will create new PhD-holders, nor does it mean that the increasing ranks of civil engineers are causing an uptick in the amount of cheese Americans are eating. We can readily discount this relationship as spurious.

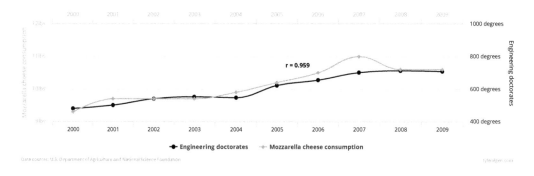

FIGURE 10.1: The high correlation between cheese and civil engineering.

We are less likely to call the relationship between education spending and inequality spurious. We did not immediately laugh off the correlation we found in Chapter 5, and there are many published academic papers that investigate the genuine relationship between them. Why? The difference between the plausibility of the two relationships goes beyond statistics. There is no believable causal story linking cheese consumption and doctoral degrees, despite the high correlation between the two. There *is* a believable causal story between inequality and education, since increasing the quality of education enhances the employment opportunities available to students after graduation.

In this chapter, we present new non-statistical language for creating, measuring, and evaluating causal stories and relationships using observational (i.e. non-experimental) data. We introduce the concept of causal directed acyclic graphs (DAGs) that allow us to formally encode our understanding of causal stories. With well-crafted DAGs, we can use a set of rules called *do*-calculus to make specific adjustments to statistical models and isolate or identify causal relationships between variables of interest.

10.2 Causation and causal graphs

Unlike correlation, which at its core is simply a mathematical formula, there is no `causation()` function in R. Causation is a question of philosophy and theory, not math. Debates over the definition of causation have been waged for thousands of years—Plato and Aristotle wrote about the metaphysics and components of causal relationships. For the sake of simplicity, in this chapter we will use a fairly universal definition: X can be said to cause Y if:

- **Association**: X and Y are *associated* with each other
- **Time ordering**: X *precedes* Y
- **Nonspuriousness**: The association between X and Y is *not spurious*

More simply, we can collapse all three of these conditions into a single definition:

> A variable X is a cause of a variable Y if Y in any way relies on X for its value.... X is a cause of Y if Y listens to X and decides its value in response to what it hears (Pearl et al., 2016, 5–6)

The concept of variables "listening to" each other simultaneously incorporates association, time ordering, and nonspuriousness. Consider the relationship between flipping a light switch and turning on a light. The act of enabling a light switch is associated with a light bulb emitting light, indicating an association between the two. A light bulb cannot emit light before it is turned on, thus ensuring correct time ordering—flipping the switch must precede the light. Finally, the association is nonspurious since there is a plausible link between the two: electrical pulses travel through wires to equipment that transforms the power to current that can ultimately power the bulb. We can also more simply say that the a light bulb "listens to" the light switch. Changes to the switch's on/off state influence the state of the bulb further down the causal chain. A light bulb listens to many other factors—electricity must flow into the building, transformers must function properly, and multiple switches might control the same light—but a single switch is definitely one of the causes of emitting light.

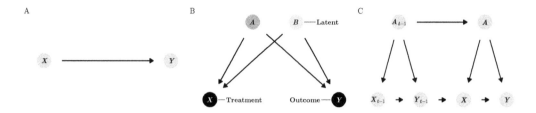

FIGURE 10.2: Various DAGs.

We can encode the philosophy or theory of a causal chain of events in a directed acyclic graph, or DAG. DAGs are graphical models of the process that generates the data and outline how X causes Y (see panel A of Figure 10.2) These graphs consist of three core elements: nodes, edges (or arrows), and direction. **Nodes** represent phenomena that have causal relationships to other elements in the system. These can be things that are measurable and could included a dataset, such as GDP, population, political party, length of time in office, income, geographic location, or socioeconomic status. Nodes do not necessarily need to be measurable—limitations in data collection or variable abstraction can make it impossible to collect reliable measurements on different phenomena. If this is the case, the node should still be included in the graph, but should be considered *unobserved* or *latent*. Nodes also represent treatment and outcome variables (see panel B in Figure 10.2). For instance, if we are interested in the causal effect of education expenditures on inequality (as in Chapter 5), education expenditures would be the *treatment* or *exposure*, and Gini-based social inequality would be the *outcome*.

Edges (or arrows) transmit associations between nodes. For instance, if a graph shows that $X \to Y$, X has a causal association with Y, or Y "listens to" or responds to X. Crucially, the absence of an edge between nodes implies that there is no causal relationship between the nodes. Causal associations between nodes can only flow in one **direction**—arrows can never go in both directions (e.g. $X \leftrightarrow Y$), and it should be impossible to return to any given node while walking through the graph (e.g. $X \to Z \to Y \to Z \to X$). Often there is a valid reason to include looping or feedback edges. For instance, education expenditures can lead to inequality, which then leads to changes in education expenditures. In these situations, rather than include bidirectional arrows, it is best to distinguish between the two periods of education expenditures, often using the subscript t for time: $X_{t-1} \to Y_{t-i} \to X_t \to Y_t$ (see panel C in Figure 10.2).

The presence and absence of nodes, arrows, and direction in a DAG communicates our theory and our philosophy of causation. Using DAGs makes it clear to your audience what your assumptions are. Linking nodes together implies that there is a statistical relationship between two phenomena, while omitting a node or an arrow implies that the node does not play any role in the causal story. If readers want to quibble with your theory of causation, they can do so easily by referring to the DAG and noting which nodes and arrows should be included or excluded.

10.3 Measuring causal effects

Since causation means that Y listens to X, measuring the effect of X on Y requires that you can manipulate X directly to see what happens to Y. We can use a special piece of mathematical notation to represent direct interventions: the *do*-operator, or $do(\cdot)$. If Y listens to X, we can say that there is some expected value of Y when we "do" X, or $E(Y|do(X))$ (read this as "the expected value of Y given that you do X").[2] We can make this notation a little easier to understand with some examples of causal questions social scientists are interested in:

- E(Votes won | do(Campaign fundraising)): The effect of "doing" (i.e. engaging in) campaign fundraising on the number of votes won in an election
- E(Inequality | do(Education expenditures)): The effect of "doing" education expenditures (i.e. spending money) on a country's level social inequality
- E(Air quality | do(Carbon tax)): The effect of "doing" (i.e. implementing) a carbon tax on a country's air quality
- E(Malaria infection rate | do(Mosquito net)): The effect of "doing" (i.e. using) a mosquito net on a village's malaria infection rate

To measure the causal effect of an intervention, we need to compare the expected value of the outcome ($E(Y)$) when we intervene and when we do not intervene with X. For example, we can compare what happens to a light bulb when we flip a switch to what happens to a light bulb when we do not flip the switch. Subtracting the difference between these two states of the world results in a causal effect:

$$\text{Causal effect of switch on light} = E(\text{Light} \mid do(\text{Switch} = \text{On})) - E(\text{Light} \mid do(\text{Switch} = \text{Off}))$$

We can write this definition more generally using a few more mathematical symbols. Here δ stands for the causal effect and the i subscript stands for a specific individual (i.e. comparing the effects in the same light bulb):

$$\delta_i = E(Y \mid do(X = 1))_i - E(Y \mid do(X = 0))_i$$

Calculating the causal effect δ_i in the physical world is fairly straightforward. Find two identical light bulbs, turn one on (do(Switch = On)), leave one off (do(Switch = Off)),

---
[2]This can also be written as $P(Y|do(X))$, of the probability distribution of Y given that you "do" X.

and compare the light that is emitted. Or, since a single light bulb is ostensibly unchanged over time, measure the change in light within the same light bulb.

In social science, however, measuring the causal effect of interventions in individual people, provinces, or countries is far more difficult. We cannot measure $E(Y|do(X = 1))_i$ in one individual and then go back in time to measure $E(Y|do(X = 0))_i$ in that same individual. This poses the *fundamental problem of causal inference*: individual-level causal effects are impossible to observe because we cannot see individual-level counterfactuals.

Instead of measuring individual-level effects, it is easier (and actually feasible) to measure the average outcomes for every unit in a group. We can find the *average treatment effect* (ATE) by calculating the difference between the expected value of the average outcome ($\bar{Y}$) for those who $do(X)$ and those who do not:

$$\delta_{\text{ATE}} = E(\bar{Y} \mid do(X = 1)) - E(\bar{Y} \mid do(X = 0))$$

Under experimental conditions, like randomized controlled trials where participants are randomly assigned to receive treatment or not, groups that $do(X = 1)$ and $do(X = 0)$ will generally be comparable because no individual willingly selected themselves in or out of the treatment. As long as the randomization is done well, the ATE that we calculate should be unbiased and accurate.

The majority of political science data, however, is observational. National GDP, election results, government spending, enrollment in primary schools, and the number of historical coups are all generated from non-experimental processes. If we are interested in the effect of being a democracy (i.e. "doing" democracy) on GDP, we could try to define the causal effect as:

$$\delta_{\text{ATE}} = E(\text{Average GDP} \mid do(\text{Democracy} = 1)) \\ - E(\text{Average GDP} \mid do(\text{Democracy} = 0))$$

However, simply subtracting the average GDP for non-democracies from the average GDP for democracies will result in a biased and incorrect estimate. Countries are not randomly assigned to be democracies. There are countless economic, historical, sociological, and political factors that influence a country's ability to $do(\text{Democracy})$, all of which ensure that there is systematic bias in a country's selection of political system. Any correlation we might find would not imply causation.

There are many econometric techniques to identify and estimate causal effects from observational data, including natural experiments, difference-in-difference analysis, regression discontinuity analysis, and instrumental variables (see Cunningham, 2018; Angrist and Pischke, 2008; Angrist and Pischke, 2015, for a complete overview of all of these methods). In general, these methods attempt to approximate treatment and control groups, organizing individuals or countries into comparable groups and removing selection bias. In addition to these approaches, we can use the logic of causal models to identify causal relationships, remove statistical confounding, and use observational data to estimate valid causal effects.

10.4 DAGs and statistical associations

Drawing nodes and edges is useful for understanding the various elements of a social phenomenon, but on their own, nodes and edges do not identify causal relationships. A causal effect is considered to be *identified* if the association between the treatment node and the outcome node is isolated and stripped of statistical associations that come from other nodes in the graph (see figure 10.3).

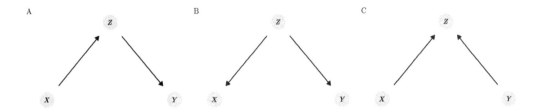

FIGURE 10.3: Basic types of relationships in DAGs.

The **direction** of arrows along the paths between nodes in a graph determines how statistical associations are transmitted between nodes. To illustrate how certain directional relationships can pass statistical associations between nodes in a graph, assume that two variables X and Y are not related to each other. We can say this with mathematical notation using the $\perp$ symbol: $X \perp Y$ means that X is independent of Y (and $X \not\perp Y$ means that the two variables are not independent). In the language of causal graphs, we can say that if X and Y are independent of each other, they are *d-separated* (the d- stands for "direction"). If X and Y are not independent of each other, they are *d-connected*—features of the directional relationships in the causal graph connect the two variables together and allow information to pass between them. Figure 10.3 shows how the inclusion of a third node Z in the pathway between X and Y changes the statistical relationship between the two and influences whether they are *d*-separated or *d*-connected.

1. **Mediators**: In panel A, X and Y are correlated with each other because of the mediating variable Z. Any changes in X will cause changes in Y downstream through Z. The path between X and Y is thus *open* and $X \not\perp Y$. X and Y are *d*-connected.
2. **Confounders**: In panel B, even though X and Y do not cause each other, Z is a common cause of both and confounds the relationship. Any changes in Z will cause changes in both X and Y, thus eliminating any statistical independence between the two. The path between X and Y is thus *open* and that again, $X \not\perp Y$. X and Y are *d*-connected.

In reality, X and Y should be independent of each other, but in the case of both mediators and confounders, X and Y are *d*-connected because Z opens up a pathway between the two and passes information between them.

One powerful element of the logic of causal diagrams is that we can control the flow of statistical information by blocking and opening pathways between nodes. For both mediating and confounding, if we can stop the association between both X and Z and Z and Y, X and Y will once again be independent of each other and will be d-separated. To do this, we can *adjust for* or *condition on* Z by removing the parts of X and Y that are explained by Z. There are many ways to adjust for variables, and examples are included later in this chapter. A basic way to think about adjusting variables is with regression: if we ran a regression and included Z as a control variable (e.g. `lm(Y ~ X + Z)`), the coefficient for Z would account for (and remove) the joint variation of Z across both X and Y. Once we adjust for Z, the path between X and Y is blocked and the two are d-separated. Mathematically, we can write this as $X \perp Y | Z$, or X is independent of Y conditional on Z.

The concept of blocking or conditioning helps us understand the final type of relationship between nodes:

3. **Colliders**: In panel C of Figure 10.3, X and Y are common causes of Z, but each are independent of each other. The path between X and Y is closed because it is blocked by Z. $X \perp Y$ and X and Y are d-separated. If we control for Z in a regression model, we inadvertently open up the pathway between X and Y and create a spurious statistical relationship between the two.

So far, we have talked about DAGs and d-separation in the abstract with X, Y, and Z. Figure 10.4 provides a more concrete causal graph that illustrates all three relationships simultaneously. Suppose we are interested in the causal effect of campaign fundraising (X) on total votes (Y). In this case, we no longer assume that X and Y are independent of each other (i.e. $X \perp Y$)—we want to measure the relationship between the two. In order to isolate that relationship, though, we must ensure that the pathway between campaign money and total votes is the only d-connected pathway in the graph. Each of the other nodes in the graph—"hired campaign manager," "candidate quality," and "won election"—pass on different types of statistical associations between money and electoral success. We can examine them in turn:

1. The relationship between money and total votes is **mediated** by hiring a campaign manager. Any changes in fundraising has an effect on the number of votes won, but fundraising also influences the chances of hiring a campaign manager, which then has an effect on total votes. The path between money and votes is thus open because of the mediating campaign manager variable.
2. The relationship between money and electoral success is **confounded** by candidate quality. High quality candidates are more likely to raise more money *and* win elections, which means that the relationship between money and electoral success is no longer isolated. If we adjust for candidate quality and compare candidates at the same quality (or control for quality in a regression, holding quality constant), we can close the pathway between money and success and isolate the path.

10.5 Backdoors and do-calculus

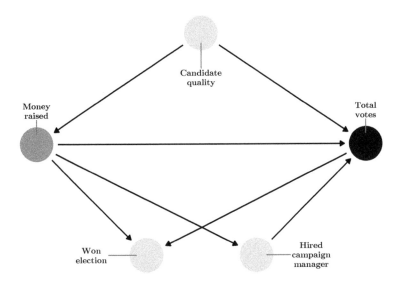

FIGURE 10.4: Simplified DAG showing the relationship between campaign spending and votes won in an election.

3. Winning an election is a **collider** on the path between money and electoral success. If we adjust for winning the election (i.e. only look at winning candidates), the actual relationship between money and electoral success will be distorted. Collider bias is a type of selection bias—by only looking at successful campaigns, we miss the relationship between money and success for candidates who did not win, and our estimates will be wrong. We should *not* control for winning an election.

Knowing the direction of the arrows between nodes in a DAG thus provides useful guidance in what to control for or adjust for. If we want to isolate the relationship between campaign fundraising and electoral success, we should control for candidate quality (since it is a confounder) and we should not control for winning the election (since it is a collider). The decision to adjust for having a campaign manager depends on our research question. If we adjust for it, we will remove the campaign manager effect from the total effect of fundraising on votes, and the remaining effect of X on Y will really be X without the campaign manager effect. If we are interested in the total effect of fundraising, including whatever effect comes from having a campaign manager, we should *not* control for having a campaign manager.

10.5 Backdoors and *do*-calculus

Following the fundamental problem of causal inference (i.e. due the the fact that we do not have a time machine), answering causal questions without an experiment appears impossible. However, if we apply a special set of logical rules called *do*-calculus to our causal graph, we can strip away any confounding relationships between our treatment and outcome nodes and isolate the causal effect between them using only observational data.

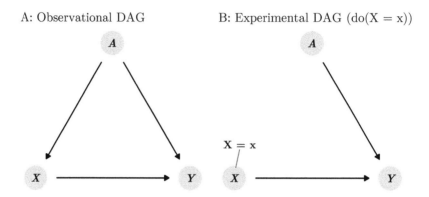

FIGURE 10.5: Edges in a DAG are deleted when using the $do(\cdot)$ operator.

The *do*-operator represents a direct intervention in a DAG and allows us to set a node to a specific value. For instance, in a randomized controlled trial, we as researchers have control over who gets assigned to treatment and control groups in the intervention. The causal effect (or δ) that we then find from the trial is the effect of X on our outcome Y when we $do(X = \text{Treatment})$. In experimental settings, a $do(\cdot)$ intervention implies that all arrows coming into the treatment node are deleted (see Figure 10.5). This ensures that X is *d*-separated from Y and the arrow between X and Y is completely isolated and identified. With observational data, however, it is not possible to estimate $E(Y|do(X))$ because as researchers, we do not have control over X. We cannot assign some countries to increase and other to decrease education spending as part of an experiment to see how social inequality changes, or assign some campaigns to spend extra effort in fundraising to see the effect of money on votes. Instead of $E(Y|do(X))$, which is our main quantity of interest, we can only estimate $E(Y|X)$, or the expected value of Y given existing levels of X (i.e. the correlation between the two). Unfortunately, the phrase that this chapter opened with—"correlation is not causation"—holds true here: $E(Y|X) \neq E(Y|do(X))$. In order to estimate a causal effect from observational data, we need to transform $E(Y|do(X))$ into something that is *do*-free, since we cannot actually undertake a *do*-intervention.

A set of three special logical rules named "*do*-calculus" allows us to do just that: through the application of different rules, we can remove $do(\cdot)$ operators from DAGs and identify causal effects with only observational data. Covering the full set of *do*-calculus rules goes beyond the scope of this chapter; consult Pearl et al. (2016), Pearl and Mackenzie (2018), Shpitser and Pearl (2008) for more details and examples.

10.5 Backdoors and do-calculus

One particular derivation of *do*-calculus rules defines a special *backdoor criterion* that allows us to remove $do(\cdot)$ operators by adjusting for confounding variables along the path between X and Y. The backdoor criterion states that the causal effect of X on Y is identifiable (i.e. can be isolated) after adjusting for the set of confounding variables Z, using the formula:

$$P(Y|do(X)) = \sum_Z P(Y|X,Z) \times P(Z)$$

For the sake of this chapter, the exact derivation and interpretation of this formula is less important than the intuition. According to this backdoor criterion, the *do*-operator on the left-hand side of the equation can be transformed into a *do*-free expression on the right-hand side, estimated with $P(Y|X,Z)$ (the distribution of Y conditioned on both X and Z) and $P(Z)$ (the distribution of Z), which are both estimable using observational data alone.

In practice, backdoor adjustment is often far more intuitive than trying to derive long formulas. Instead, it is possible to find and adjust for confounding backdoor paths graphically using a DAG following this process:

1. List all the pathways between X and Y, regardless of the direction of the arrows.
2. Identify any pathways that have arrows pointing backwards towards X.
3. The nodes that point back to X are confounders and therefore open up backdoor paths. These need to be adjusted for.

We can apply this process to the DAG in Figure 10.4. We are interested in the effect of campaign fundraising on the number of votes won in an election, or $E(\text{Votes won} \mid do(\text{Campaign fundraising}))$. Because we did not experimentally assign some campaigns to raise money and others to not, we can only work with observational data, leaving us with just the correlation between campaign fundraising and votes, or $E(\text{Votes won} \mid \text{Campaign fundraising})$. If we look at the causal graph are four paths between "Money raised" and "Total votes":

- Money raised → Total votes
- Money raised ← Candidate quality → Total votes
- Money raised → Hired campaign manager → Total votes
- Money raised → Won election ← Total votes

In the second path, "Candidate quality" points backward into "Money raised" and is a confounder that opens up a backdoor path between fundraising and votes. The first, third, and fourth paths only have right-point arrows and introduce no confounding. If we adjust for quality and hold it constant, we ensure that the relationship between fundraising and votes is *d*-separated from all other nodes and is therefore isolated and identified.

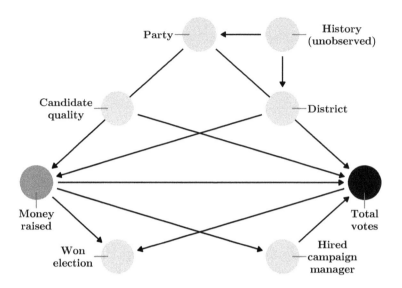

FIGURE 10.6: More complicated DAG showing the relationship between campaign spending and votes won in an election.

The same logic applies to more complex DAGs as well. Consider Figure 10.6, which is an expanded version of Figure 10.4 with three new nodes: the political party of the candidate, the district the candidate runs in, and the unobserved history of both the district and the party, which influences party decision-making and district demographic dynamics. We are interested in identifying or isolating the path between fundraising and total votes, but there are confounding paths that d-separate the causal pathway we care about. We can close these backdoors to isolate the causal effect. First, we list all the paths between "Money raised" and "Total votes":

- Money raised → Total votes
- Money raised → Hired campaign manager → Total votes
- Money raised → Won election ← Total votes
- Money raised ← Candidate quality → Total votes
- Money raised ← District → Total votes
- Money raised ← Party → Total votes
- Money raised ← District ← History → Party → Total votes
- Money raised ← Party ← History → District → Total votes

Of these eight possible paths, the last five have arrows pointing leftward into "Money raised" from three unique nodes: candidate quality, party, and district. We should therefore adjust for quality, party, and district to close these backdoor paths and ensure that the connection between money and votes is identified. Importantly, note that history, which is not measurable, is also a confounder. However, because we close off party and district through adjustment, history does not statistically confound money and votes. Even though it is latent and unmeasurable, downstream nodes that *are* measurable allow us to close its backdoor effects. Also note that "Won election" is a collider, not a confounder, and should not be adjusted to avoid creating spurious connections between money and votes, and "Hired campaign manager" is a mediator and should not be adjusted if we are interested in the total effect of money on votes.

The backdoor criterion is not the only method for finding which nodes need to be adjusted. Another common method is the front door criterion, which relies on *d*-separated mediator variables to identify the relationship between X and Y (see Rohrer (2018), Pearl et al. (2016), Pearl and Mackenzie (2018), Elwert (2013), and Hernán and Robbins (2020) for details and examples). Complex DAGs that do not easily fit either the backdoor or front door criteria can use the application of the rules of *do*-calculus to find appropriate and feasible adjustment sets. Special algorithms have been developed to work through *do*-calculus rules to determine if an effect is identifiable—the `causaleffect` package in R includes functions for running these algorithms.

Importantly, it is not always possible to identify causal effects in a DAG. If there is no way to translate $P(Y|do(X))$ to a *do*-free expression using *do*-calculus, then it is impossible to isolate and identify the causal relationship using observational data.

10.6 Drawing and analyzing DAGs

Since DAGs are a collection of nodes and arrows, they are simple to draw. I recommend sketching them by hand on paper or on a whiteboard when you are first mapping out your causal theory, and then transferring the handwritten draft to a computer. While it is possible to use many different computer programs to draw DAGs, including Microsoft PowerPoint or Adobe Illustrator, it is best to use graphing software created with DAGs in mind, such as DAGitty or the `ggdag` package in R.

10.6.1 Drawing DAGs with DAGitty

DAGitty (https://www.dagitty.net) is an in-browser graphical editor for creating and analyzing causal diagrams. By pointing and clicking, you can create nodes and link them together with edges. You can also assign specific nodes to be treatment, outcome, and latent/unobserved variables, which each have their own coloring scheme. You can export DAGs as PNG, JPG, and SVG files for inclusion in other documents. Figure 10.7 shows an example of a causal graph made in DAGitty. Consult the online manual[3] for more details about DAGitty's features, or spend a few minutes playing around to get used to adding and connecting nodes.

---

[3] See http://dagitty.net/manual-3.x.pdf

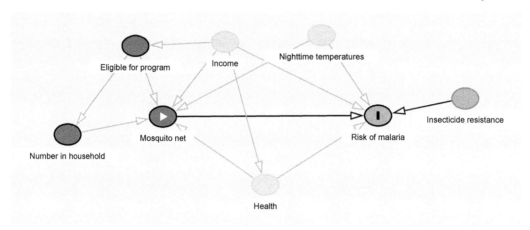

FIGURE 10.7: Example DAG built in DAGitty.

10.6.2 Finding adjustment sets with DAGitty

DAGitty automatically colors nodes and pathways by whether or not they are open. For instance, in Figure 10.7, the nodes for nighttime temperatures, income, and health conditions are all red, indicating that they confound the relationship between treatment and outcome. You can specify that these nodes will be adjusted and clear up the path.

In the right sidebar, DAGitty includes a list of nodes that need to be adjusted for in order to identify the causal effect. If the effect is identifiable, it will list all the minimally sufficient sets of nodes; if it is not identifiable, DAGitty will tell you. For the DAG in Figure 10.7, the minimally sufficient adjustment set includes nighttime temperatures, income, and underlying health conditions (see Figure 10.8).

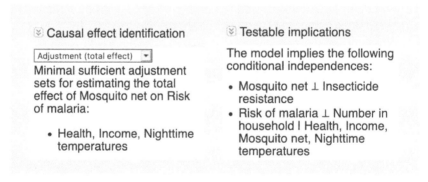

FIGURE 10.8: Minimal sufficient adjustment sets and testable implications for a DAGitty model.

The right sidebar also includes a list of all the testable implications of the DAG. Recall that after adjustment, certain nodes become d-separated and thus do not

10.6 Drawing and analyzing DAGs

pass statistical associations between each other. Based on this DAG, after adjusting for nighttime temperatures, income, and underlying health conditions, some of the following statements should be true:

- Income ⊥ Insecticide resistance: Income should be independent of insecticide resistance
- Income ⊥ Nighttime temperatures: Income should be independent of nighttime temperatures
- Number in household ⊥ Underlying health conditions | Income: The number of people in a household should be independent of underlying health conditions across similar levels of income

Each of these statements is testable with with actual data. If you have columns in a dataset for these different nodes, you can use R to check the correlations between them (i.e. `cor(data$income, data$temperature)` or `lm(number_houshold ~ health_conditions + income)`) and see if they are actually independent of each other.

10.6.3 Drawing DAGs with R

The `ggdag` R package allows you to use `ggplot2` to create and analyze DAGs with R. The documentation for the package is full of helpful examples of the full range of the package's functions. Below are a few examples of the most common things you can do with it.

In general, you create a DAG object with `dagify()` and plot it with `ggdag()` or `ggplot()`. The syntax for creating a DAG in `dagify()` is similar to what you use when creating models with `lm()` or `glm()`: build systems of formulas that indicate relationships between the nodes. For instance, in the DAG below, y is caused by x, a, and b (y ~ x + a + b), and x is caused by a and b (x ~ a + b), which makes a and b confounders.

```
library(ggdag)

simple_dag <- dagify(
  y ~ x + a + b,
  x ~ a + b,
  exposure = "x",
  outcome = "y"
)

# theme_dag() puts the plot on a white background without axis labels
ggdag(simple_dag) +
  theme_dag()
```

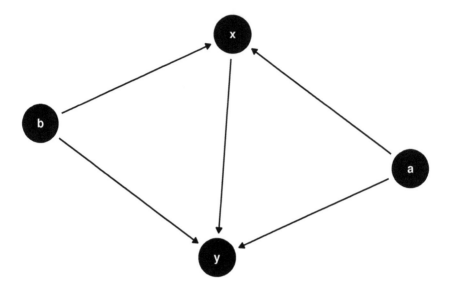

FIGURE 10.9: Basic DAG built with `ggdag()`.

Setting x and y as the exposure and outcome variables is optional if you want a simple graph, but if you do set them, you can color the points by node status:

```
ggdag_status(simple_dag) +
  theme_dag()
```

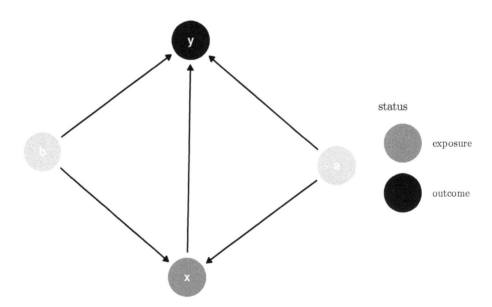

FIGURE 10.10: DAG with nodes colored by status.

10.6 Drawing and analyzing DAGs

Notice how the layout is different in both of those graphs. By default, ggdag() positions the nodes randomly every time using a network algorithm. You can change the algorithm by using the layout argument: ggdag(simple_dag, layout = "nicely"). You can see a full list of possible algorithms by running ?layout_tbl_graph_igraph in the console.

Alternatively, you can specify your own coordinates so that the nodes are positioned in the same place every time. Do this with the coords argument in dagify():

```
simple_dag_with_coords <- dagify(
  y ~ x + a + b,
  x ~ a + b,
  exposure = "x",
  outcome = "y",
  coords = list(x = c(x = 1, a = 2, b = 2, y = 3),
                y = c(x = 2, a = 1, b = 3, y = 2))
)

ggdag_status(simple_dag_with_coords) +
  theme_dag()
```

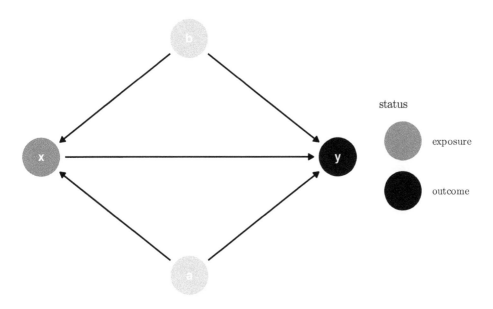

FIGURE 10.11: DAG with manual coordinates.

The variable names you use do not have to be limited to just x, y, and other lowercase letters. You can any names you want, as long as there are no spaces.

```
dag_with_var_names <- dagify(
  outcome ~ treatment + confounder1 + confounder2,
  treatment ~ confounder1 + confounder2,
  exposure = "treatment",
  outcome = "outcome"
)


ggdag_status(dag_with_var_names) +
  theme_dag()
```

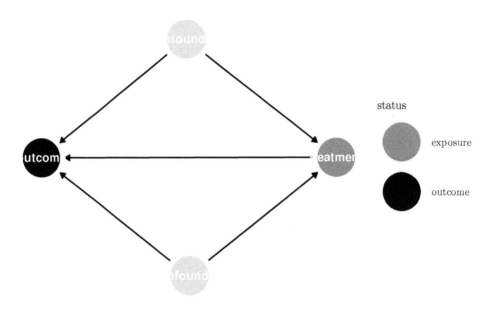

FIGURE 10.12: DAG with node names.

However, unless you use very short names, it is likely that the text will not fit inside the nodes. To get around this, you can add labels to the nodes using the `labels` argument in `dagify()`. Plot the labels by setting `use_labels = "label"` in `ggdag()`. You can turn off the text in the nodes with `text = FALSE` in `ggdag()`.

```
simple_dag_with_coords_and_labels <- dagify(
  y ~ x + a + b,
  x ~ a + b,
  exposure = "x",
  outcome = "y",
  labels = c(y = "Outcome", x = "Treatment",
             a = "Confounder 1", b = "Confounder 2"),
  coords = list(x = c(x = 1, a = 2, b = 2, y = 3),
                y = c(x = 2, a = 1, b = 3, y = 2))
)
```

10.6 Drawing and analyzing DAGs

```
ggdag_status(simple_dag_with_coords_and_labels,
             use_labels = "label", text = FALSE) +
  guides(fill = FALSE, color = FALSE) +   # Disable the legend
  theme_dag()
```

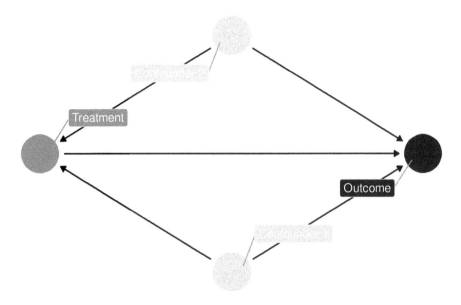

FIGURE 10.13: DAG with node labels.

10.6.4 Finding paths and adjustment sets with R

R can also perform analysis on DAG objects. For example, we can find all the testable implications from the DAG using the impliedConditionalIndependencies() function from the dagitty package. For this simple DAG, there is only one: a should be independent of b. If we had a dataset with columns for each of these variables, we could check if this is true by running cor(a, b) to see if the two are related.

```
library(dagitty)
```

```
impliedConditionalIndependencies(simple_dag)
## a _||_ b
```

We can also find all the paths between x and y using the paths() function from the dagitty package. We can see that there are three open paths between x and y:

```
paths(simple_dag)
## $paths
## [1] "x -> y"       "x <- a -> y"  "x <- b -> y"
```

```
## 
## $open
## [1] TRUE TRUE TRUE
```

The first open path is fine—we want a single *d*-connected relationship between treatment and outcome—but the other two indicate that there is confounding from a and b. We can see what each of these paths are with the `ggdag_paths()` function from the `ggdag` package:

```
ggdag_paths(simple_dag_with_coords) +
  theme_dag()
```

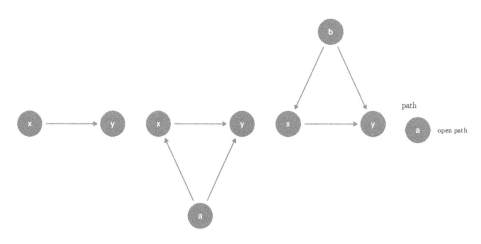

FIGURE 10.14: All possible paths between x and y.

Instead of listing out all the possible paths and identifying backdoors by hand, you can use the `adjustmentSets()` function in the `dagitty` package to programmatically find all the nodes that need to be adjusted. Here we see that both a and b need to be controlled for to isolate the x -> y relationship.

```
adjustmentSets(simple_dag)
##  { a, b }
```

You can also visualize the adjustment sets with `ggdag_adjustment_set()` in the `ggdag` package. Make sure you set `shadow = TRUE` to draw the arrows coming out of the adjusted nodes—by default, those are not included.

```
ggdag_adjustment_set(simple_dag_with_coords, shadow = TRUE) +
  theme_dag()
```

10.6 Drawing and analyzing DAGs

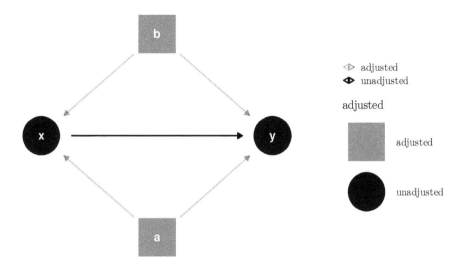

FIGURE 10.15: Adjustment set for DAG.

R will find minimally sufficient adjustment sets, which includes the fewest number of adjustments needed to close all backdoors between x and y. In this example DAG there was only one set of variables (a and b), but in other situations there could be many possible sets, or none if the causal effect is not identifiable.

Exercise 10A. In Chapter 5, you used multiple regression to estimate the determinants of inequality in Latin America and the Caribbean (Huber et al., 2006). For this exercise, you will draw a DAG that models the causal effect of *ethnic diversity* on *social inequality*. Do the following:

1. List all the variables you used in that chapter (GDP, foreign direct investment, health expenditure, etc.) and anything else that seems relevant to social inequality.

2. Draw an initial DAG by hand on paper or a whiteboard and carefully consider the causal relationships between all the different nodes.

3. Draw the DAG with DAGitty. Assign inequality as the outcome and ethnic diversity as the exposure. If any of your nodes are unobserved, assign them to be latent. Determine what nodes need to be adjusted for.

4. Draw the DAG in R with `ggdag()`.

10.7 Making adjustments

Throughout this chapter, we have talked about adjusting for confounders to close backdoor paths, but thus far we have not explored how to actually make these adjustments. There is no one correct method to adjust for nodes. Mathematically, adjustment means removing the variation that comes from confounders out of treatment and control nodes. For instance, in the DAG in Figure 10.4, we remove the effect of candidate quality from money raised, remove the effect of candidate quality from total votes, and then compare the unconfounded effect of money raised on total votes. We can account for the candidate quality effect by stratifying our sample into high and low quality candidates, by running a regression that includes candidate quality as a covariate, by finding matched pairs in the data that have similar values of quality, or by weighting observations by quality. In this section, we will walk through three common ways of making adjustments: multiple regression, matching, and inverse probability weighting.

10.7.1 Simulated mosquito net data

Instead of using an actual political science dataset, we will explore different methods of making adjustments using synthetic data that I generated about an international development program designed to reduce the risk of malaria using mosquito nets. The same methods and principles apply to analysis using real data, but it is impossible to know the true preexisting causal effect in real observational data, so there is no way to compare estimates to the truth. Because this data is simulated, however, we know the (hypothetical) truth—using mosquito nets *causes* malaria risk to decrease by 10 points, on average.

Researchers are interested in whether using mosquito nets decreases an individual's risk of contracting malaria. They have collected data from 1,752 households in an unnamed country and have variables related to environmental factors, individual health, and household characteristics. Additionally, this country has a special government program that provides free mosquito nets to households that meet specific requirements: to qualify for the program, there must be more than 4 members of the household, and the household's monthly income must be lower than $700 a month. Households are not automatically enrolled in the program, and many do not use it. The data is not experimental—researchers have no control over who uses mosquito nets, and individual households make their own choices over whether to apply for free nets or buy their own nets, as well as whether they use the nets if they have them.

```
mosquito_dag <- dagify(
  malaria_risk ~ net + income + health + temperature + resistance,
  net ~ income + health + temperature + eligible + household,
  eligible ~ income + household,
```

10.7 Making adjustments

```
    health ~ income,
  exposure = "net",
  outcome = "malaria_risk",
  coords = list(x = c(malaria_risk = 7, net = 3, income = 4, health = 5,
                      temperature = 6, resistance = 8.5, eligible = 2,
                      household = 1),
                y = c(malaria_risk = 2, net = 2, income = 3, health = 1,
                      temperature = 3, resistance = 2, eligible = 3,
                      household = 2)),
  labels = c(malaria_risk = "Risk of malaria", net = "Mosquito net",
             income = "Income", health = "Health",
             temperature = "Nighttime temperatures",
             resistance = "Insecticide resistance",
             eligible = "Eligible for program",
             household = "Number in household")
)

ggdag_status(mosquito_dag, use_labels = "label", text = FALSE) +
  theme_dag()
```

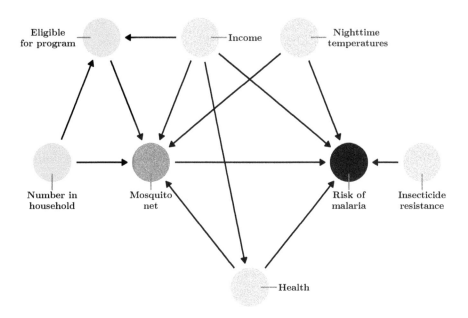

FIGURE 10.16: DAG for the effect of a hypothetical mosquito net program on malaria risk.

The causal graph in Figure 10.16 outlines the complete relationship between mosquito net use and risk of malaria. Each node in the DAG is a column in the dataset collected by the researchers, and includes the following:

- Malaria risk (`malaria_risk`): The likelihood that someone in the household will be infected with malaria. Measured on a scale of 0–100, with higher values indicating higher risk.
- Mosquito net (`net` and `net_num`): A binary variable indicating if the household used mosquito nets.
- Eligible for program (`eligible`): A binary variable indicating if the household is eligible for the free net program.
- Income (`income`): The household's monthly income, in US dollars.
- Nighttime temperatures (`temperature`): The average temperature at night, in Celsius.
- Health (`health`): Self-reported healthiness in the household. Measured on a scale of 0–100, with higher values indicating better health.
- Number in household (`household`): Number of people living in the household.
- Insecticide resistance (`resistance`): Some strains of mosquitoes are more resistant to insecticide and thus pose a higher risk of infecting people with malaria. This is measured on a scale of 0–100, with higher values indicating higher resistance.

According to the DAG, malaria risk is caused by income, temperatures, health, insecticide resistance, and mosquito net use. People who live in hotter regions, have lower incomes, have worse health, are surrounded by mosquitoes with high resistance to insecticide, and who do not use mosquito nets are at higher risk of contracting malaria than those who do not. Mosquito net use is caused by income, nighttime temperatures, health, the number of people living in the house, and eligibility for the free net program. People who live in areas that are cooler at night, have higher incomes, have better health, have more people in the home, and are eligible for free government nets are more likely to regularly use nets than those who do not. The DAG also shows that eligibility for the free net program is caused by income and household size, since households must meet specific thresholds to qualify.

We can load the data and use `glimpse()` to see the first few observations of every column:

```
library(tidyverse)


library(politicalds)
data("mosquito_nets")


glimpse(mosquito_nets)
## Rows: 1,752
## Columns: 10
## $ id           <dbl> 1, 2, 3, 4, 5, 6, 7, 8, 9, 10, 11, 12, 13, 14, ...
## $ net          <lgl> TRUE, FALSE, FALSE, TRUE, FALSE, FALSE, TRUE, F...
## $ net_num      <dbl> 1, 0, 0, 1, 0, 0, 1, 0, 0, 0, 0, 1, 1, 0, 0, 1,...
## $ malaria_risk <dbl> 33, 42, 80, 34, 44, 25, 19, 35, 32, 40, 30, 14,...
```

```
## $ income      <dbl> 781, 974, 502, 671, 728, 1050, 1146, 1093, 1037...
## $ health      <dbl> 56, 57, 15, 20, 17, 48, 65, 75, 60, 36, 75, 62,...
## $ household   <dbl> 2, 4, 3, 5, 5, 1, 3, 5, 3, 3, 6, 3, 4, 3, 1, 5,...
## $ eligible    <lgl> FALSE, FALSE, FALSE, TRUE, FALSE, FALSE, FALSE,...
## $ temperature <dbl> 21, 26, 26, 21, 19, 25, 27, 30, 28, 21, 17, 19,...
## $ resistance  <dbl> 59, 73, 65, 46, 54, 34, 45, 65, 55, 54, 33, 39,...
```

10.7.2 Verify the conditional independencies

Before proceeding with identifying the causal relationship between mosquito net use and malaria risk, we should first check to see if the relationships defined by our DAG reflect the reality of the data. Recall that *d*-separation implies that nodes are statistically independent of each other and do not transfer associational information. If you draw the DAG in Figure 10.16 in DAGitty, or if you run `impliedConditionalIndependencies()` in R, you can see a list of all the implied conditional independencies.

```
impliedConditionalIndependencies(mosquito_dag)
## eligible _||_ health | income
## eligible _||_ malaria_risk | health, income, net, temperature
## eligible _||_ resistance
## eligible _||_ temperature
## health _||_ household
## health _||_ resistance
## health _||_ temperature
## household _||_ income
## household _||_ malaria_risk | health, income, net, temperature
## household _||_ resistance
## household _||_ temperature
## income _||_ resistance
## income _||_ temperature
## net _||_ resistance
## resistance _||_ temperature
```

In the interest of space, we will not verify all these implied independencies, but we can test a few of them:

- Health ⊥ Household members: Health should be independent of the number of people in each household. In the data, the two variables should not be correlated. This is indeed the case:

```
cor(mosquito_nets$health, mosquito_nets$household)
## [1] 9.8e-05
```

- Income ⊥ Insecticide resistance: Income should be independent of insecticide resistance. This is again true:

```
cor(mosquito_nets$income, mosquito_nets$resistance)
## [1] 0.014
```

- Malaria risk ⊥ Household members | Health, Income, Bet net use, Temperature: Malaria risk should be independent of the number of household members given similar levels of health, income, mosquito net use, and nighttime temperatures. We cannot use `cor()` to test this implication, since there are many variables involved, but we can use a regression model to check if the number of household members is significantly related to malaria risk. It is not significant ($t = -0.17$, $p = 0.863$), which means the two are independent, as expected.

```
lm(malaria_risk ~ household + health + income + net + temperature,
   data = mosquito_nets) %>%
  broom::tidy()
## # A tibble: 6 x 5
##   term          estimate std.error statistic  p.value
##   <chr>            <dbl>     <dbl>     <dbl>    <dbl>
## 1 (Intercept)     76.2      0.966      78.9   0.
## 2 household       -0.0155   0.0893     -0.173 8.63e- 1
## 3 health           0.148    0.0107     13.9   9.75e-42
## # ... with 3 more rows
```

After checking all the other conditional dependencies, we can know if our DAG captures the reality of the full system of factors that influence mosquito net use and malaria risk. If there are substantial and significant correlations between nodes that should be independent, there is likely an issue with the specification of the DAG. Return to the theory of how the phenomena are generated and refine the DAG more.

10.7.3 Find the adjustment set

There is a direct path between mosquito net use and the risk of malaria, but the effect is not causally identified due to several other open paths. We can either list out all the paths and find which open paths have arrows pointing backwards into the mosquito net node (run `paths(mosquito_dag)` to see these results), or we can let R find the appropriate adjustment sets automatically:

```
adjustmentSets(mosquito_dag)
## { health, income, temperature }
```

Based on the relationships between all the nodes in the DAG, adjusting for health, income, and temperature is enough to close all backdoors and identify the relationship

10.7 Making adjustments

between net use and malaria risk (see Figure 10.17). Importantly, we do not need to worry about any of the nodes related to the government program for free nets, since those nodes are not d-connected to malaria risk. We only need to worry about confounding relationships.

We can confirm this graphically with `ggdag_adjustment_set()`:

```
ggdag_adjustment_set(mosquito_dag, shadow = TRUE,
                     use_labels = "label", text = FALSE)
```

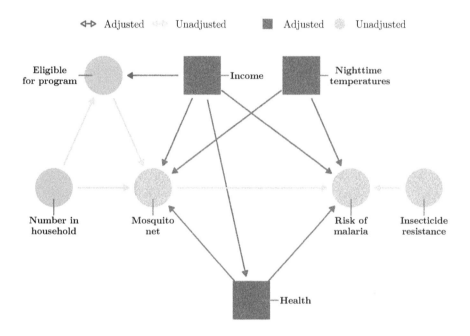

FIGURE 10.17: Adjustment set to identify the relationship between mosquito net use and malaria risk.

10.7.4 Naive unadjusted estimate

As a baseline for the other adjustment approaches we will try, we can first see what the relationship between mosquito net use and malaria risk is in the absence of any adjustment. If we create a boxplot of the distribution of malaria risk across people who do and do not use mosquito nets, we see that the average risk is substantially lower among those who use nets (see Figure 10.18).

```
ggplot(mosquito_nets, aes(x = net, y = malaria_risk)) +
  geom_boxplot()
```

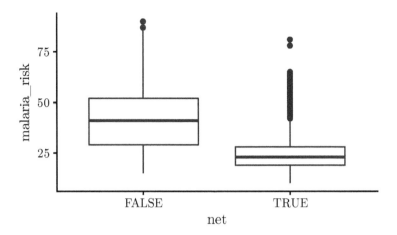

FIGURE 10.18: Distribution of malaria risk across those who did and did not use mosquito nets.

We can run a simple regression model to measure the exact average difference:

```
model_naive <- lm(malaria_risk ~ net, data = mosquito_nets)
```

```
texreg::screenreg(model_naive)
## 
## =========================
##                Model 1   
## -------------------------
## (Intercept)    41.94 ***
##                (0.40)   
## netTRUE       -16.33 ***
##                (0.65)   
## -------------------------
## R^2            0.27     
## Adj. R^2       0.27     
## Num. obs.      1752     
## =========================
## *** p < 0.001; ** p < 0.01; * p < 0.05
```

Based on this model, it appears that using a mosquito net is associated with a decrease in 16 points in malaria risk. However, this is *not* the causal effect. This is an instance where correlation does not equal causation. Other factors like income, health, and temperatures confound the relationship between net use and risk.

10.7.5 Regression

Because you are already familiar with multiple regression models from Chapter 5, one quick and easy way to try to adjust for confounders is to include them as covariates in a linear regression. At first glance this makes intuitive sense—the whole goal of

10.7 Making adjustments

adjustment is to compare treatment and outcome nodes at the same values of the various confounders, and the whole purpose of multiple regression is to explain variation in the outcome by holding different explanatory variables constant. **However, making confounding adjustments with linear regression will result in properly identified causal relationships only under very specific circumstances.** For regression-based adjustment to work, the relationships between all the treatment, outcome, and confounding nodes *must be linear,* which is difficult to test and verify with real observational data. It is almost always better to use one of the other adjustment techniques described below (matching or inverse probability weighting), since those methods do not rely on the assumption of linearity.

With that *large* caveat, we can adjust for our backdoors by including them in a regression model:

```
model_regression <- lm(malaria_risk ~ net + income + temperature + health,
                      data = mosquito_nets)
```

```
texreg::screenreg(model_regression)
## 
## ==========================
##                 Model 1
## --------------------------
## (Intercept)    76.16 ***
##                 (0.93)
## netTRUE        -10.44 ***
##                 (0.26)
## income         -0.08 ***
##                 (0.00)
## temperature     1.01 ***
##                 (0.03)
## health          0.15 ***
##                 (0.01)
## --------------------------
## R^2             0.89
## Adj. R^2        0.88
## Num. obs.       1752
## ==========================
## *** p < 0.001; ** p < 0.01; * p < 0.05
```

Based on these results, using a mosquito net *causes* a decrease of 10.44 points in malaria risk, on average. Note that because we have adjusted for confounders, we can now justifiably use causal language instead of merely talking about associations.

10.7.6 Matching

The main issue with using observational rather than experimental data is that people who used a mosquito net did so without being assigned to a treatment group. Individual characteristics led people to self-select into treatment, which makes the people who used mosquito nets fundamentally different from those who did not. Thanks to the DAG, we know many of the factors that caused people to choose to use nets: income, health, and temperature. If we could lump observations together that are similar to each other in income, health, and temperature, but differ in their use of mosquito nets, we can simulate experimental treatment and control groups and arguably calculate a more accurate causal effect, since we are finding the estimate across comparable groups.

Chapter 6 explored the idea of matching to identify relevant qualitative case studies in mixed methods research. There you created propensity scores that estimated the likelihood of treatment, and then identified cases with similar (or wildly different) propensity scores. While matching with propensity scores is popular, it can cause problems when you use it for causal identification (see King and Nielsen (2019)). There are alternatives, though. One common technique is to find matches that minimize the distance between the different confounders for each observation.

To illustrate this concept very briefly, consider a causal model where age and education are the only confounders of the relationship between a treatment and an outcome. Some of the observations self-selected into the treatment, while the rest did not. Panel A in Figure 10.19 shows the distribution of self-selected treatment across different values of age and education. There are some noticeable patterns: only one treated observation has more than 25 years of education, and only one has less than 15 years. If we want to create synthetic treatment and control groups from this data, it would not make sense to compare the single treated highly educated observation with untreated observations with low levels of education, since the two observations have such different levels of education. Ideally, we want to find the closest and most similar untreated highly educated observation and use that as the comparison with the single treated highly educated observation.

If we can find a collection of untreated observations that are close in distance to the treated observations, we can create a balanced matched set of treated and untreated observations. Panel B in Figure 10.19 shows each of the closest pairs of treated and untreated observations. We can discard any unmatched untreated observations and only use the matched set to make our causal inference, since we now ostensibly have comparable treatment and control groups. There are multiple ways to measure this distance—Mahalanobis and Euclidean distance are common, but not the only methods. There are also multiple methods to define pairs. In Figure 10.19, each treated observation is matched with one unique untreated observation, but it is also possible to allow doubling and tripling of matches. The example in Figure 10.19 minimizes the distance between two dimensions of confounders (age and education), but you can use as many confounders as necessary to create matched pairs.

We can use matching to adjust for the confounders in our mosquito net and malaria risk DAG. The `matchit()` function in the `MatchIt` R package provides many different matching methods, including nearest-neighbor Mahalanobis matching. There is no one best matching method, and you should play around with the different options for

10.7 Making adjustments

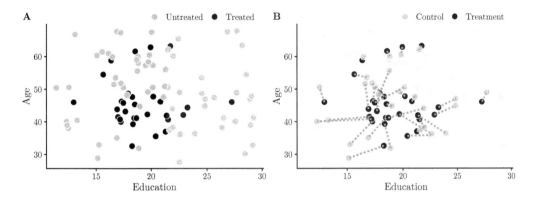

FIGURE 10.19: Matching based on nearest neighbor Mahalanobis distance.

the `method` and `distance` arguments. The `replace = TRUE` argument allows treated observations to be matched with more than one untreated observation.

The `matchit()` function uses familiar formula-based syntax for defining how to match the assignment to mosquito nets. Because our adjustment set includes income, temperature, and health, we use all three to create the matches.

```
library(MatchIt)

matched <- matchit(net ~ income + temperature + health,
                   data = mosquito_nets,
                   method = "nearest", distance = "mahalanobis",
                   replace = TRUE)

matched

##
## Call:
## matchit(formula = net ~ income + temperature + health, data =
##         mosquito_nets, method = "nearest", distance = "mahalanobis",
##         replace = TRUE)
##
## Sample sizes:
##           Control Treated
## All          1071     681
## Matched       439     681
## Unmatched     632       0
## Discarded       0       0
```

According to the output from `matched`, the 681 treated observations (i.e. those who used mosquito nets) were matched with 439 untreated observations (i.e. people who are similar in income, temperature, and health, but did not use nets). 632 untreated observations were not matched and will be discarded. Had we used `replace = FALSE`, there would have been an equal number of treated and untreated observations; there are fewer untreated observations here because some are doubled or tripled up.

We can create a new data frame based on this matching with the `match.data()` function. Notice how there are now only 1,120 rows instead of 1,752, since we discarded the unmatched observations. Also notice that there is a new column named `weights`. The `matchit()` function assigns matched pairs of observations different weights depending on how close or distant the matches are in an attempt to control for variation in distance. We can use these weights in our regression model to improve our estimation of the causal effect.

```
mosquito_nets_matched <- match.data(matched)
```

```
glimpse(mosquito_nets_matched)
## Rows: 1,120
## Columns: 12
## $ id           <dbl> 1, 3, 4, 5, 6, 7, 8, 11, 12, 13, 14, 16, 17, 20...
## $ net          <lgl> TRUE, FALSE, TRUE, FALSE, FALSE, TRUE, FALSE, F...
## $ net_num      <dbl> 1, 0, 1, 0, 0, 1, 0, 0, 1, 1, 0, 1, 1, 1, 0, 0,...
## $ malaria_risk <dbl> 33, 80, 34, 44, 25, 19, 35, 30, 14, 31, 34, 54,...
## $ income       <dbl> 781, 502, 671, 728, 1050, 1146, 1093, 889, 1133...
## $ health       <dbl> 56, 15, 20, 17, 48, 65, 75, 75, 62, 42, 53, 29,...
## $ household    <dbl> 2, 3, 5, 5, 1, 3, 5, 6, 3, 4, 3, 5, 2, 6, 4, 3,...
## $ eligible     <lgl> FALSE, FALSE, TRUE, FALSE, FALSE, FALSE, FALSE,...
## $ temperature  <dbl> 21, 26, 21, 19, 25, 27, 30, 17, 19, 28, 24, 26,...
## $ resistance   <dbl> 59, 65, 46, 54, 34, 45, 65, 33, 39, 37, 53, 55,...
## $ distance     <dbl> NA, NA, NA, NA, NA, NA, NA, NA, NA, NA, NA, NA,...
## $ weights      <dbl> 1.00, 0.64, 1.00, 0.64, 1.29, 1.00, 1.93, 0.64,...
```

Finally, we can run a regression using the matched data:

```
model_matched <- lm(malaria_risk ~ net, data = mosquito_nets_matched,
                    weights = weights)
```

```
texreg::screenreg(model_matched)
##
## =========================
##                 Model 1
## -------------------------
## (Intercept)      36.09 ***
##                  (0.60)
## netTRUE         -10.49 ***
##                  (0.76)
## -------------------------
## R^2              0.14
## Adj. R^2         0.14
## Num. obs.     1120
## =========================
## *** p < 0.001; ** p < 0.01; * p < 0.05
```

Based on these results, using a mosquito net *causes* a decrease of 10.49 points in malaria risk, on average. Again, we can use causal language now because we adjusted for confounders when matching, thus identifying the causal path between mosquito net use and malaria risk.

10.7.7 Inverse probability weighting

One disadvantage to using matching is that we throw away a lot of information—unmatched observations get discarded, and our sample size can shrink significantly. Matching also tends to be highly discrete, since each treated observation must be paired with one (or more) untreated observations. Look back at Panel B in Figure 10.19 and notice that some untreated observations are actually very close to treated observations, but still get discarded because they were beat out by observations that have a slightly smaller distance.

Instead of throwing away potentially useful data, we can use other methods to create matches that are less discrete and more informative. One common method in epidemiology and biostatistics is inverse probability weighting (IPW). Under IPW, each observation is assigned a weight based on how well its actual assignment to treatment matches the predicted probability of treatment, and those weights are then used in a regression model to estimate the causal effect of the treatment on the outcome.

We can illustrate this process with the education and age example from Figure 10.19. Instead of matching, we use logistic regression to create propensity scores for treatment:

```
data("edu_age") # from the "politicalds" package


edu_age <- edu_age %>%
  mutate(treatment = factor(treatment))

model_treatment <- glm(treatment ~ education + age, data = edu_age,
                       family = binomial(link = "logit"))

edu_age_propensities <- broom::augment_columns(
  model_treatment, edu_age, type.predict = "response"
) %>%
  rename(propensity = .fitted)
```

We can look at a few rows in the data to see these propensity scores. Person 59 had a 14% chance of self-selecting into treatment given their education and age, but they ended up not doing the treatment, as predicted. Person 27, on the other hand, also had a 14% chance of choosing treatment and they did, which is a fairly unlikely outcome. That choice is unexpected!

```
edu_age_propensities %>%
  select(id, treatment, education, age, propensity) %>%
  slice(59, 27)
## # A tibble: 2 x 5
##      id treatment education   age propensity
##   <dbl> <fct>         <dbl> <dbl>      <dbl>
## 1    59 Control        28.9  47.0      0.147
## 2    27 Treatment      21.7  63.2      0.148
```

With matching, we were interested in pairing unexpected and expected observations—recall that we needed to find a highly educated untreated observation to match up with the single highly educated treated observation. We do the same thing here. Observations that are unlikely to receive treatment and then do not receive treatment follow our expectations and should carry less weight. In contrast, observations that are unlikely to receive treatment and then *do* should carry more weight. Inverse probability weighting allows us to assign a numerical value to the unexpectedness of observations. We use the following formula for calculating the inverse probability weights for an average treatment effect, where Treatment is a binary 0/1 variable:[4]

$$\frac{\text{Treatment}}{\text{Propensity}} + \frac{1 - \text{Treatment}}{1 - \text{Propensity}}$$

We can add inverse probability weights to our predicted propensity scores with `mutate()`. Compare the `ipw` values for Person 59 and 27. Remember that Person 59 had a low probability of selecting into treatment and they did not, as expected. Their inverse probability weight is only 1.17. Person 27, on the other hand, had a low probability of being in treatment and yet they beat the odds and joined treatment, against expectations. Accordingly, they have a high inverse probability weight of 6.78.

```
edu_age_ipw <- edu_age_propensities %>%
  mutate(ipw = (treatment_num / propensity) +
           (1 - treatment_num) / (1 - propensity))

edu_age_ipw %>%
  select(id, treatment, education, age, propensity, ipw) %>%
  slice(59, 27)
## # A tibble: 2 x 6
##      id treatment education   age propensity   ipw
##   <dbl> <fct>         <dbl> <dbl>      <dbl> <dbl>
## 1    59 Control        28.9  47.0      0.147  1.17
## 2    27 Treatment      21.7  63.2      0.148  6.78
```

---

[4]There are many other versions of inverse probability weights that are aimed at estimating other causal quantities, such as the average treatment on the treated effect, the average treatment among the evenly matchable effect, and the average treatment among the overlap population effect. See this blog post by Lucy D'Agostino McGowan for more details: https://livefreeordichotomize.com/2019/01/17/understanding-propensity-score-weighting/

10.7 Making adjustments

Since every observation in the dataset has an inverse probability score, we do not need to throw any data away. Instead, we can weight each observation by its inverse probability score. Figure 10.20 shows the full dataset with points sized by their IPW. Observations that meet expectations receive less weight than observations that behave contrary to expectations.

```
ggplot(edu_age_ipw, aes(x = education, y = age,
                        color = treatment, size = ipw)) +
  geom_point()
```

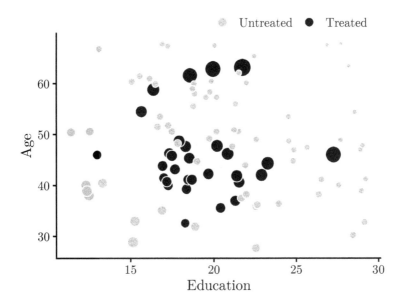

FIGURE 10.20: Observations sized by inverse probability weights.

We can adjust for the confounders in our mosquito net and malaria risk example using inverse probability weighting. First, we use logistic regression to predict the propensity to use a mosquito net using income, temperature, and health. Then we use the propensity scores to calculate the inverse probability weights using this formula:

$$\frac{\text{Mosquito net}}{\text{Propensity}} + \frac{1 - \text{Mosquito net}}{1 - \text{Propensity}}$$

```
model_mosquito_net <- glm(net ~ income + temperature + health,
                          data = mosquito_nets,
                          family = binomial(link = "logit"))

mosquito_nets_ipw <- broom::augment_columns(model_mosquito_net,
                                            mosquito_nets,
                                            type.predict = "response") %>%
```

```
  rename(propensity = .fitted) %>%
  mutate(ipw = (net_num / propensity) + (1 - net_num) / (1 - propensity))
```

Now that we have inverse probability weights, we can use them in a regression:

```
model_ipw <- lm(malaria_risk ~ net, data = mosquito_nets_ipw,
                weights = ipw)

texreg::screenreg(model_ipw)
##
## ========================
##                Model 1
## ------------------------
## (Intercept)    39.68 ***
##                (0.47)
## netTRUE       -10.13 ***
##                (0.66)
## ------------------------
## R^2             0.12
## Adj. R^2        0.12
## Num. obs.    1752
## ========================
## *** p < 0.001; ** p < 0.01; * p < 0.05
```

Based on this IPW model, using a mosquito net *causes* a decrease of 10.13 points in malaria risk, on average. Once again, we can safely use causal language because we identified the causal path between mosquito net use and malaria risk by accounting for confounders in the inverse probability weights.

10.7.8 Comparing all the methods

Now that we have run several regression models that adjust for confounders in different ways, we can compare the results all together. The naive estimate of 16 appears to definitely be an overestimate: after adjusting with regression, matching, and inverse probability weighting, the causal effect of using a mosquito net on the risk of malaria is consistently around 10. Assuming our DAG is correct, we successfully found a causal effect from non-experimental, observational data.

```
texreg::screenreg(list(model_naive, model_regression,
                       model_matched, model_ipw),
                  custom.model.names = c("Naive", "Regression",
                                         "Matching", "IPW"))
##
```

10.7 Making adjustments

```
## ==========================================================
##                 Naive       Regression    Matching      IPW
## ----------------------------------------------------------
## (Intercept)    41.94 ***    76.16 ***    36.09 ***    39.68 ***
##                 (0.40)       (0.93)       (0.60)       (0.47)
## netTRUE       -16.33 ***   -10.44 ***   -10.49 ***   -10.13 ***
##                 (0.65)       (0.26)       (0.76)       (0.66)
## income                      -0.08 ***
##                              (0.00)
## temperature                  1.01 ***
##                              (0.03)
## health                       0.15 ***
##                              (0.01)
## ----------------------------------------------------------
## R^2            0.27          0.89         0.14         0.12
## Adj. R^2       0.27          0.88         0.14         0.12
## Num. obs.      1752          1752         1120         1752
## ==========================================================
## *** p < 0.001; ** p < 0.01; * p < 0.05
```

Exercise 10B: In Exercise 10A, you drew a DAG that modeled the causal relationship between ethnic diversity and social inequality. In this exercise you will use the adjustment set from that DAG to attempt to estimate the causal effect. Do the following:

1. Load the `welfare` dataset from the book's package:

   ```
   library(politicalds)
   data("welfare")
   ```

   Your outcome variable is the Gini index (`gini`). Your treatment variable is ethnic diversity (`ethnic_diversity`), a binary variable that is 1 if between 20–80% of the population is ethnically diverse, and 0 if not.

2. Use the DAG you made previously to determine the minimum sufficient adjustment set. What nodes need to be adjusted for to ensure the pathway between ethnic diversity and inequality is identified?

3. Build a naive and incorrect correlation-is-not-causation model to test the baseline relationship between diversity and inequality (i.e. `lm(gini ~ ethnic_diversity, data = welfare)`). How is diversity associated with inequality? Why is this estimate not causal?

4. Use multiple regression to close the backdoors. Include the variables from your adjustment set as explanatory variables in a regression model.

5. Use matching to close the backdoors. Use the variables from your adjustment set to match observations' assignment to treatment, then use the matched observations in a regression model.

6. Use inverse probability weighting to close the backdoors. Use the variables from your adjustment set to generate propensity scores for assignment to treatment, then create inverse probability weights and use those weights in a regression model.

7. How do these adjusted effects compare to the naive model? How confident would you be claiming that these are causal effects? Why? What could you do to improve your causal identification?

10.8 Caveats

DAGs are powerful tools for incorporating the theory and philosophy of your causal model into the statistical estimation of the relationship between treatment and control. However, drawing a DAG does not automatically grant you the ability to speak with causal language. Causal graphs do not solve the fundamental problem of causal inference—they do not create a time machine that allows you to see what outcomes would be in a counterfactual world. The core assumption when using DAGs is that it is possible to completely identify and isolate a causal relationship using only observable data. Critics often look down on this selection-on-observables approach. If you omit an important node because it is not observable, your causal estimate will be incorrect and biased regardless of whether you use matching, inverse probability weighting, or any other method for isolating relationships between nodes.

Nodes in DAGs can represent actual observable variables in a dataset, but that is not always the case. Recall that causal graphs encode our philosophy of the data generating process—each node represents some phenomenon that has some influence somewhere in the chain of events that ultimately causes the treatment and the outcome. Nodes do not need to be measurable. In Figure 10.6 we included a node for the history of the electoral district, which influences both party dynamics and district characteristics. It is impossible to distill the entire history of a district into a single number in a dataset. We could attempt to do so by splitting the "history" node into smaller (and more measurable) components, such as the party of prior candidates, district boundary decisions, or the wealth of party leaders and supporters, but even then, we need to be comfortable declaring that these smaller nodes represent a district's entire history. That is a difficult assumption.

Contrary to criticisms of the selection-on-observables approach inherent in causal graphs, the presence of unobservable nodes in a DAG does not prevent us from using them to isolate and identify causal effects. Because of the logic of *d*- separation and the way we have drawn the arrows in the DAG, we can exclude the unobservable history of a district in Figure 10.6 because its node is blocked by the nodes for district and party. If we believe that our DAG is correct and that these two nodes are indeed sufficient for blocking all the unmeasured variation from the history node, we can safely proceed with estimating the causal effect. If a district's unobservable history passes statistical associations to campaign fundraising or total votes through other unblocked pathways, we will not be able to identify the causal effect with observational data.

In the end, your causal estimates are only as good as your DAG. If you accidentally omit confounding nodes, or if you cannot measure confounding nodes, or if you include nodes that are colliders, your results will be wrong. Moreover, the relationships between observed and unobserved nodes might be too complex to be solved using *do*-calculus adjustment. Unless you are completely confident that your DAG reflects the actual data generating process and the underlying theory of the the phenomenon you are studying and that the adjustments you make are sufficient for identifying the causal relationship, avoid placing too much confidence in the results.

Part III

Applications

11
Advanced Political Data Management

Andrés Cruz and Francisco Urdinez[1]

Suggested readings

- Graham, B. A. T. and Tucker, J. R. (2019). The international political economy data resource. *The Review of International Organizations*, 14(1):149–161.

- Lall, R. (2016). How Multiple Imputation Makes a Difference. *Political Analysis*, 24(4):414–433.

Packages you need to install

- `tidyverse` (Wickham, 2019), `politicalds` (Urdinez and Cruz, 2020), `skimr` (Waring et al., 2020), `countrycode` (Arel-Bundock, 2020), `stringdist` (van der Loo, 2019), `naniar` (Tierney et al., 2020), `mice` (van Buuren and Groothuis-Oudshoorn, 2020), `remotes` (Hester et al., 2020), `inexact` (Cruz, 2020).

11.1 Introduction

In this chapter, we will address two common problems that arise when using data from different sources. The first of them is to join datasets, a process that is often addressed by using one or identification variables (for example, the name of a country, the official code of a subnational unit, a code created by yourself). In some cases, when the identification variables are well standardized, the process becomes simple. For example, in official governmental datasets, countries have a unique code. In all of them Brazil will be "BRA", and Chile will be "CHI", which makes the work in R easy.

However, in the real world, different dataset sources are not meant for working together. After looking at a tool for standardizing country codes, `countrycode`, we will check a general solution, the *fuzzy* or inaccurate joining of data. Consider the following scenario: your main dataset has a variable called `country`, where Brazil is coded

---

[1]Institute of Political Science, Pontificia Universidad Católica de Chile. E-mail: arcruz@uc.cl and furdinez@uc.cl.

in uppercase as "BRAZIL". Then, you want to join your data with data from the World Bank, where the information of Brazil is labeled as "Brazil". There are extreme cases, such as the difference between "Venezuela" and its official name, "República Bolivariana de Venezuela" (Bolivarian Republic of Venezuela). You will probably face these types of differences in the countries of your dataset. How can we solve this puzzle in a quick and effective way?

The second problem is that, in most cases, our datasets contain missing values. This occurs for multiple reasons: coding error, governments that do not register information, data to be collected, etc. In the subsection, you will learn how R registers and works with missing values, besides some tools that allow exploring these in your datasets. One specific topic that we will address is that of imputations. When missing values are present in a regression, R simply drops the observation that has the missing value, either in its dependent variable, independent variable or controls (this elimination is called *listwise deletion*). Imagine that we want to compare ten countries of Latin America in the evolution of their unemployment rates between 2008 and 2018, and yet one of them has no data for the triennial 2011-2013. Can we, perhaps, fill those values by "guessing" the unobserved values? This process receives the name of imputation.

Deciding whether to impute is a researcher's decision. The imputation will be adequate depending on whether the data is missing in a **random** way or not. This dilemma is present when the data is used for graphing and we do not want it to be incomplete. Additionally, in some advanced models, such as the spatial ones, having missing values can inhibit us from using commands. For these cases, we can consider imputing data. Doing this has a cost since, like every solution, it is not perfect.

Throughout this chapter we will use the dataset of international treaties created by Carsten Schulz[2] based on United Nation's existing international treaties repository[3]. This repository hosts all the international treaties concluded between states, with their texts and information on the signatories. The dataset of the example is simplified, since it only has four international treaties instead of the hundreds that are used by Schulz for studying the reasons behind the decision of a country to join an international agreement.

Let's start by loading the dataset from our package, `politicalds`:

```
library(tidyverse)
```

```
library(politicalds)
data("treaties")
```

Now, the dataset has been loaded into our R session.

```
ls()
## [1] "treaties"
```

---

[2] See https://www.carstenschulz.eu/
[3] See https://treaties.un.org/

11.2 Merging datasets

The unit of analysis of the dataset is treaty-country. Each observation contains information about the process of domestic incorporation of the treaty at a national level, that is, of its signature and ratification. These actions are categorized under the `action_type` variable, which is accompanied by the `action_year` variable. The latter registers the year in which each action took place. Let's glimpse at the dataset with `glimpse()`:

```
glimpse(treaties)
## Rows: 248
## Columns: 5
## $ treaty_name   <chr> "Comprehensive Nuclear-Test-Ban Treaty", "Comp...
## $ adoption_year <dbl> 1996, 1996, 1996, 1996, 1996, 1996, 1996, 1996...
## $ country_name  <chr> "Antigua and Barbuda", "Antigua and Barbuda", ...
## $ action_type   <chr> "Ratification", "Signature", "Ratification", "...
## $ action_year   <dbl> 2006, 1997, 1998, 1996, 2007, 2005, 2008, 2008...
```

Our dataset only contains information for 31 American countries and their responses (signature, ratification) to four relevant treaties from the 1990s.

- The Comprehensive Nuclear-Test–Ban Treaty (1996).

- The Kyoto Protocol of the United Nations Framework Convention on Climate Change (1997).

- The Rotterdam Convention on the Prior Informed Consent Procedure for Certain Hazardous Chemicals and Pesticides in International Trade (1998)

- The Roma Statue, the constitutive act of the International Criminal Court (1998).

Thus, for each treaty we have 62 observations (31 countries, 2 possible answers for each):

```
treaties %>%
  count(treaty_name)
## # A tibble: 4 x 2
##   treaty_name                                                         n
##   <chr>                                                           <int>
## 1 Comprehensive Nuclear-Test-Ban Treaty                              62
## 2 Kyoto Protocol to the United Nations Framework Convention on Cli~  62
## 3 Rome Statute of the International Criminal Court                   62
## # ... with 1 more row
```

11.2 Merging datasets

A common exercise in observational studies is to merge datasets from different sources. Let's suppose that, for example, we have the following summary on how many treaties of the sample have been signed or ratified by each country:

```
summary_treaties <- treaties %>%
  group_by(country_name) %>%
  summarize(
    # we will find the ones that are not missing with !is.na()
    sum_signed = sum(action_type == "Signature" & !is.na(action_year)),
    sum_ratif  = sum(action_type == "Ratification" & !is.na(action_year))
  )

summary_treaties
## # A tibble: 31 x 3
##   country_name         sum_signed sum_ratif
##   <chr>                     <int>     <int>
## 1 Antigua and Barbuda           3         3
## 2 Argentina                     4         4
## 3 Bahamas (the)                 2         1
## # ... with 28 more rows
```

By exploring the dataset, an interesting case is that of United States, who signed the four treaties but did not ratify them:

```
summary_treaties %>%
  filter(country_name == "United States of America (the)")
## # A tibble: 1 x 3
##   country_name                    sum_signed sum_ratif
##   <chr>                                <int>     <int>
## 1 United States of America (the)           4         0
```

It would be interesting to explore the relationship between countries' treaty responses and some other variables about them. As an example, we will load a panel dataset with information about per capita GDP with exchange parity, according to the World Bank.

```
data("americas_gdp_pc")
```

We have information for the 31 countries of interest between 1996 and 1998:

```
americas_gdp_pc %>%
  count(country_name)
## # A tibble: 31 x 2
##   country_name            n
##   <chr>               <int>
## 1 Antigua and Barbuda     3
## 2 Argentina               3
## 3 Bahamas (the)           3
## # ... with 28 more rows

americas_gdp_pc %>%
  count(year)
```

11.2 Merging datasets

```
## # A tibble: 3 x 2
##    year      n
##   <int>  <int>
## 1  1996     31
## 2  1997     31
## 3  1998     31
```

Based on this dataset, we obtain average change of the per capita GDP with purchasing power parity for the 1996-1998 period:

```
summary_gdp <- americas_gdp_pc %>%
  group_by(country_name) %>%
  summarize(avg_gdp_pc = mean(gdp_pc))
```

Now we have two summary datasets, `summary_treaties` and `summary_gdp`, both at a country level. How can we merge their information? In this particular case, both have 31 rows (observations). One option to avoid is to simply paste one dataset next to the other (for example, with the `bind_cols()` function). In some cases, this option might be a good idea, but is usually a risky choice: it is hard to know if both datasets are ordered in the exact same way required for joining them correctly, especially when the number of observations is large. Thus, we usually want to guide the merge by one or more identification variables present in both datasets. In this case, the `country_name` column can guide the merge: we want to add information about the GDP to our dataset if and only if there is an exact *match* between their country names. In code, we can add new variables to our dataset from another with `left_join()`[4]:

```
summary_complete <- left_join(x = summary_treaties, y = summary_gdp,
                              # we can explicitly provide the id
                              # variable name:
                              by = "country_name")
```

This code will deliver the same result using pipes:

```
summary_complete <- summary_treaties %>%
  left_join(summary_gdp, by = "country_name")
```

```
summary_complete
## # A tibble: 31 x 4
##   country_name        sum_signed sum_ratif avg_gdp_pc
##   <chr>                    <int>     <int>      <dbl>
```

---

[4]This way of joining datasets, known as `left join`, is particularly common. In some specific cases, we might need other types of merges, more advanced. We recommend you check the prolific summary written by Jenny Bryan (https://stat545.com/join-cheatsheet.html).

```
## 1 Antigua and Barbuda            3         3      18965.
## 2 Argentina                      4         4      15362.
## 3 Bahamas (the)                  2         1      29530.
## # ... with 28 more rows
```

It is also possible to join datasets using more than one identification variable. For example, we could join our original datasets, `treaties` and `americas_gdp_pc`, both with the country-year observational unit. However, there is a particularity. In both cases, the country variable is named `country_name`, but the name of the year variable is not the same: in the first one, the name is `adoption_year` (for the original date of the treaty), and in the latter is `year`. We can easily provide this information to `left_join()` with the `by = ` argument:

```
treaties_with_gdp <- treaties %>%
  left_join(americas_gdp_pc,
            by = c("country_name", "adoption_year" = "year"))
```

Thus, to the original "treaties" dataset, which had 248 observations and 5 columns, we added the additional column with information about per capita GDP with exchange rate parity for each country-year participating in the treaty:

```
treaties_with_gdp
## # A tibble: 248 x 6
##    treaty_name   adoption_year country_name  action_type  action_year gdp_pc
##    <chr>                 <dbl> <chr>         <chr>              <dbl>  <dbl>
## 1 Comprehensiv~          1996 Antigua and~  Ratificati~         2006 18408.
## 2 Comprehensiv~          1996 Antigua and~  Signature           1997 18408.
## 3 Comprehensiv~          1996 Argentina     Ratificati~         1998 14557.
## # ... with 245 more rows
```

Exercise 11A. Download the World Economics and Politics (WEP) Dataverse[5] and choose ten country-year variables, including both institutional and economic characteristics of states as new variables to be added to `treaties_with_gdp` and merge them. Was it easy to find the unique identifiers (codes, names)?

11.2.1 Standardizing country codes

You can notice that the country names in the datasets we are working with are very particular. For example, Bolivia is "Bolivia (Plurinational State of)" and Dominican

---
[5]See https://ncgg.princeton.edu/wep/dataverse.html

11.2 Merging datasets

Republic is "Dominican Republic (the)". These are their official names to the international community. We will probably want to join them with data from other sources that do not have the same names: we can have Bolivia, Estado Plurinacional de Bolivia,Plurinational State of Bolivia, etc. For these cases, which are very common for those who study comparative politics and international relations, the best option is to have *standardized codes*, something that we can also find for subnational divisions and other types of analysis units.

Most of the coding reduces substantially the names of the countries to a handful of characters and/or numbers. In general, these come from international organizations or big academic projects that support them. The countrycode package allows us to transform different codifications and standards with ease, which helps us in further operations with the datasets.

```
library(countrycode)
```

After loading the package, we can check in the help file ?codelist the different codifications supported. Following this, using our initial variable and the countrycode() function, we will create a variable with standardized codes. iso2c and iso3c are among the most common codifications, which reduce the countries to 2 or 3 characters, according to ISO (International Organization of Standardrization). Meanwhile, cown and imf use numeric codifications relatively common in social sciences, which come from the Correlates of War project and the International Monetary Fund, respectively[6].

```
summary_complete_with_codes <- summary_complete %>%
  mutate(
    country_iso2c = countrycode(country_name, origin = "country.name",
                                destination = "iso2c"),
    country_iso3c = countrycode(country_name, origin = "country.name",
                                destination = "iso3c"),
    country_cown  = countrycode(country_name, origin = "country.name",
                                destination = "cown"),
    country_imf   = countrycode(country_name, origin = "country.name",
                                destination = "imf")
)
```

```
summary_complete_with_codes %>%
  select(starts_with("country_"))
## # A tibble: 31 x 5
##    country_name        country_iso2c country_iso3c country_cown country_imf
##    <chr>               <chr>         <chr>                <dbl>       <dbl>
## 1 Antigua and Barbu~ AG            ATG                     58         311
## 2 Argentina          AR            ARG                    160         213
```

---

[6]countrycode supports many other codifications used in our discipline, some more complete than others. Two of them, which are prevalent in political science, are those of Polity IV and the Varieties of Democracy (V-Dem) project. We recommend you check the help file ?codelist for more information.

```
## 3 Bahamas (the)       BS           BHS                    31         313
## # ... with 28 more rows
```

Having a dataset with at least one standardization code is particularly useful, as it often reduces the friction of merging our data with other. `countrycode`, then, is a good tool for pre-processing datasets with country information before merging them.

11.3 Fuzzy or inexact join of data

Although `countrycode` helps enormously in some cases, it is not always flexible enough. Sometimes, instead of needing to translate between codifications, we just do not have a standard to guide the merging, it is all a mess. Think of people's names: in a dataset of ex-presidents of Brazil, the same person can be coded as Lula, Luiz Inácio Lula da Silva, Lula da Silva. What if the person changed his/her name or is known by a nickname? Chilean legislator José Manuel Ismael Edwards Silva is known as Rojo Edwards. These real-world problems can cause huge headaches to those working with political data. Fuzzy merge is the solution for these situations.

Let's load a dataset equal to our previous `summary_gdp`, but with messy names in the countries:

```
data("summary_messy")
```

```
unique(summary_messy$country_name)
## [1] "Antigua & Barbuda" "Argentina"           "Bahamas"
## [4] "Barbados"           "Belize"              "Bolivia"
## [7] "Brasil"             "Canada"
##  [ reached getOption("max.print") -- omitted 23 entries ]
```

The names in this dataset do not correspond to any standardization, rather they are the result of a manual codification. A clear example is the abbreviation of `Dominican Rep`. What would happen if we were to merge our `summary_treaties` dataset with a muddled dataset such as this one?

```
summary_treaties %>%
    left_join(summary_messy, by = "country_name")
## # A tibble: 31 x 4
##    country_name         sum_signed sum_ratif avg_gdp_pc
##    <chr>                     <int>     <int>      <dbl>
## 1 Antigua and Barbuda            3         3         NA
```

```
## 2 Argentina                          4         4      15362.
## 3 Bahamas (the)                      2         1         NA
## # ... with 28 more rows
```

As you can observe, the merging fails for multiple observations because no exact match is found for the identification variable. That is the case of "Antigua and Barbuda" in the original dataset, and "Antigua & Barbuda" in the additional one. R considers they are different units. Intuitively, these values are alike, but R cannot guess or assume without our precise instructions.

Luckily, we have a solution for these types of cases, as well as with similar situations for politician's names, political parties, companies or regions. The initial intuition is as follows: The text strings `Antigua and Barbuda` and `Antigua & Barbuda` are akin. They share letters, they have similar longitudes, etc. Both have more in common than, for example, the pair "Antigua and Barbuda" and "Argentina". Computer scientists, acknowledging this fact, have developed multiple algorithms for assigning a difference score between two text strings, whatever they may be. Most of these algorithms assign a value of 0 when the strings are equal, and then it increases according to their dissimilarity.

Based on the above, we could calculate the distances between all possible pairs of strings in our two identification variables. Then, we could retrieve the lowest values for making pairs, and finally, we could perform a *fuzzy join*. As an example, we will use the `stringdist` package to generate a matrix with all these distances for our two datasets:

```
library(stringdist)
stringdistmatrix(summary_treaties$country_name,
                 summary_messy$country_name,
                 # we will use the default algorithm, named "osa" or
                 # "optimal string alignment"
                 method = "osa",
                 useNames = T)
```

```
##                      Antigua & Barbuda Argentina Bahamas Barbados
## Antigua and Barbuda                  3        15      16       15
## Argentina                           14         0       9        9
## Bahamas (the)                       15        12       6       10
## Barbados                            13         9       4        0
```

By looking at the first row, we can see that the string `Antigua and Barbuda` gets a score of 3 compared to `Antigua & Barbuda`, while it receives a score of 15 compared to `Argentina`.

Although the algorithms available in `stringdist` are robust, they are not infallible, and human supervision is often necessary.

For performing a *fuzzy join* we will make use of a work-in-progress package developed by Andrés Cruz, `inexact`, which combines the algorithms of `stringdist` with human supervision. It allows to make a human-supervised fuzzy join.

After installing `inexact` (`remotes::install_github("arcruz0/inexact")`), we can use it in the `Addins` tab of RStudio, or by running the following command:

```
inexact::inexact_addin()
```

A window will appear as the one in Figure 11.1, in which we will have to provide the characteristics of the merge we are trying to perform.

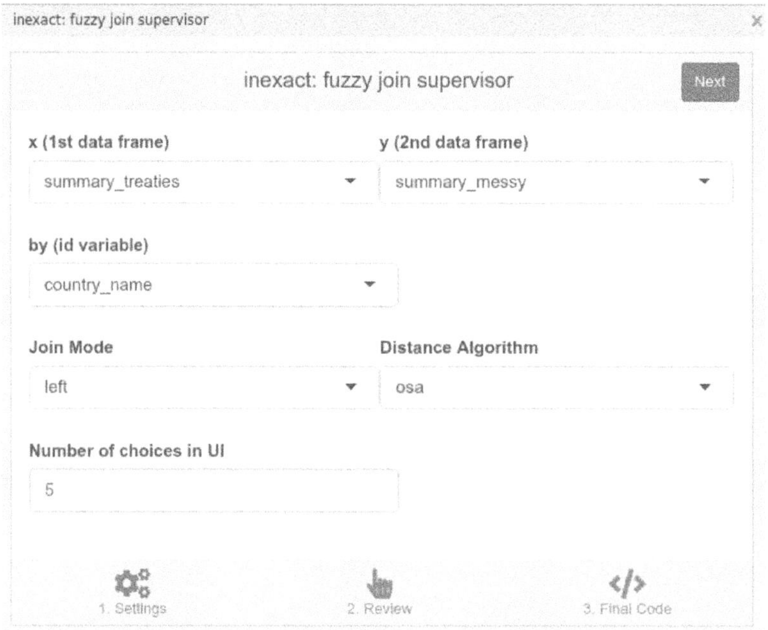

FIGURE 11.1: First panel of `inexact`, where you need to select the merging options.

After clicking `Next`, a supervision window of `inexact` will appear, as we can observe in Figure 11.2. Only the pairs that have a distance greater than 0 (that is, the imperfect ones) will appear in this window. By default, these will be ordered from the most conflicting to the least conflicting, in terms of the calculated distances. You can check each pair and modify them manually according to your knowledge. In this case, one of the six default pairs of the algorithm is wrong: it paired `Bolivia (Plurinational State Of)` with `United States`. We can correct this easily by selecting `Bolivia` in the upper right list (see Figure 11.2).

Lastly, after clicking `Next`, we will arrive at a final window, where a code is presented that allows us to make the *fuzzy* join, just as Figure 11.3 shows.

The code and its results are presented below. Note how the argument `custom_match` = is the one that allows `inexact::inexact_join()` to modify the default pairs of

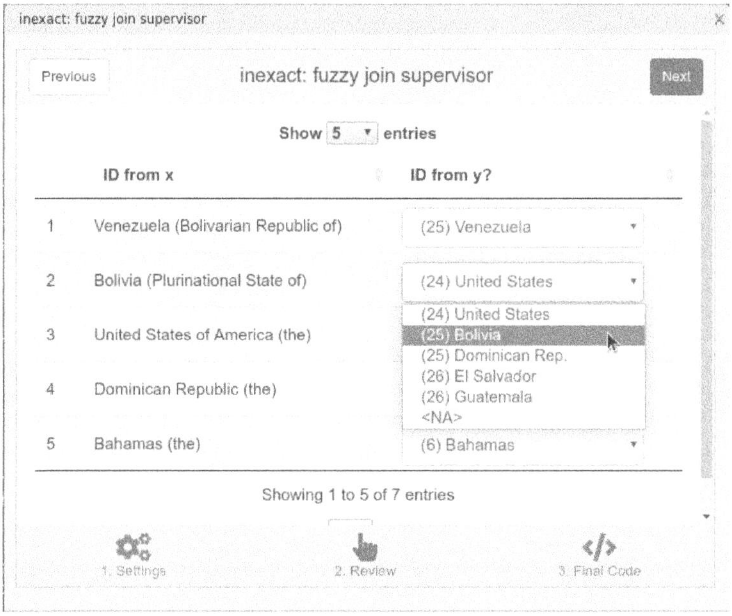

11.3 Fuzzy or inexact join of data

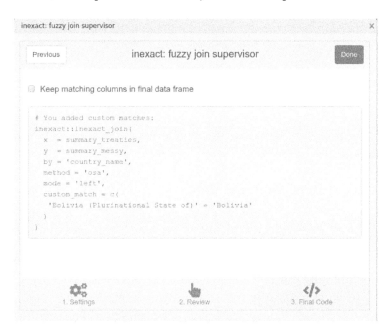

FIGURE 11.2: Second panel of `inexact`, where we supervise the default pairs.

FIGURE 11.3: Third panel of `inexact`, with the final code.

the algorithm. Now we can merge both datasets perfectly, even though they do not have completely standardized variables in common!

```
inexact::inexact_join(
  x = summary_treaties,
  y = summary_messy,
```

```
    by = 'country_name',
    method = 'osa',
    mode = 'left',
    custom_match = c(
      'Bolivia (Plurinational State Of)' = 'Bolivia'
    )
  )
## # A tibble: 31 x 4
##   country_name         sum_signed sum_ratif avg_gdp_pc
##   <chr>                     <int>     <int>      <dbl>
## 1 Antigua and Barbuda           3         3     18965.
## 2 Argentina                     4         4     15362.
## 3 Bahamas (the)                 2         1     29530.
## # ... with 28 more rows
```

Exercise 11B. In the following dataset you have names and information of Argentinean deputies in 2019[7]. Download it and merge it with the following dataset[8] which contains information of the commissions each deputy takes part of. Use unique codes and names to merge both variables and compare results.

11.4 Missing values' management

We have already encountered missing values (NA's) in some occasions, including the introductory section. In the R context, these are special values that can assume vectors which designate that there is missing information. However, missing values are not only a programming quirk. Taking them loosely may imply serious biases in our investigation. Thus, we will approach the subject in greater detail. We will discuss imputations, a possible technique for amending biases produced by missing values.

11.4.1 Types of missing values

Although R simply registers NA in the missing cell, it is important to theoretically understand the type of missing value we are dealing with. In short, the imputation

---

[7]See https://datos.hcdn.gob.ar/dataset/a80e0fa7-d73a-4ed1-9dec-80465e368951/resource/169de2eb-465f-4007-a4c2-39a5ba4c0df3/download/diputados1.8.csv

[8]See https://datos.hcdn.gob.ar/dataset/b238e6ab-691a-4e64-91d9-1445c4506ef4/resource/237f85b1-0b99-42e4-9b64-31a3935870a2/download/bloques_integracion2.5.csv

11.4 Missing values' management

literature identifies four different types. Missing values can be **structural**, *missing completely at random* (**MCAR**), *missing at random* (**MAR**) and *missing not at random* (**MNAR**).

first, we will examine the case of *structural* missing data. Let's look at the following example of our treaties' dataset:

```
treaties %>%
  filter(country_name == "United States of America (the)" &
         treaty_name == "Comprehensive Nuclear-Test-Ban Treaty") %>%
  select(country_name, action_type, action_year)
## # A tibble: 2 x 3
##   country_name                   action_type  action_year
##   <chr>                          <chr>              <dbl>
## 1 United States of America (the) Ratification          NA
## 2 United States of America (the) Signature           1996
```

The `action_year` variable codes the year when, in this case the United States, took action over the Comprehensive Nuclear-Test–Ban Treaty (1996). The United States signed the treaty in 1996, but never ratified it, as we previously observed. In other words, the year of ratification is simply not existent for the United States, as might be the case of other countries. Thus, missing data is structural when data is missing because it is non-existent. Secondly, we have data missing completely at random (**MCAR**) if there was any randomization process at work when the data was generated. For example, if we have asked eight questions to each person at random in a ten question survey. The two missing questions for each person are not explained by variables related to the surveyed (their ideology, age, gender, religion, etc), because by design we chose to randomly not ask them. Other cases of data missing completely at random can occur without the interference of the design, but because of unplanned events that affect our process of data collection.

On the other hand, our data could be missing at random (**MAR**), in addition to the rest of the variables that will allow us to model the distribution of missing cells. For example, let's think of the Composite Index of National Capability (CINC) by the Correlates of War (COW) project. This index is a *proxy* of the national power just as the realism school understands it. It combines six indicators of hard power: primary energy consumption, total population, urban population, steel and iron production, military expenditure and the number of military troops. The index varies from 0 to 1, as it represents each country's share of total global power in a given year. Let's suppose that we have twenty missing values for some countries-years. We probably have other variables for describing these countries-years that theoretically correlate with the index. We could make a model for trying to **guess** our missing values through the information we do observe. This is what we call an `imputation`.

Finally, our data could be missing not randomly (**MNAR**). This is possibly the most frequent scenario in which data is incorrectly imputed in political science, when in fact it should not. Non-random missing values are those whose missing condition is correlated to another variable in such a way that there is a *pattern* in the missing data. When facing a scenario in which missing data is present, we must carefully consider which variables may be explaining this missing data, and whether a selection bias is present. For example, if we use data from the World Bank about GDP growth, it is possible that poor countries do not present any data. This is because the quality of the measurements in these countries is not optimal. It may also be that measurements have not been made if the country is going through a difficult time in its economy, which could have negative effects on their national statistics. For example, since 2016 the World Bank does not report inflation rates for Venezuela. Ideally, it is necessary to correct this bias through a co-variable in the model by using selection models (unfortunately, we do not address them in the book).

After explaining the example case and presenting the descriptive statistics for missing values, we will make two scenarios for imputations. The first one is an imputation for descriptive purposes (creation of graphs). The second one, perhaps more helpful, will be for correcting possible bias in the regression model.

11.4.2 Description of missing values in the dataset

Let's start by loading some data on state capacity gathered from the Correlates of War project, which we previously mentioned. Our sample has only two countries, United States and China, with annual data from 1860 to 2012. Thus, the observational unit of the dataset is country-year.

```
library(politicalds)
data("cinc_index")
```

We have two variables with standardized codes of countries, `country_iso3c` (categorical) and `country_cown` (numeric), in addition to `year`, which denotes the year. Following this, `mil_expenditures` is the military expenditure of the country, `mil_personnel` is the size of the army in active men, `iron_steel` denotes the production of iron and steel, `energy_cons` corresponds to energy consumption, `population` is the total population, and `urban_population` is the urban population. Lastly, `capabilities_index` is the index composed by the material capacities we are interested in. This reflects the share of global power that each country possesses. If we observe the value of the two most powerful countries in the world, we will notice that China surpassed the United States a couple of years ago:

```
cinc_index %>%
  filter(year == 2012) %>%
  select(country_iso3c, capabilities_index)
## # A tibble: 2 x 2
##   country_iso3c capabilities_index
##   <chr>                      <dbl>
## 1 USA                         13.9
## 2 CHN                         21.8
```

11.4 Missing values' management

In International Relations literature, there is an existing debate about the extent of this index in relation to the real share of total global power of the countries, which you can observe in Chan (2005) and Xuetong (2006).

Let's start describing the missing values of the dataset. Using the quick summary of `skimr::skim()`, we observe in Figure 11.4 that `mil_expenditures` and `urban_population` are the two variables in which these are present, having 49 and 68 values missing, respectively.

```
> skimr::skim(cinc_index)
-- Data Summary ------------------------
                           Values
Name                       cinc_index
Number of rows             306
Number of columns          10

Column type frequency:
  character                1
  numeric                  9

Group variables            None

-- Variable type: character ------------------------------------------
  skim_variable n_missing complete_rate min max empty n_unique whitespace
1 country_iso3c         0             1   3   3     0        2          0

-- Variable type: numeric --------------------------------------------
  skim_variable      n_missing complete_rate    mean       sd    p0     p25      p50      p75     p100 hist
  country_cown               0             1     356    355.      2       2      356      710      710
  year                       0             1    1936     44.2  1860    1898     1936     1974     2012
  mil_expenditures          49         0.840 62097165. 130157898.  2660   64307  5830000 47808000 693600000
  mil_personnel              0             1    1719    1545.     28     470     1501     2580.   12123
  iron_steel                 0             1   50800.  95841.      0    50.5   14508    80568.   731040
  energy_cons                0             1     900.   1138.      0    25.3     332.    1437.    5334.
  population                 0             1  405424. 352084.  31513  128376   330135   507110.  1377065
  urban_population          68         0.778    81.6    109.    2.64   19.2     42.6     92.5     613.
  capabilities_index         0             1    16.1    5.58    7.74   11.9     14.8     18.5     38.4
```

FIGURE 11.4: Skim of our dataset.

The `naniar` package contains visual tools for helping us to better understand the possible patterns in our missing values. A way of visually exploring the latter comes with the function `gg_miss_var()`:

```
library(naniar)
```

```
gg_miss_var(cinc_index)
```

Similarly, we can only present the percentages with the argument `show_pct = T` (see Figure 11.6):

```
gg_miss_var(cinc_index, show_pct = T)
```

Where are our missing values? A visual inspection of the dataset, for example, with `View()`, allows us to get an idea in small and medium-size datasets. In any case, the `gg_mis_fct()` function of `naniar` is often useful, which allows us to separate the NA's according to a categorical variable of our dataset. Let's do it by country:

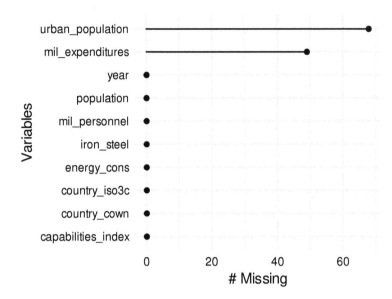

FIGURE 11.5: Missing values per variable in our dataset.

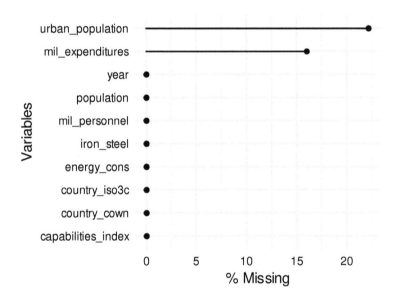

FIGURE 11.6: Missing values per variable in our dataset, expressed as a percentage of observations.

```
gg_miss_fct(x = cinc_index, fct = country_iso3c)
```

In Figure 11.7 we rapidly observe that all of our missing values correspond to China. In contrast, only complete observations are present for the United States.

It can also be interesting to explore the relationship between the missing values of a variable and a numeric column of the dataset, instead of a categorical. For this slightly more complex operation, `naniar` provides the `bind_shadow()` function, which

11.4 Missing values' management

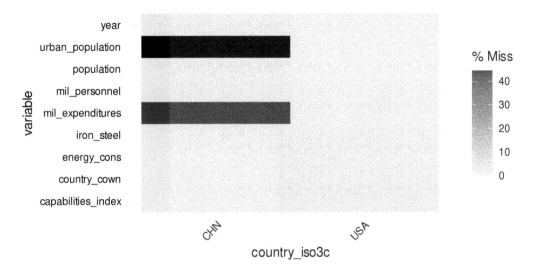

FIGURE 11.7: Alternative plot. Missing values per variable in our dataset, expressed as a percentage of observations.

generates `shadow` variables from those of our dataset. These variables contain only the binary information of missing or not missing values for each observation.

```
cinc_index_miss_plot <- cinc_index %>%
  bind_shadow(only_miss = T)
```

With this new dataset we can generate graphs with `ggplot2` that can help us to explore the distribution of missing values (according to our new variables from the `shadows`) in relation to the year (`year` variables).

```
ggplot(cinc_index_miss_plot,
       aes(x = year, fill = urban_population_NA)) +
  geom_histogram(binwidth = 1) +
  # the rest is to make the plot more readable:
  scale_x_continuous(breaks = seq(1860, 2020, 10)) +
  scale_y_continuous(breaks = c(1, 2)) +
  scale_fill_manual(values = c("lightgray", "black")) +
  theme(axis.text.x = element_text(angle = 90, hjust = 1))


ggplot(cinc_index_miss_plot,
       aes(x = year, fill = mil_expenditures_NA)) +
  geom_histogram(binwidth = 1) +
  # the rest is to make the plot more readable:
  scale_x_continuous(breaks = seq(1860, 2020, 10)) +
  scale_y_continuous(breaks = c(1, 2)) +
  scale_fill_manual(values = c("lightgray", "black")) +
  theme(axis.text.x = element_text(angle = 90, hjust = 1))
```

FIGURE 11.8: Diagnose to answer to the question: 'In which years are there NAs for urban population?'

FIGURE 11.9: Diagnose to answer to the question: 'In which years are there NAs for military expenditure?'

In the case of `urban_population`, the urban population of the country, the missing values are clearly gathered in the first years of the dataset. The same applies for `mil_expenditures`, although occasionally there is complete data for a few more years than in `urban_population`.

11.5 Imputation of missing values

Let's observe at a scatterplot for the evolution of the Composite Index of National Capability of China and the United States (see Figure 11.10):

```
ggplot(cinc_index) +
  geom_point(aes(x = year, y = capabilities_index,
                 group = country_iso3c, color = country_iso3c)) +
  scale_x_continuous(breaks = seq(1860, 2020, 10)) +
  theme(axis.text.x = element_text(angle = 90, hjust = 1))
```

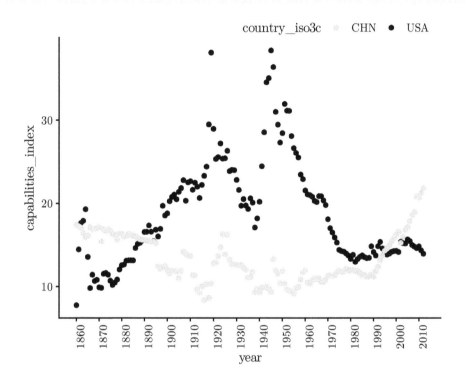

FIGURE 11.10: Instead of looking at the index of material capacities as a line we see each year as a dot to better observe the missing values.

How would the chart look if the United States did not have data between 1950 and 1970?

```
cinc_index2 <- cinc_index %>%
  mutate(capabilities_index = if_else(
    country_iso3c == "USA" & year %in% 1950:1970,
    NA_real_, capabilities_index
  ))
```

Then, we observe the chart 11.11:

```
ggplot(cinc_index2) +
  geom_point(aes(x = year, y = capabilities_index, group = country_iso3c,
             color = country_iso3c)) +
  scale_x_continuous(breaks = seq(1860, 2020, 10)) +
  theme(axis.text.x = element_text(angle = 90, hjust = 1))
```

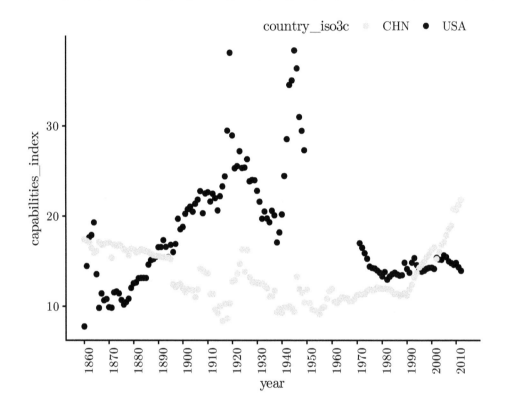

FIGURE 11.11: The figure is missing data for the United States between 1950 and 1970.

Now, it is important to think about the type of missing values we have. Let's suppose that there were no declared motives for assuming the missing values in the dataset, thus, they are missing at random. For doing the imputation we will use `mice` (which stands for Multivariate Imputation via Chained Equations). This is one of many options that exist for the ecosystem of R packages, which also includes the popular package `Amelia`[9], a software created by political scientists Gary King, James Honaker and Matthew Blackwell. We chose to exemplify with `mice`, since it is the one which has remained most updated. Recall that these packages assume that your missing data is MAR.

---

[9]See https://gking.harvard.edu/amelia

11.5 Imputation of missing values

The package is designed so as we can indicate how many imputations we want made for each missing value (with the m = option, which by default makes 5 imputations), and allows us to work both with cross-sectional data (such as surveys) and with panel data, as is our case with the CINC index, which contain data from many years for each country. After comprehending the structure of the missing values in the dataset and exposing the assumptions about their randomness, the second step for the imputation is to select the variables to be used. You just need the id of each observation (`country_iso3c` or `country_cown`), the temporal variable (`year`) and the variables that will be used for modeling the values. Our dataset does not have extra variables that we would like to eliminate, but this is a great idea if you want to apply these steps in your own examples.

We will call the imputation `imputed_init`. For executing it, we need to indicate `mice` some arguments in the command: Since our data is in a panel format we need to inform which is our unit variable and which is our time variable. We do so by using the `predictor()` function. `country_cown` is our ID numerical variable, and `year` our time variable.

```
library(mice)
```

```
imputed_init <- mice(cinc_index2, maxit = 0)
## Warning: Number of logged events: 1

predictor <- imputed_init$predictorMatrix
# set country_cown as ID variable for 2l.norm:
predictor[, "country_cown"] <- -2
# set year as temporal variable (random effects):
predictor[, "year"] <- 2
```

Also we need to specify by which method the imputation will be carried out using the `method=` option. `pmm` (Predictive Mean Matching) works for numeric variables, `logreg` (Logistic Regression) works for binary variables and `polyreg` (Bayesian polytomous regression) is recommeded for factor variables. In this case, since the CINC index is a continious variable, we will use `pmm`. We also define a two-level normal model using `"2l.lmer"`, and you want to set it as `"2l.bin"` when imputing binary variables, as this option imputes by a two-level logistic model.

```
method_1 <- imputed_init$method
method_1[which(method_1 == "pmm")] <- "2l.lmer"
```

Once the imputation is made we store it as `imputed_mice`. The imputation is relatively hidden within the list of objects we created, so we need to assign it a name directly.

```
imputed_mice <- mice(cinc_index2, m = 5, seed = 1,
                     method = method_1, predictorMatrix = predictor)
```

We can always give a glance to our newly imputed data, stored within `imp`:

```
imputed_mice$imp$capabilities_index
## # A tibble: 21 x 5
##      `1`   `2`   `3`   `4`   `5`
##    <dbl> <dbl> <dbl> <dbl> <dbl>
## 1  20.5  21.8  19.4  26.2  28.6
## 2  23.4  26.8  24.2  22.2  22.4
## 3  15.6  33.4  17.5  25.7  25.3
## # ... with 18 more rows
```

Let's observe visually where the imputed values are located in the distribution of the `capabilities_index` variable.

```
stripplot(imputed_mice, capabilities_index, pch = 20)
```

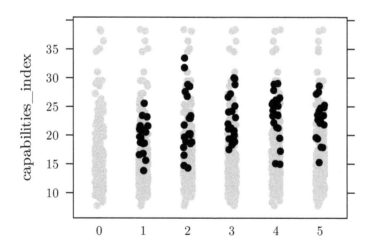

FIGURE 11.12: Visual representation of imputed values in `capabilities_index`.

To work with the imputations we want to extract them from our object `imputed_mice`. Recall we asked for five imputations `m = 5` which means we need to extract each of them separately. We do so using the function `complete()` within `mice`. Because `complete()` is a function which is also available in other packages, we make sure R uses the right one by setting it as `mice::complete`, which means "use the `complete()`

11.5 Imputation of missing values

function *within* mice package". We will join our five "complete" datasets, one after another, with the `bind_rows()` command:

```
complete_data1 <- mice::complete(imputed_mice, 1)
complete_data2 <- mice::complete(imputed_mice, 2)
complete_data3 <- mice::complete(imputed_mice, 3)
complete_data4 <- mice::complete(imputed_mice, 4)
complete_data5 <- mice::complete(imputed_mice, 5)
```

After generating five new datasets, we will join them together one after another with the `bind_rows()` command so we combine the five imputations into a single one:

```
complete_data_all <- bind_rows(
  complete_data1 %>% mutate(num_imp = 1),
  complete_data2 %>% mutate(num_imp = 2),
  complete_data3 %>% mutate(num_imp = 3),
  complete_data4 %>% mutate(num_imp = 4),
  complete_data5 %>% mutate(num_imp = 5),
) %>%
  select(num_imp, everything()) %>%
  mutate(source = "Specific Imp.") %>%
  filter(country_iso3c == "USA" & year %in% 1950:1970)
```

To analyze the imputed data in a two-way graph, we will calculate the average of the imputations.

```
avg_imp <- complete_data_all %>%
  group_by(country_iso3c, country_cown, year) %>%
  summarize(capabilities_index = mean(capabilities_index)) %>%
  ungroup() %>%
  mutate(source = "Average Imp.") %>%
  filter(country_iso3c == "USA" & year %in% 1950:1970)
```

The figure displays in clear values the individual imputations, and in a darker tone the mean value of the five imputations:

```
ggplot(mapping = aes(x = year, y = capabilities_index,
                    group = country_iso3c, color = country_iso3c)) +
  geom_point(data = cinc_index2) +
  # add imputed data and its average values:
  geom_point(data = complete_data_all, color = "darkgray") +
  geom_point(data = avg_imp, color = "black") +
  # add vertical lines for emphasis:
  geom_vline(xintercept = c(1950, 1970), linetype = "dashed") +
  scale_x_continuous(breaks = seq(1860, 2020, 10)) +
  scale_color_manual(values = c("lightgray", "black")) +
```

```
theme(axis.text.x = element_text(angle = 90, hjust = 1))
## Warning: Removed 21 rows containing missing values (geom_point).
```

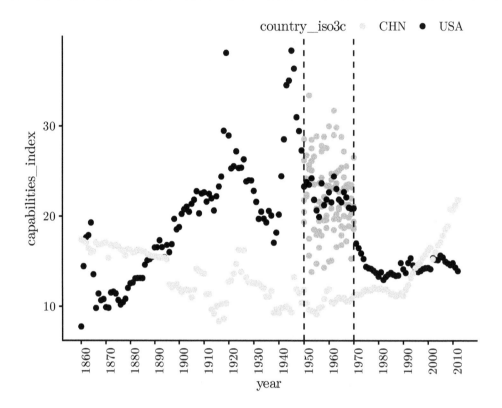

FIGURE 11.13: This figure displays the four imputations we made with mice (lighter), in addition to its average (darker).

It is important to remark that executing imputations for descriptive studies is different from the exercise of doing imputations for regressions (see Figure 11.14).

11.5.1 Regressions after imputing data

Unlike the process for visualizing data, when imputing data that will be used in the analysis of regressions, what is recommended is to not average the imputations but to average the regression coefficients obtained with each imputation executed. The process is as follows (see Figure 11.15):

First, let's observe how the regression looks for the United States with the incomplete data. We will explain the CINC index of each country by the energy_cons and urban_population variables.

11.5 Imputation of missing values

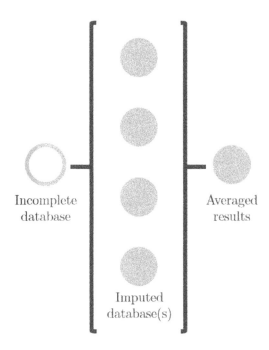

FIGURE 11.14: The process of imputations for visual analysis is done after averaging the imputations asked to 'mice'.

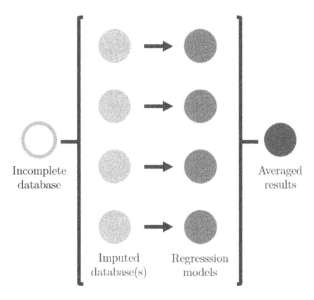

FIGURE 11.15: The process of imputations for regression analysis consists of averaging the coefficients of the imputed data, not averaging the imputed datasets.

```r
model_incomplete <- lm(capabilities_index ~ energy_cons + urban_population,
                       data = cinc_index2)

texreg::screenreg(model_incomplete)
## 
## ============================
##                   Model 1   
## ----------------------------
## (Intercept)        15.32 ***
##                    (0.54)   
## energy_cons         0.00 ** 
##                    (0.00)   
## urban_population   -0.01 *  
##                    (0.00)   
## ----------------------------
## R^2                 0.04    
## Adj. R^2            0.03    
## Num. obs.         217       
## ============================
## *** p < 0.001; ** p < 0.01; * p < 0.05
```

We are using the datasets "complete_data" from 1 to 5, which we previously made. We do a regression for each five imputations and compare three of them with the model with missing values.

```r
model_imp1 <- lm(capabilities_index ~ energy_cons + urban_population,
                 data = complete_data1)

model_imp2 <- lm(capabilities_index ~ energy_cons + urban_population,
                 data = complete_data2)

model_imp3 <- lm(capabilities_index ~ energy_cons + urban_population,
                 data = complete_data3)

model_imp4 <- lm(capabilities_index ~ energy_cons + urban_population,
                 data = complete_data4)

model_imp5 <- lm(capabilities_index ~ energy_cons + urban_population,
                 data = complete_data5)

# we'll only show the first three regressions:

model_list <- list(model_incomplete, model_imp1, model_imp2, model_imp3)

texreg::screenreg(
  model_list,
  custom.model.names = c("Incomp. m.", "Imp. m.1", "Imp. m.2", "Imp. m.3")
)
```

11.5 Imputation of missing values

```
## 
## ================================================================
##                    Incomp. m.   Imp. m.1   Imp. m.2   Imp. m.3
## ----------------------------------------------------------------
## (Intercept)        15.32 ***    15.02 ***  15.87 ***  15.26 ***
##                    (0.54)       (0.37)     (0.46)     (0.39)
## energy_cons         0.00 **      0.00 ***   0.00 ***   0.00 ***
##                    (0.00)       (0.00)     (0.00)     (0.00)
## urban_population   -0.01 *      -0.01 **   -0.01 **   -0.01 **
##                    (0.00)       (0.00)     (0.00)     (0.00)
## ----------------------------------------------------------------
## R^2                 0.04         0.06       0.07       0.07
## Adj. R^2            0.03         0.06       0.07       0.06
## Num. obs.         217          306        306        306
## ================================================================
## *** p < 0.001; ** p < 0.01; * p < 0.05
```

When running the regressions, `mice` makes it again very easy to fit a a model to each of the imputed dataset and then pool the results together:

```
imputed_mice_form <- with(
  imputed_mice,
  lm(capabilities_index ~ energy_cons + urban_population)
)

imputed_model_pooled <- summary(pool(imputed_mice_form))
```

```
imputed_model_pooled
## # A tibble: 3 x 6
##   term             estimate std.error statistic    df  p.value
##   <fct>               <dbl>     <dbl>     <dbl> <dbl>    <dbl>
## 1 (Intercept)      15.4       0.583      26.5   14.3  1.42e-13
## 2 energy_cons       0.00146   0.000358    4.07  58.7  1.40e- 4
## 3 urban_population -0.0103    0.00355    -2.90 214.   4.09e- 3
```

Displaying the results with `texreg` will be slightly more difficult, as it is not implemented by default. We will create a custom specification with `texreg::createTexreg()`, manually providing the coefficients, SEs, R^2, etc.

```
tr_imputed_model_pooled <- texreg::createTexreg(
  # coefficient names:
  coef.names = as.character(imputed_model_pooled$term),
  # coefficients, SEs and p-values:
  coef = imputed_model_pooled$estimate,
  se = imputed_model_pooled$std.error,
  pvalues = imputed_model_pooled$p.value,
```

```
    # R^2 and number of observations:
    gof.names = c("R^2", "Num. obs."),
    gof = c(pool.r.squared(imputed_mice_form)[1, 1], nrow(imputed_mice$data)),
    gof.decimal = c(T, F)
)
```

Then we can employ the regular `texreg::screenreg()` with out new object:

```
texreg::screenreg(tr_imputed_model_pooled)
##
## ============================
##                    Model 1
## ----------------------------
## (Intercept)        15.44 ***
##                    (0.58)
## energy_cons         0.00 ***
##                    (0.00)
## urban_population   -0.01 **
##                    (0.00)
## ----------------------------
## R^2                 0.07
## Num. obs.           306
## ============================
## *** p < 0.001; ** p < 0.01; * p < 0.05
```

Finally, we achieved what we were looking for. A regression that has more observations, thanks to the process of imputation. In this case, the difference between observations is small, since we went from 12165 to 12187, with almost no difference in the coefficients. When the percentage of missing values is larger, you will notice that the work of imputation makes a difference in the results.

```
texreg::screenreg(list(model_incomplete, tr_imputed_model_pooled),
                  custom.model.names = c("Incomp. m.", "Pooled imp. m."))
##
## =============================================
##                    Incomp. m.   Pooled imp. m.
## ---------------------------------------------
## (Intercept)        15.32 ***    15.44 ***
##                    (0.54)       (0.58)
## energy_cons         0.00 **      0.00 ***
##                    (0.00)       (0.00)
## urban_population   -0.01 *      -0.01 **
##                    (0.00)       (0.00)
## ---------------------------------------------
## R^2                 0.04         0.07
## Adj. R^2            0.03
## Num. obs.           217          306
```

11.5 Imputation of missing values

```
## ===================================================
## *** p < 0.001; ** p < 0.01; * p < 0.05
```

We hope you found the chapter useful! If you are interested in learning more about working with missing values, we recommend you read James Honaker and Gary King's text (Honaker and King, 2010).

Exercise 11C. Instead of 5 imputations, repeat the exercise of using imputed data for regression models, but this time use 10 imputations.

12

Web Mining

Gonzalo Barría[1]

Suggested readings

- Calvo, E. (2015). *Anatomía Política de Twitter En Argentina: Tuiteando #Nisman*. Capital Intelectual, Buenos Aires, Argentina.

- Steinert-Threlkeld, Z. C. (2018). *Twitter as Data*. Cambridge University Press, Cambridge.

Packages you need to install

- `tidyverse` (Wickham, 2019), `glue` (Hester, 2020), `rvest` (Wickham, 2019), `rtweet` (Kearney, 2020).

12.1 Introduction

Information on the Internet is growing exponentially every day. If the problem of political science in the 20th century was the lack of data to test hypotheses, the 21st century presents another challenge: information is abundant and within reach, but you have to know how to collect and analyze it. This huge amount of data is generally available but in an unstructured way so when you come across the information you will have two main challenges: to extract it and then to sort it out (i.e. to leave it as *tidy data*).

One of the most used techniques to extract information from websites is the technique known as *web scraping*. Web scraping is becoming an increasingly popular technique in data analysis because of its versatility in dealing with different websites. As we can see in the following graph, Google searches for the term "web scraping" have grown steadily year by year since 2004:

---

[1] Institute of Political Science, Pontificia Universidad Católica de Chile. E-mail: ghbarria@uc.cl. Twitter: @GonzaloBarria.

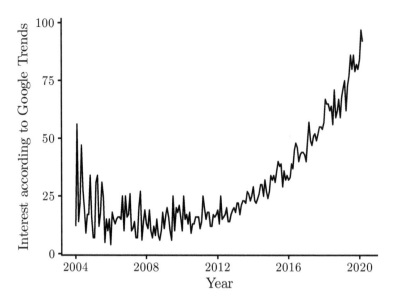

FIGURE 12.1: Searches for web scraping in Google.

This consists of obtaining unstructured data (HTML tags) from a web page that we usually then transform into a structured format of columns and rows, which can be accessed and used more easily. It allows us to obtain data from non-traditional sources (practically any website!)

- The information we can obtain is the same as what we could do manually (copying and pasting to an Excel document, for example), but we can automate very tedious tasks.

What could we use web scraping for? As a practical example, we are going to perform two simple data extraction exercises. But before that it is good to be familiar with the sources of information and the multiple uses that can be given to the extracted data. Some other possible applications for which you can use web scraping are

1. For example, they can be used to classify products or services to create recommendation engines, to obtain text data, like in Wikipedia, to make systems based on Natural Language Processing.

2. Generate data from image tags, from websites like Google, Flickr, etc. to train image classification models.

3. Consolidate data from social networks: Facebook and Twitter, to perform sentiment or opinion analysis.

4. Extract comments from users and e-commerce sites such as Alibaba, Amazon or Walmart.

12.2 Ways to do web scraping

We can scrape data, that is obtain it, in different ways:

1. Copy and paste it: obviously this need to be done by a human, it is a slow and inefficient way to get data from the web.

2. Use of APIs: API means *application programming interface*. Websites such as Facebook, Twitter, LinkedIn, among others, provide a public and/or private API, which can be accessed using programming, to retrieve data in the desired format. An API is a set of definitions and protocols used to develop and integrate application software. To put it simply, it is a code that tells applications how they can communicate with each other. APIs allow your products and services to communicate with others, without the need to know how they are implemented. This simplifies application development and saves time and money.

3. DOM Analysis: DOM stands for *Document Object Model* and is essentially a platform interface that provides a standard set of objects to represent HTML and XML documents. The DOM allows dynamic access through programming to access, add and dynamically change structured content in documents with languages such as JavaScript. Through some programs it is possible to retrieve dynamic content, or parts of websites generated by scripts from a client. DOM objects model both the browser window and the history, the document or web page, and all the elements that the page itself may have, such as paragraphs, divisions, tables, forms and their fields, etc. Through the DOM you can access, by means of JavaScript, to any of these elements, that is to say to their corresponding objects to alter their properties or to invoke their methods. However, through the DOM, any element of the page is available to Javascript programmers, to modify them, delete them, create new elements and place them on the page.

12.2.1 Robot exclusion standard

Before getting into the practice of *web scraping* we have to better understand what it is and how the *robots.txt* file present in most websites works. This file contains the so called *robot exclusion standard* which is a series of instructions specially directed to programs that seek to index the content of these pages (for example the Google bot that "saves" new pages that are created on the Internet). This method is used to prevent certain bots that analyze Internet sites from adding "unnecessary" information to search results. A *robots.txt* file on a web page will function as a request that certain bots ignore specific files or directories in their search. This is important when we talk about *web scraping* as it is always advisable to check the *robots.txt* file of a website before starting the scraping as it may include information we will need later.

An example of how one of these files is can be found in *robots.txt* file from Google
https://www.google.com/robots.txt

```
User-agent: *
Disallow: /search
Allow: /search/about
Allow: /search/static
Allow: /search/howsearchworks
Disallow: /sdch
Disallow: /groups
Disallow: /index.html?
Disallow: /?
Allow: /?hl=
Disallow: /?hl=*&
Allow: /?hl=*&gws_rd=ssl$
```

FIGURE 12.2: Glance at Google's robot exclusion standard.

In this image we find the expression *User-agent:* * This allows all the robots to access the files that are stored in the main code of the web page since the wildcard * *means "ALL". Below we see that bots are not allowed to index or visit websites of the type* /index.html?* or */groups* for example. If we find the expression *Disallow:* / it means that you are denied access to all bots (the wildcard / is automatically applied to all files stored in the root directory of the website).

12.3 Web scraping in R

We'll leave by loading the packages we need. You are already familiar with some of them as `ggplot2` which is part of the `tidyverse`. The `rvest` package is the one that will allow us to do the scraping of data with R. It's essentially a library that allows us to bring in and manipulate data from a web page, using HTML and XML. Finally, the `glue` package is designed to make it easier to interpolate (paste) data into *strings*.

```
library(tidyverse)
library(glue)
library(rvest)
```

To "read" data from different websites we will need the help of an open source tool, a plugin called "selectorgadget". This is used to extract information from a website. In this case we'll use it to select and highlight the parts of the website that we want to extract.

More information can be found at this link `https://selectorgadget.com/` and here `https://cran.r-project.org/web/packages/rvest/vignettes/selectorgadget.html`

12.3.1 Applied example: the Organization of American States (OAS) press releases

As a first example we will do a web scraping of a static site. That is, a web site that has text in HTML and that doesn't change. Let's suppose that we are working on a project of diplomacy and international relations in which we must systematize information on the interaction among Latin American countries. One of the most useful repositories to start this kind of research is the website of the Organization of American States (OAS). It can be found at the following address: `https://www.oas.org/es/`.

This site offers information that is very relevant for political data analysts since it is unstructured data that allows to identify, for example, country networks, alliances, sources of conflict and issues that foreign ministries find relevant.

Before starting with the *web scraping* itself we will analyze the *robots-txt* file of the OAS website `http://oas.org/robots.txt`

```
User-agent: *
Crawl-delay: 3
Disallow: /teldir/
Disallow: /agdocs/
Disallow: /america_viva/
Disallow: /Assembly2001/
Disallow: /catedraOLD/
Disallow: /documents/
Disallow: /directory_missions/
Disallow: /ezine/
Disallow: /FilmFestival/
Disallow: /foodfestival/
```

FIGURE 12.3: Glance at the robot file of OAS's website.

We find that some directories are prohibited from being indexed but all bots and users are allowed to visit the site. The most important thing is the expression *Crawl-delay: 3* that basically tells us that for every request made by a robot it is advisable to wait 3 seconds between one query and another so as not to saturate the site which can result in it blocking you.

In this particular example we want to extract the titles of the press releases found on this website. For simplicity it is recommended to use Google Chrome (`https://chrome.google.com/`) and the Selector Gadget extension for this search engine.[2]

---

[2] you can install it from this link `https://chrome.google.com/webstore/detail/selectorgadget/mhjhnkcfbdhnjickkkdbjoemdmbfginb`.

The Selector Gadget extension also works in Firefox and Safari (as well as in other browsers), you simply have to drag the marker to your bookmarks bar.

First we load the page as an object so that we can then read it as an html file. In this case we want the titles of the press releases of the Organization of American States (OAS) for October 2019, in English.

```
dir.create("webs") # create a "webs/" folder to store the file
download_html(url = "https://www.oas.org/en/media_center/press_releases.asp
         ?nMes=10&nAnio=2019",
         file = "webs/pr_oas_10_2019.html")
```

12.3.2 Load the html to work with R

Once we downloaded the data we import it in R using `read_html()`:

```
web_pr_10_2019 <- read_html("webs/pr_oas_10_2019.html")
```

12.3.3 Extract the information with `html_nodes()` + `html_text()`

Next, we open the page in Google Chrome where the first thing we will see is the site with the news. There we click on the Chrome extension of the Selector Gadget as indicated in the figure below (it's the small magnifying glass in the upper right corner).

FIGURE 12.4: Screenshot of OAS press release website in Spanish.

12.3 Web scraping in R

The first step is to find the CSS selector that contains our information. In most cases we will just use SelectorGadget[3]. Let's start with the titles. As we can see in the figure 12.5, we select the part of the website we want to extract. In this case we want the titles of the press releases that after being selected are highlighted in yellow. You will notice that when you click on one of them, appears a message ".itemmenulink" in the empty space that is in the Selector. These are the characters designated for the titles in this website

FIGURE 12.5: Titles of the press releases that we want to extract.

Next we create an object to read this information as an html object in R. We will call it `titles_web_pr_10_2019`. Inside the function `html_nodes()` we add the character ".itemmenulink" since that represents the titles according to the selector.

```
titles_web_pr_10_2019 <- web_pr_10_2019 %>%
  html_nodes(".itemmenulink") %>%
  html_text()
```

This is how we got the headlines for that whole website.

Now we will review the first 10 results to see how the data was sorted. The idea later on would be to transform this object into a dataframe. We can also extract the content of each news item but for that we would need the URL in English (or whatever language you prefer) for each press release.

```
head(titles_web_pr_10_2019, n = 3)
```

---

[3]See https://selectorgadget.com/

```
## [1] "Security Ministers of the Americas Adopted Recommendations of Quito "
## [2] "OAS General Secretariat commences the electoral integrity analysis a"
## [3] "Statement of the Ministers Responsible for Public Security in the Am"
```

To extract the links from an element, instead of its text, we must replace `html_text()` with `html_attr("href")`:

```
links_web_pr_10_2019 <- web_pr_10_2019 %>%
  html_nodes(".itemmenulink") %>%
  html_attr("href") %>%
  str_c("https://www.oas.org/en/media_center/", .)


links_web_pr_10_2019
```

```
## [1] "https://www.oas.org/en/media_center/press_release.asp?sCodigo=E-092/"
## [2] "https://www.oas.org/en/media_center/press_release.asp?sCodigo=E-091/"
## [3] "https://www.oas.org/en/media_center/press_release.asp?sCodigo=S-023/"
## [4] "https://www.oas.org/en/media_center/press_release.asp?sCodigo=S-022/"
## [5] "https://www.oas.org/en/media_center/photonews.asp?sCodigo=FNE-97927"
## [6] "https://www.oas.org/en/media_center/press_release.asp?sCodigo=E-090/"
## [7] "https://www.oas.org/en/media_center/press_release.asp?sCodigo=S-021/"
## [8] "https://www.oas.org/en/media_center/press_release.asp?sCodigo=E-089/"
##  [ reached getOption("max.print") -- omitted 7 entries ]
```

Now we can create a *data frame* with all the information:

```
df_web_pr_10_2019 <- tibble(
  title = titles_web_pr_10_2019,
  link  = links_web_pr_10_2019
)


df_web_pr_10_2019
## # A tibble: 15 x 2
##    title                                link
##    <chr>                                <chr>
## 1 Security Ministers of the Americas ~  https://www.oas.org/en/media_cent~
## 2 OAS General Secretariat commences t~  https://www.oas.org/en/media_cent~
## 3 Statement of the Ministers Responsi~  https://www.oas.org/en/media_cent~
## # ... with 12 more rows
```

Exercise 12A. Get the date of each OAS press release for October 2019. Call the vector "web_date_releases_10_2019".

Exercise 12B. Get the news titles from the website of *The Economist*'s international section: https://www.economist.com/international/?page=1

12.3.4 Iterations

Iterations allow us to repeat the same operation for a set of elements. Let's analyze the syntax in the following example:

```
print(1)
```

[1] 1

```
walk(.x = 1:10,
     .f = ~ {
       print(.x)
     })
```

[1] 1 [1] 2 [1] 3 [1] 4 [1] 5 [1] 6 [1] 7 [1] 8 [1] 9 [1] 10

We could download several pages from OAS website with an iteration. For example, let's download the press releases for all months from 2017 to 2019. To do so, we use the `cross_df()` function to create a *data.frame* that contains all the possible combinations between months and years:

```
iteration_df <- cross_df(list(month = 1:12, year = 2017:2019))
```

Then, we use the `walk2()` function that will receive two arguments that will iterate, the months and years.

```
walk2(.x = iteration_df$month,
      .y = iteration_df$year,
      .f = ~ {
        Sys.sleep(2) # stops
        download_html(
          url  = glue("https://www.oas.org/en/media_center/press_releases.
                      asp?nMes={.x}&nAnio={.y}"),
          file = glue("webs/pr_oas_{.x}_{.y}.html"))
      })
```

Our next steps will show you how to process these websites (in this case we will do 36 URLs, but it could be many more) to create a unique dataset. We will use custom functions and iterations in the process.

12.3.5 Custom functions (recipes)

What we did so far to get our *data frame* with information from OAS's website can be summarized in the following steps:

1. Download the site

2. Upload the html to R
3. Extract the vectors for title and link using selectors
4. Create the *data frame* with these vectors

This can be thought of as a "recipe", which should work for any file. Thus, we will create it as a custom function for you to work with:

```
f_process_site <- function(file){
  web <- read_html(file, encoding = "UTF-8")

  titles <- web %>%
  html_nodes(".itemmenulink") %>%
  html_text()

  links <- web %>%
  html_nodes(".itemmenulink") %>%
  html_attr("href") %>%
  str_c("https://www.oas.org/en/media_center/", .)

  df_info <- tibble(
    title = titles,
    link  = links
  )

  return(df_info) # what returns/delivers the function
}
```

Now the function takes any file as an argument and gives us what we expect:

```
f_process_site(file = "webs/pr_oas_10_2017.html")
## # A tibble: 15 x 2
##    title                                link
##    <chr>                                <chr>
## 1 OAS Electoral Observation Mission t~  https://www.oas.org/en/media_cent~
## 2 New Ambassador of Mexico to the OAS~  https://www.oas.org/en/media_cent~
## 3 Chief of OAS Mission Arrives in Nic~  https://www.oas.org/en/media_cent~
## # ... with 12 more rows
```

Thus, we can iterate this function in our 36 files to create a complete dataset. In this case we won't use `walk()`, but `map_dfr()` –this function expects each iteration to return a *data frame*, and (below) pastes them in order with `bind_rows()`.

```
files <- list.files("webs/", full.names = T)
files

## [1] "webs//pr_oas_1_2017.html"  "webs//pr_oas_1_2018.html"
```

```
## [3] "webs//pr_oas_1_2019.html"  "webs//pr_oas_10_2017.html"
## [5] "webs//pr_oas_10_2018.html" "webs//pr_oas_10_2019.html"
## [7] "webs//pr_oas_11_2017.html" "webs//pr_oas_11_2018.html"
##  [ reached getOption("max.print") -- omitted 28 entries ]

df_web_pr_2017_2019 <- map_dfr(.x = files, .f = ~ {f_process_site(.x)})
df_web_pr_2017_2019
## # A tibble: 517 x 2
##    title                                 link
##    <chr>                                 <chr>
## 1 OAS observes conclusion of the elec~ https://www.oas.org/en/media_cent~
## 2 OAS Deploys Observation Mission for~ https://www.oas.org/en/media_cent~
## 3 OAS Electoral Observation Mission i~ https://www.oas.org/en/media_cent~
## # ... with 514 more rows
```

12.4 Using APIs and extracting data from Twitter

Twitter is a social network founded in 2006 that allows users to interact with each other by sending messages of no more than 280 characters. It is widely used by public officers and politicians especially during campaign periods. With new data analysis techniques the study of users' interaction on Twitter has become relevant, for example, to measure the issues people are talking about (*trending topics*) and whether the opinions about certain person or issue are positive or negative. For example, Ernesto Calvo wrote *Political Anatomy of Twitter in Argentina: Tweeting #NISMAN* in which he identifies through the study of tweets over the causes of the death of former prosecutor Alberto Nisman that they reflected (and were very much correlated with) political cleavages of government-oppostition. Based on the Nisman case, Ernesto Calvo analyzes the tweets of Argentine users and shows that polarization in Twitter is explained by a mix of political beliefs, trolls and smartphone brands.

If you have read the previous section, you already know the R packages that will be used in this section of the book. To do the extraction of data from Twitter the best option in hand is `rtweet`, which allows to access for free to the Twitter's API to download information of users, trending topics and hashtags. To extract data from Twitter with R it is recommended to consult *Twitter as Data*, which contains some standardized routines to download data from this platform.

12.4.1 Some background on APIs

Application Programming Interfaces (APIs) are a set of protocols and functions that govern certain interactions between web applications and users.

APIs are similar to web browsers but serve different purposes:

- Web browsers play back content from browsers
- APIs allow you to manipulate and organize data

In order for the public APIs to be used, many sites only allow authorized users (e.g. those who have an account on the platform). This is the case of Twitter, Facebook, Instagram and Github.

Although these APIs are widely known, it is not superfluous to mention that some were created by the same R community specialized in political data. For example, the lobbyR package created by Daniel Alcatruz[4] allows loading and structuring data from the lobby API found in the Lobby Law platform (https://www.leylobby.gob.cl/) implemented by the Government of Chile, which allows consultations on hearings in certain public services and state agencies such as Congress and municipalities. Another package that deserves to be mentioned is inegiR created by Eduardo Flores that allows interaction with the API of INEGI (National Institute of Statistics and Geography of Mexico) for specific consultations.[5]

12.4.2 Extract the data

The first thing not to depend on the official Twitter API is to have your own Twitter account that you can create at (https://twitter.com/i/flow/signup). Then you proceed to load the rtweet package that will allow you to extract data from Twitter. We will show you this routine and we will extract data from different users and hashtags.

```
library(rtweet)
```

For example, we can obtain the account IDs that the United Nations follows in its Spanish account. By default the function shows you up to 5000 users.

```
## Get the account IDs that the UN follows in Spanish
friends_ONU_es <- get_friends("ONU_es")
friends_ONU_es
```

```
## # A tibble: 173 x 2
##   user    user_id
##   <chr>   <chr>
## 1 ONU_es  2574658173
## 2 ONU_es  851590790
## 3 ONU_es  1176867875744649216
## # ... with 170 more rows
```

---

[4] See https://github.com/Dalcatruz/lobbyR
[5] See http://enelmargen.org/ds/inegiR/vignette_spa.html

12.4 Using APIs and extracting data from Twitter

To find out more about these users we can use the `lookup_users()` function

```
info_friends_ONU_es <- lookup_users(friends_ONU_es$user_id)

info_friends_ONU_es
```

```
## # A tibble: 173 x 90
##   user_id status_id created_at          screen_name text   source
##   <chr>   <chr>     <dttm>              <chr>       <chr>  <chr>
## 1 257465~ 12312639~ 2020-02-22 17:06:14 ConsueloVi~ Ante~  Twitt~
## 2 851590~ 12303084~ 2020-02-20 01:49:17 ONU_Cuba    Las ~  Twitt~
## 3 117686~ 12312560~ 2020-02-22 16:34:50 AcnurMexico En r~  Twitt~
## # ... with 170 more rows, and 84 more variables
```

We can also obtain information from the followers through the function `get_followers()`. As the UN Twitter account in Spanish has more than one million followers, it is more comfortable to do an analysis of a portion of them. To get users' IDs of all 48.8 million followers of @ONU_es, you only need two things:

- A stable internet connection
- Time - approximately five and a half days

As, an example, we will only get 200 users. Notice the `retryonratelimit = T` argument, which would help us if we were trying to obtain data from a great number of accounts.

```
followers_ONU_es <- get_followers("ONU_es",
                        n = 200, # we'll only get 200 followers
                        retryonratelimit = T)
```

Here we get the information of the sample of users who follow the account @ONU_es:

```
info_followers_ONU_es <- lookup_users(followers_ONU_es$user_id)

info_followers_ONU_es
```

```
## # A tibble: 200 x 90
##   user_id status_id created_at          screen_name text   source
##   <chr>   <chr>     <dttm>              <chr>       <chr>  <chr>
## 1 422091~ 12312538~ 2020-02-22 16:26:06 fbaraza27   Esta~  Twitt~
## 2 586633~ 11364218~ 2019-06-05 23:57:37 carranzasn~ @Iva~  Twitt~
## 3 154435~ 12314260~ 2020-02-23 03:50:21 Chile20Love @Tdn~  Twitt~
## # ... with 197 more rows, and 84 more variables
```

12.4.2.1 Searching for specific tweets

Now we're ready to look for specific tweets. Let's look for all the tweets with the hashtag "#Piñera", in reference to the Chilean president. Remember that a hashtag is a label that corresponds to a word preceded by the symbol "#". It is mainly used to find tweets with similar content, since the purpose is for both the system and the user to identify it quickly. We'll use the function `rtweet::search_tweets()`. This function requires the following arguments:

q: the word you want to look up. n: the maximum number of tweets you want to extract. You can request a maximum of 18,000 tweets every 15 minutes.

Your particular analysis might benefit from some other argument(s), which you can check with the `?search_tweets` help command.

An important aspect that we have to clarify regarding the queries we make to the Twitter API is that the results we obtained will not be able to be completely replicated. They will certainly change, since the free version of Twitter allows us to make queries in a finite period of time in the past (for example a week) so the results of the queries will change over time. If you are interested in a certain agenda or hashtag, a good advice is to regularly download them and store them in a hard drive. You won't be able to go back to them in the future.

```
# Look for 1000 tweets with the hashtag #piñera
pinera_tweets <- search_tweets(q = "#piñera",
                               n = 1000)
# We see the first 3 rows
head(pinera_tweets, n = 3)
```

```
## # A tibble: 3 x 90
##    user_id  status_id  created_at          screen_name  text  source
##    <chr>    <chr>      <dttm>              <chr>        <chr> <chr>
## 1  117214~  12314261~  2020-02-23 03:50:41 elOliverde~  "Des~ Twitt~
## 2  395223~  12314259~  2020-02-23 03:50:11 carlosv733~  "Bri~ Twitt~
## 3  192719~  12314257~  2020-02-23 03:49:23 frantoro70   "Bri~ Twitt~
## # ... with 84 more variables
```

For information from users who are tweeting about #piñera at the time this chapter is being written we could use `lookup_users()`, like we did before:

```
lookup_users(pinera_tweets$user_id)
```

Tweet queries can be made in several ways depending on what you are looking for:

Search for a keyword

```
query_1 <- search_tweets(q = "piñera", n = 20)
```

Search for a phrase

```
query_2 <- search_tweets(q = "piñera economia", n = 20)
```

12.4 Using APIs and extracting data from Twitter

Search for multiple keywords

```
query_3 <- search_tweets(q = "piñera AND bachelet", n = 20)
```

By default `search_tweets()` returns 100 tweets. To get more you need to specifiy a larger n with the `n =` argument.

Exercise 12C. Instead of using AND, use OR between the different search terms. Both searches will return very different tweets.

Search for any mention of a list of words

```
query_4 <- search_tweets("bolsonaro OR piñera OR macri", n = 20)
```

Exercise 12D. Search for Spanish tweets that are not retweets.

Specifies the language of tweets and excludes retweets

```
query_5 <- search_tweets("piñera", lang = "es", include_rts = FALSE, n=20)
```

12.4.3 Downloading features related to Twitter users

12.4.3.1 Retweets

A retweet is when you or someone else shares a tweet so that your followers can see it. It's similar to the "share" feature on Facebook. Let's do the same routine as before but this time ignoring the retweets. We do it with the `include_rts =` argument defined as `FALSE` (F). We can then get tweets and retweets in a separate dataframe

---

Note: The `include_rts =` option is very useful if you are studying the patterns of viralization of certain hashtags.

```
# Look for 1000 tweets with the hashtag #piñera, but ignore retweets
pinera_tweets_no_rts <- search_tweets("#piñera",
                                      n = 1000,
                                      include_rts = F)
# See the first three rows
head(pinera_tweets_no_rts, n = 3)

## # A tibble: 3 x 90
##   user_id  status_id  created_at          screen_name text  source
##   <chr>    <chr>      <dttm>              <chr>       <chr> <chr>
## 1 628103~  12314218~  2020-02-23 03:33:42 Patho1980   "La ~ Twitt~
## 2 297832~  12314206~  2020-02-23 03:28:58 FLOYD3178   "un ~ Twitt~
## 3 119235~  12314125~  2020-02-23 02:56:39 leandrofie~ "Ins~ Twitt~
## # ... with 84 more variables
```

Now let's see who's tweeting about the hashtag "#piñera":

```
head(unique(pinera_tweets$screen_name))
## [1] "elOliverde18"     "carlosv73348197" "frantoro70"
## [4] "CarlosC77161692" "Pedro07316240"   "ojeda_prufer"
```

We can also use the `search_users()` function to explore which users are tweeting using a particular hashtag. This function extracts a data.frame of the users and information about their accounts.

```
# Which users are tweeting about #piñera?
pinera_users <- search_users("#piñera",
                             n = 1000)
# see the first 3 users
head(pinera_users, n = 2)

## # A tibble: 2 x 90
##   user_id  status_id  created_at          screen_name text  source
##   <chr>    <chr>      <dttm>              <chr>       <chr> <chr>
## 1 136235~  12308574~  2020-02-21 14:10:50 sebastianp~ Anoc~ Twitt~
## 2 216071~  11645976~  2019-08-22 17:58:01 mnegropine~ Feli~ Twitt~
## # ... with 84 more variables
```

Let's learn more about these people. Where are they from? As we see, there are 403 unique places so the graph we use to plot the information is not addecuate to visualize it.

```
# How many places are represented?
length(unique(pinera_users$location))
## [1] 403
```

12.4 Using APIs and extracting data from Twitter

Let's order by frequency the most named locations and plot them. To do this we use `top_n()` which will extract the locations with at least 20 users associated to it

```
pinera_users %>%
  count(location, sort = TRUE) %>%
  mutate(location = reorder(location, n)) %>%
  top_n(20) %>%
  ggplot(aes(x = n, y = location)) +
  geom_col() +
  labs(x = "Frequency", y = "Location")
```

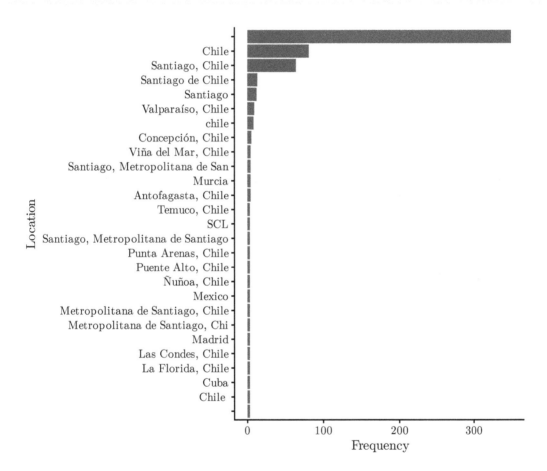

FIGURE 12.6: Sample of users tweeting about Piñera – unique locations.

It is better if we remove the blank locations to better see the pattern more clearly. In general, the geolocation feature of Twitter is quite mediocre because it is voluntary and few people have it activated in their mobile phones. So what we see is people who want us to know their location, meaning that there is a bias in these results.

```
pinera_users %>%
  count(location, sort = T) %>%
  mutate(location = reorder(location, n)) %>%
  na.omit() %>% filter(location != "") %>%
  top_n(20) %>%
  ggplot(aes(x = n, y = location)) +
  geom_col() +
  labs(x = "Frequency", y = "Location")
```

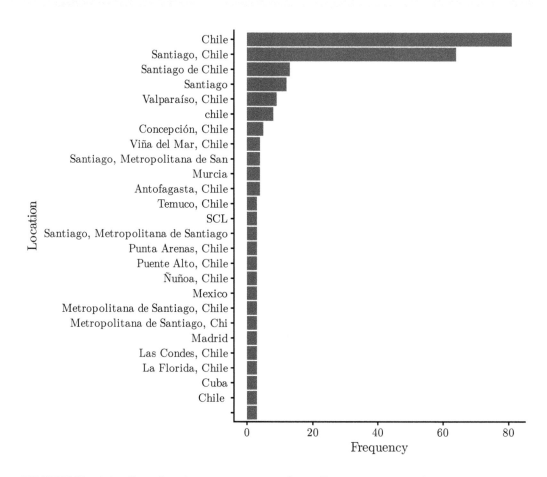

FIGURE 12.7: Sample of users tweeting about Piñera – unique locations.

Finally we repeat the exercise using hashtags that refer to Michelle Bachelet, Chile's former president:

```
## Search for tweets that use the hashtag #bachelet
bachelet_tweets <- search_tweets(
  "#bachelet", n = 1000, include_rts = F
)
```

12.4 Using APIs and extracting data from Twitter

How has been the time trend of this hashtag in the last days?

```
## plot a series of tweets in time
bachelet_tweets %>%
  ts_plot("3 hours") +
  labs(
    x = "", y = "Frequency",
    caption = "Source: Data collected from the Twitter API."
  )
```

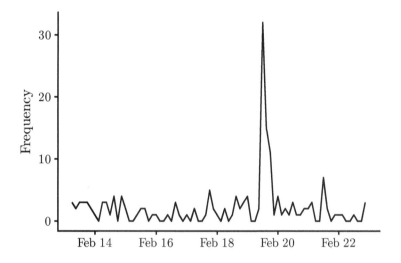

Source: Data collected from the Twitter API.

FIGURE 12.8: #Bachelet frequency in tweets from the last nine days. Tweet count added using 3-hour intervals.

We hope you have found this chapter useful. In Chapter 13 we will show you how to further explore data from Twitter once you have downloaded it.

13

Quantitative Analysis of Political Texts

Sebastián Huneeus[1]

Suggested readings

- Salganik, M. J. (2017). *Bit by Bit: Social Research in the Digital Age.* Princeton University Press, Princeton, NJ.

- Silge, J. and Robinson, D. (2017). *Text Mining with R: A Tidy Approach.* O'Reilly, Sebastopol, CA.

- Steinert-Threlkeld, Z. C. (2018). *Twitter as Data.* Cambridge University Press, Cambridge.

Packages you need to install

- `tidyverse` (Wickham, 2019), `politicalds` (Urdinez and Cruz, 2020), `lubridate` (Spinu et al., 2020), `skimr` (Waring et al., 2020), `ggwordcloud` (Le Pennec and Slowikowski, 2019), `tidytext` (Robinson and Silge, 2020), `stopwords` (Benoit et al., 2020), `quanteda` (Benoit et al., 2020), `quanteda.textmodels` (Benoit et al., 2020), `qdapRegex` (Rinker, 2017), `stm` (Roberts et al., 2019), `tidystm` (Johannesson, 2020), `remotes` (Hester et al., 2020).

This chapter is divided into three sections, which employ different strategies to analyze textual data from Twitter. Subsection 13.1 covers text analysis exploration, Subsection 13.2.3 deals with Wordfish (a technique to position texts along an axis), while Subsection 13.3 covers structural topic models (STM), which helps us discover underlying themes in text data.

In the study of contentious politics, #olafeminista, #metoo, #blacklivesmatter and #niunamenos are hashtags that went viral and tell a rich story of the social media activism and protest. We will use as a case study a protest cycle called Ola Feminista (Feminist Wave), which occurred in Chile from May to June of 2018. The feminist protest cycle denounced structural gender inequalities and started as a mobilization of students in Santiago, and gradually grew expanding to broader demands from feminist and worker organizations across Chile.[2]

---

[1]Institute of Political Science, Pontificia Universidad Católica de Chile. E-mail: lshuneeus@uc.cl.
[2]For a deeper understanding on the extent of this movement, we recommend the book edited by journalist Faride Zerán, "Mayo Feminista: La rebelión contra el patriarcado" (2018).

In the first half of this chapter, you will learn how to use basic descriptive statistics to understand how policy makers use Twitter. We will analyze how deputies in Chile made use of gender related hashtags during the #olafeminista (feminist wave). We will analize simple frequency variations in the usage of hashtags to adrresss different levels of engagement with the online debate around gender and feminist issues.

In the second half or the chapter, you will learn how to use Wordfish and Structural Topic Modeling (STM), two recent natural language processing (NLP) techniques used in political science for unsupervised text mining. While Wordfish will let us position the political coalitions along a left-right ideological axis, the STM will let us indentify the most regular topics –or groups of words– and see how these topics are correlated to a variable of interest. In our example we will explore the correlation between the gender of the deputy and the use of certain hashtags. All in all, these techniques are a great tool to gain knowledge on how coalitions and policy makers are digitally engaged in a political conversation.

In this chapter, you will use an original dataset with identification variables for the deputies, such as name and last name, district, political party, age, among others. The identification variables were extracted from the official web page of the House of Representatives (*Cámara de Diputados*)[3]. For the data extraction from Twitter we used the `rtweet` package, which allows us to freely access Twitter's API for downloading information by users, dates and hashtags (see Chapter 12).

In this type of analysis, the most difficult task is gathering and cleaning the dataset to make it look "tidy". Fortunately, Twitter's API and packages such as `rtweet` or `twitter` are very helpful in managing the downloaded information in a simple and orderly way.[4]

13.1 Analysis of political hashtags

What are hashtags (#) and how are they related to politics? Hashtags are texts that connect users in a digital conversation. Analyzing them helps us to understand how, when and by whom these conversations are taking place. Also, hashtags could help political movilization. Indeed, there is literature that addresses social protest through the viralization of hashtags, like the recent works that study the hashtag-feminism (Trott, 2018) and hashtags of racial minority activism, such as #blacklivesmatter (Ince et al., 2017).[5]

---

[3]See https://www.camara.cl/camara/deputys.aspx#tab

[4]To learn more about some great research strategies in social sciences using Twitter data we recommend the chapter "Observing Behavior" of the book "Bit by Bit" by Matthew Salganik (2017, ch. 2).

[5]An essential read about how a Latin-American political crisis can be studied through Twitter is *Political Anatomy of Twitter in Argentina* (Anatomía Política de Twitter en Argentina) (2015) on the structure of the digital network generated after the death of prosecutor Alberto Nisman in 2015.

13.1 Analysis of political hashtags

In general, using a hashtag indicates interest in a topic, independently of one being in favor or against it. Therefore, this first exercise is not intended to measure levels of support, but instead, this analysis allows us to identify those representatives discussing gender issues. The organization of the section is as follows. The first part is the exploration of the dataset and a bivariate descriptive analysis of the # frequencies. In the second part, we make a comparison by gender. Thirdly, we compare the use of hashtags by political coalition. In the fourth part, we will see the weekly variation in the use of some #. In the fifth section, this temporal variation will be separated by gender.

13.1.1 Twitter data exploration

```
library(tidyverse)
library(tidytext)
```

```
library(politicalds)
data("poltweets")
```

After loading the dataset `poltweets`, we explore it with `skimr::skim(poltweets)`, as shown in Figure 13.1. With this function you can do a quick exploration of the size of the dataset, the number of observations and variables, and the type of variables (character, integer, factor, etc). We can also look at the number of missing values, the number of categories or values that the factor variable assumes (`n_unique`), as well as the dispersion statistics for quantitative variables (min, max, quantiles, mean and standard deviation). `skimr::skim()` is a good first step that will allow us to diagnose

```
> skimr::skim(poltweets)
-- Data Summary ------------------------
                           Values
Name                       poltweets
Number of rows             18658
Number of columns          8

Column type frequency:
  character                7
  POSIXct                  1

Group variables            None

-- Variable type: character -------------------------------------------------
  skim_variable n_missing complete_rate min  max empty n_unique whitespace
1 status_id             0             1   1    5     0    18658          0
2 screen_name           0             1   7   15     0      122          0
3 names                 0             1   4   17     0       98          0
4 lastname              0             1   3   16     0      112          0
5 text                  0             1   1  524     0    18623          0
6 coalition             0             1   2    3     0        3          0
7 gender                0             1   4    6     0        2          0

-- Variable type: POSIXct ---------------------------------------------------
  skim_variable n_missing complete_rate min                 max                 median              n_unique
1 created_at            0             1 2018-04-01 00:00:53 2018-06-30 23:58:27 2018-05-17 08:46:47    18658
```

FIGURE 13.1: Skim of our dataset.

the data we will work with. In the case of our dataset `poltweets`, we see that there are 7 variables of the "character" type and one of the "POSIXct" type, which also helps working with dates.

Look how 75.4% of the rows in the dataset correspond to tweets by congressmen. We also notice thata there are are 29 congresswomen and 93 congressmen in the dataset.

```
# among tweets:
poltweets %>% count(gender) %>% mutate(freq = n / sum(n))
## # A tibble: 2 x 3
##   gender     n  freq
##   <chr>  <int> <dbl>
## 1 Female  4595 0.246
## 2 Male   14063 0.754

# deputies' characteristics:
poltweets %>%
   distinct(screen_name, gender)%>%
   count(gender)
## # A tibble: 2 x 2
##   gender     n
##   <chr>  <int>
## 1 Female    29
## 2 Male      93
```

After we loaded the dataset and took a quick view to the size and variables included, we must extract the hashtags from the tweets using the `unnest_tokens()` function of `tidytext`, creating a tokenized data frame with one row per hashtag. We then just filter all the rows starting with a hashtag (#), leaving us with a one-hashtag-per-row data frame.

```
poltweets_hashtags <- poltweets %>%
  unnest_tokens(output = "hashtag", input = "text", token = "tweets") %>%
  filter(str_starts(hashtag, "#"))
```

We want to see the differences in how representatives, parties and coalitions engage in the gender political debate. To do so, we create a new dummy variable that takes value "1" each time the character string variable matches any of the regular expresions like "femi", "niunamenos", "aborto", "mujer" and "genero":

```
poltweets_hashtags <- poltweets_hashtags %>%
   mutate(fem_hashtag = case_when(str_detect(hashtag, "femi") ~ 1,
                                  str_detect(hashtag, "niunamenos") ~ 1,
                                  str_detect(hashtag, "aborto") ~ 1,
                                  str_detect(hashtag, "mujer") ~ 1,
                                  str_detect(hashtag, "genero")~ 1,
```

13.1 Analysis of political hashtags

```
                                 TRUE ~ 0)) %>%
  mutate(fem_hashtag = as.character(fem_hashtag))
```

We see that this is a good measure for capturing gender and feminist related hashtags. Notice that only 4.1% of the rows contain a hashtag related to gender under this criteria and that the three most frequent hashtags are #aborto3causales, #interpelacionaborto3causales and #leydeidentidaddegeneroahora.[6]

```
poltweets_hashtags %>% count(fem_hashtag) %>% mutate(freq = n / sum(n))
## # A tibble: 2 x 3
##   fem_hashtag      n   freq
##   <chr>        <int>  <dbl>
## 1 0            10423 0.958
## 2 1              455 0.0418
```

```
poltweets_hashtags %>%
  filter(fem_hashtag == "1") %>%
  count(hashtag) %>%
  arrange(-n)
## # A tibble: 63 x 2
##   hashtag                              n
##   <chr>                            <int>
## 1 #aborto3causales                    98
## 2 #interpelacionaborto3causales       64
## 3 #leydeidentidaddegeneroahora        61
## # ... with 60 more rows
```

13.1.2 Visual diagnosis

Let's make some bivariate analysis by grouping by the number of tweets per month, coalition and gender (Figures 13.2, 13.3, and 13.4).

13.1.3 Most-used hashtags

To see which are the most frenquently used hashtags, we order them from the most to the least frequent. We see that the most used hashtag is #cuentapublica

---

[6]The hashtag #aborto3causales (#abortionunder3causes) is related to a long-standing national debate: Between 1989 and 2017 Chile had one of the most restrictive abortion policies in the world, criminalizing its practice without exception. Since 2017, abortion in Chile is legal in the following cases: when the mother's life is at risk, when the fetus will not survive the pregnancy, and during the first 12 weeks of pregnancy (14 weeks if the woman is under 14 years old) in the case of rape. On February 12th, 2018, the Technical Norm was promulgated on the Official Journal of the 21.030 Law, which decriminalized abortion for the three causes. The norm was threatened with repeal and modification, which generated a substantial amount of controversy.

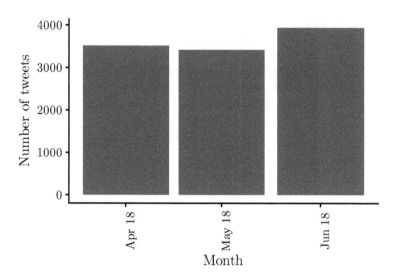

FIGURE 13.2: Total number of tweets by month.

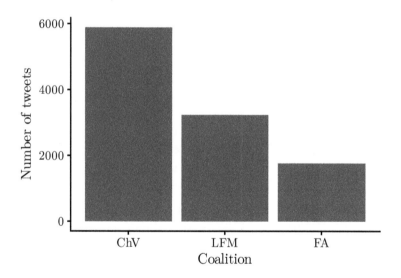

FIGURE 13.3: Total number of tweets by coalition.

(#stateoftheunion), linked to the Public Account Speech of May 21st delivered by President Sebastian Piñera to the National Congress. The hashtag `#aborto3causales` (#abortionunder3causes) is the only gender related hashtag in the ranking, with 98 mentions. This hashtag refers to the discussion about the abortion law. Since 2017, abortion in Chile is legal in the following cases: when the mother's life is at risk, when the fetus will not survive the pregnancy, and during the first 12 weeks of pregnancy (14 weeks if the woman is under 14 years old) in the case of rape. On February 12th,

13.1 Analysis of political hashtags

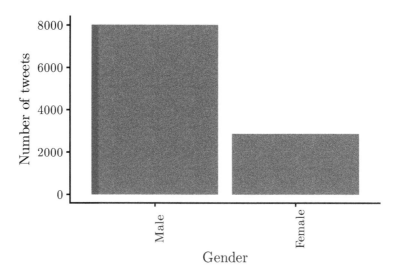

FIGURE 13.4: Total number of tweets by gender.

2018, the Technical Norm was promulgated on the Official Journal of the 21.030 Law, which decriminalized abortion for these three causes. The norm was threatened with repeal and modification, which generated a substantial amount of controversy.

```
poltweets_hashtags %>%
  count(hashtag, fem_hashtag) %>%
  arrange(-n) %>%
  slice(1:20)
## # A tibble: 20 x 3
##   hashtag        fem_hashtag     n
##   <chr>          <chr>       <int>
## 1 #cuentapublica 0             381
## 2 #valdiviacl    0             323
## 3 #losrios       0             232
## # ... with 17 more rows
```

13.1.4 Wordclouds

A quick and intuitive way of representing word frequencies are wordclouds. These graphical representations allow to place at the center and with large letters the cases that have greater frequencies. For that, we use the `ggwordcloud` package for the 35 most common hashtags. After creating a counting dataset, we'll employ the `geom_text_wordcloud()` with the "label" and "size" and "color" aesthetic mappings. In the visual inspection we see the three most used gender hashtags: #aborto3causales, #leydeidentidaddegeneroahora and #interpelacionaborto3causales (Figure 13.5).

```
library(ggwordcloud)

data_hashtags_wordcloud <- poltweets_hashtags %>%
  count(hashtag, fem_hashtag) %>%
  arrange(-n) %>%
  slice(1:35)


ggplot(data_hashtags_wordcloud,
       aes(label = hashtag, size = n, color = fem_hashtag)) +
  geom_text_wordcloud() +
  scale_size_area(max_size = 8) + # we set a maximum size for the text
  theme_void()
```

#interpelacionlarrain   #interpelacionaborto3causales
#calama #diputadoenterreno   #chilecuentacontigo   #atacama
#chiloe #altohospicio   #tiemposmejores   #aricayparinacota
#baltolutudiputado #chile #valdiviacl   #antofagasta   #tarapaca
#coquimbo #iquique #cuentapublica   #arica #laserena
#semanadistrital   #losrios   #elranco   #puertomontt   #llamioned
#diputadohugorey   #chilelohacemostodos   #aborto3causales   #riohueno
#losniñosprimero   #piñerapresidente   #vanrysselberghediputado   #diputadosrn
#leydeidentidaddegeneroahora   #osorno

FIGURE 13.5: Wordcloud of the most common hashtags.

13.1.5 Wordclouds by groups

Using the `facet_wrap()` function, wordclouds can be split by variables of interest. Classifiying by gender and coalition, we immediately see how hashtags such as #olafeminista (#feministwave), #agendamujer (#womenagenda) and #educacionnosexista (#sexisteducation) appear only among congresswomen Twitter accounts. When faceting by coalitions, we realize that the tweets from the Frente Amplio (FA) use a high proportion of gender related hashtags, whereas the oficialist coalition Chile Vamos (ChV) uses no hashtag at all (see Figures 13.6 and 13.7).

13.1 Analysis of political hashtags

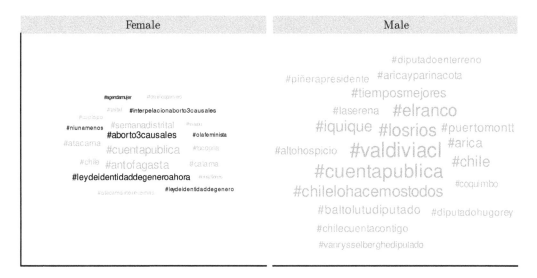

FIGURE 13.6: Wordclouds by gender.

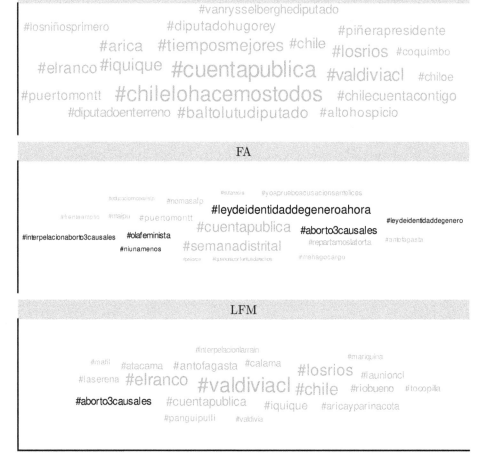

FIGURE 13.7: Wordclouds by coalition.

```
ggplot(poltweets_hashtags %>%
         count(hashtag, gender, fem_hashtag) %>%
         arrange(-n) %>%
         group_by(gender) %>%
         slice(1:20),
       aes(label = hashtag, size = n, color = fem_hashtag)) +
  geom_text_wordcloud() +
  scale_size_area(max_size = 6) +
  facet_wrap(~gender)


ggplot(poltweets_hashtags %>%
         count(hashtag, coalition, fem_hashtag) %>%
         arrange(-n) %>%
         group_by(coalition) %>%
         slice(1:20),
       aes(label = hashtag, size = n, color = fem_hashtag)) +
  geom_text_wordcloud() +
  scale_size_area(max_size = 6) +
  facet_wrap(~coalition, nrow = 3)
```

13.1.6 Barplots

Now we will rank the frequency of hashtags by gender. We will generate this graph in two steps, first we create a table with the 15 most used hashtags among women and men. Then, we will create a bar chart by adding the `geom_col()` argument to the `ggplot()` function. As a result, we see the hashtag #aborto3causales (#abortionunder3causes) and #leydeidentidaddegeneroahora (#genderidentitylawnow) appear only congresswomen accounts, whereas none of these gender related hashtags appear in masculine accounts (Figure 13.8).

```
plot_15 <- poltweets_hashtags %>%
  group_by(gender) %>%
  count(hashtag, fem_hashtag) %>%
  arrange(-n) %>%
  slice(1:15)


ggplot(data    = plot_15,
       mapping = aes(x = n, y = reorder_within(hashtag, n, gender),
                     fill = fem_hashtag)) +
  geom_col()+
  labs(x = "Frequency", y = "", fill = "Feminist hashtag") +
  facet_wrap(~gender, scales = "free", nrow = 2) +
  scale_y_reordered()
```

13.1 Analysis of political hashtags

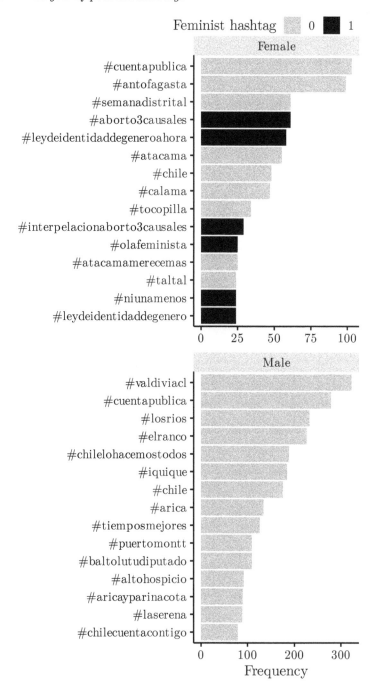

FIGURE 13.8: Most frequent hashtags by congresswomen (April 1 - June 30).

Now we calculate and plot the statistic tf-idf, intended to measure how important a word is to a document in a collection of documents. This statistic is a combination of term frequency (tf) and the term's inverse document frequency (idf), which decreases the weight for commonly used words and increases the weight for words that are not used very much in the entire collection of documents (see Figure 13.9). We see that, when separating by groups, two hashtags with the highest statistic tf-idf in the Frente Amplio are gender related (#leydeidentidaddegeneroahora).

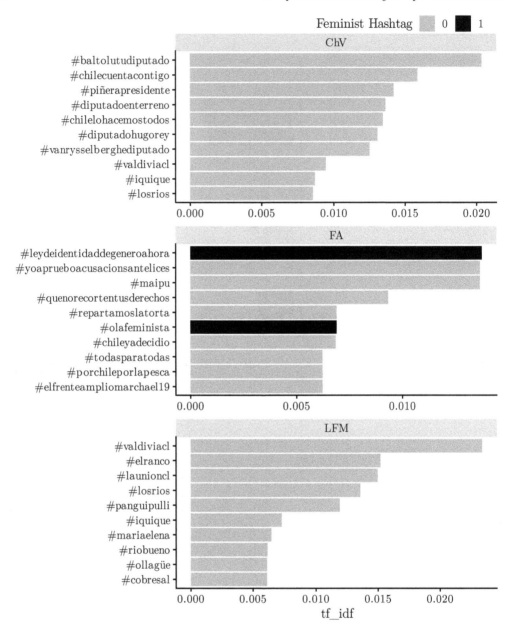

FIGURE 13.9: tf-idf statistic, intended to measure how important a word is to a document.

```
hash_tf_idf <- poltweets_hashtags %>%
  # calculate tf-idf:
  count(coalition, hashtag, fem_hashtag, sort = T) %>%
  bind_tf_idf(term = hashtag, document = coalition, n = n) %>%
  # get 10 most distinctive hashtags per coalition:
  arrange(-tf_idf) %>%
  group_by(coalition) %>%
  slice(1:10)
```

13.1 Analysis of political hashtags

```
ggplot(data    = hash_tf_idf,
       mapping = aes(x = tf_idf,
                     y = reorder_within(hashtag, tf_idf, coalition),
                     fill = fem_hashtag)) +
  geom_col() +
  labs(x = "tf_idf", y = "", fill = "Feminist Hashtag") +
  facet_wrap(~coalition, nrow = 3, scales = "free") +
  scale_y_reordered()
```

13.1.7 Temporal variation in the use of hashtags

Certain hashtags may increase or decrease in its use through time, depending on the political context. We will explore the weekly frequency of two most frequent hashtags in our example. Using the `lubridate` package, which works with data in a date format, we can look for time trends. In our dataset we have one variables with a date: `created_at`. Using this variable, we can confirm there was a peak in tweets between the 27th of May and the 2nd of June (see Figure 13.10).

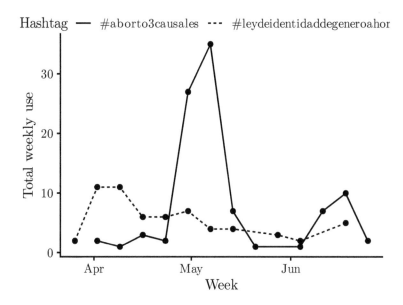

FIGURE 13.10: Temporal variation in the usage of hashtags.

```
hashtags_weekly <- poltweets_hashtags %>%
  mutate(week = floor_date(created_at, "week", week_start = 1)) %>%
  filter(hashtag %in% c("#aborto3causales",
                        "#leydeidentidaddegeneroahora")) %>%
  count(hashtag, week)
```

```
ggplot(data    = hashtags_weekly,
       mapping = aes(x = week, y = n,
                    linetype = hashtag, group = hashtag)) +
  geom_point() +
  geom_line() +
  labs(x = "Week", y = "Total weekly use", linetype = "Hashtag")
```

13.1.8 To sum up

Hashtags can tell you a lot about a political debate. We could verify some evident differences in the use of gender "#". Congresswomen used far more hashtags such as #olafeminista (#feministwave) and #agendamujer (#womenagenda) than their male counterparts. Regarding the coalitions, the left-wing ones (Frente Amplio and La Fuerza de la Mayoría) used them more. Regarding the temporal variation, the greater intensity of mentions on gender issues occurred during the week of the 14-20th of May, the week before the Public Account Speech (21st of May), which also fitted with a manifestation in various cities of the country. We observe that, in relative terms, congresswomen were almost five times more interested in the feminist movement, since they used the hashtag #agendamujer 5 times more than their male counterparts during the week of the 14-20th of May.

What did you learn in this section? We showed you how to use Twitter to analyzing political phenomena. Once you have your own dataset you can follow our step by step analysis. This would be useful as your starting point for explanatory designs that inquire on the causes of political alignment in different agendas.

13.2 Wordfish

In this section of the chapter, we will employ two NLP techniques commonly used in political science for unsupervised text mining: Wordfish and Structural Topic Models (STM). Both text processing models allow us to summarize a lot of different documents in a fast and economical way and can complement other descriptive measures like word and hashtags frequencies. As they are unsupervised, the classifications will be done without using any previous coding or dictionary (Welbers et al., 2017). This has the advantage of saving work on manual coding, as well as avoiding the coder's own bias. Another advantage is that they are not dependent on the source language, i.e. in principle they can be used in any language. Both methods use the "bag of words" approach, since the order of the words within a text does not alter the analysis. The parameters estimated by each algorithm can then be plotted with `ggplot2`, which facilitates the visual interpretation of the results (Silge and Robinson, 2017).

13.2 Wordfish

The organization of the section has the following structure. First, we will do a little data cleaning, like removing stopwords (with a stopword dictionary, in this case, incorporated into Quanteda), strange characters and numbers. Then, we will apply the Wordfish algorithm on the tweets of the Chilean Members of Parliament during the same period as the previous section. In the second part, we will do a topic modeling with STM on the same corpus.

13.2.1 Inspection and data cleaning

We load again the `poltweets` dataset and notice now that it contains a set of variables which are necessary for text analysis. Now we will use the entire tweets, not just the tokens. We will also require a variables `status_id` necessary to match each tweet to who tweeted it.

```
library(quanteda) # dfm and corpus
library(quanteda.textmodels) # wordfish
library(qdapRegex) # remove non ascii characters
```

Let's always start by doing a quick scan of the data, as we did in the previous section. The descriptive analysis allows us to summarize basic characteristics of the dataset, such as the type of variables and the number of characters per observation, the amount of missing data and the range of text units contained in each variable. We explore the character variable that contains the tweets. By using the `glimpse()` command we have a preview of each variable, specifically of its type and a preview of the first observations.

```
glimpse(poltweets)
## Rows: 18,658
## Columns: 8
## $ status_id   <chr> "1", "2", "3", "4", "5", "6", "7", "8", "9", "10...
## $ screen_name <chr> "vladomirosevic", "vladomirosevic", "vladomirose...
## $ names       <chr> "VLADO", "VLADO", "VLADO", "VLADO", "VLADO", "VL...
## $ lastname    <chr> "MIROSEVIC", "MIROSEVIC", "MIROSEVIC", "MIROSEVI...
## $ created_at  <dttm> 2018-04-11 16:19:43, 2018-04-11 16:17:56, 2018-...
## $ text        <chr> "@MarioEyza1 @bcnchile @IntendenciaXV @Insulza @...
## $ coalition   <chr> "FA", "FA", "FA", "FA", "FA", "FA", "FA", ...
## $ gender      <chr> "Male", "Male", "Male", "Male", "Male", "Male", ...
```

13.2.2 Preprocessing

Before applying the algorithm, we must pre-process the texts. This means using regular expressions to make the text cleaner. We will use regular expressions to remove strange characters, usernames, URLs, emojis and switch everything to lowercase.

```
# function to remove accents
f_remove_accent <- function(x){
  x %>%
    str_replace_all("á", "a") %>%
    str_replace_all("é", "e") %>%
    str_replace_all("í", "i") %>%
    str_replace_all("ó", "o") %>%
    str_replace_all("ú", "u") %>%
    str_replace_all("ñ", "n") # also replace "ñ", a common letter in Spanish
}

# now pre-process the dataset:
poltweets <- poltweets %>%
  mutate(text = text %>%
           # delete user names (which start with @):
           str_remove("\\@[[:alnum:]]+") %>%
           # delete URLs:
           str_remove_all("http[\\w[:punct:]]+") %>%
           # all text to lowercase:
           str_to_lower() %>%
           # remove special characters:
           str_remove_all("[\\d\\.,_\\@]+") %>%
           f_remove_accent() %>%
           # remove emojis
           rm_non_ascii()
  )
```

Once the text is clean, we want to group it according to the variable for comparison. As we are interested in obtaining the estimates at the coalition level, we group the texts by coalition. Now each coalition is a document in the dataset. When ordering by coalitions, you should place the factor levels in a way that it resembles a left-right axis:

```
by_coalition <- poltweets %>%
  group_by(coalition) %>%
  summarize(text = str_c(text, collapse = " ")) %>%
  ungroup() %>%
  # reorder the variable:
  mutate(coalition = fct_relevel(as.factor(coalition), "FA", "LFM", "ChV"))
```

For modeling with Quanteda we transform the dataset first to Corpus format, and then to Document-feature Matrix (DFM) format. This means transforming each documents in rows and "features" as columns. We make the transformation of the dataset grouped by coalitions to Corpus format and then to DFM. In addition, we take advantage of using a command that will help eliminate numbers, punctuations, symbols and stopwords (conjuctions, articles, etc.):

13.2 Wordfish

```
# Corpus format
poltweets_corpus <- corpus(by_coalition)

# DFM format
poltweets_dfm <- dfm(poltweets_corpus,
                    remove_numbers = T, remove_punct = T,
                    remove_symbols = T, remove = stopwords("spa"))
```

Using `dfm_trim()`, we eliminate words with frequency equal to or less than the 5th percentile and those words with a frequency equal or greater than the 95th percentile. In this way, we eliminate unusual words that are located at the extremes of the frequency distribution that can bias the results of the algorithm.

```
poltweets_dfm_trimmed <- dfm_trim(
  poltweets_dfm,
  min_docfreq = 0.05, max_docfreq = 0.95,
  docfreq_type = "quantile" # min 5% / max 95%
)
```

13.2.3 Wordfish

Wordfish is an algorithm that allows one-dimensional scaling of a set of texts (Slapin and Proksch, 2008). That is, to order in a one-dimensional axis the documents from how similar they are to each other in the use of certain keywords. The classification is carried out by establishing the frequency of word use. This modeling assumes that the number of times a word is said in a document follows a Poisson distribution. This model is extremely simple since the number of times a word will appear is estimated from a single parameter λ, which is both the mean and the variance of the Poisson probability distribution.

The distribution is as follows (Proksch and Slapin, 2010):

$$Wordcount_{ij} \sim Poisson(\lambda_{ij})$$

where

$$\lambda_{ij} = exp(\alpha_i + \psi_j + \beta_j * \omega_i)$$

The count of a word j for the document i follows a Poisson distribution with parameter $\dot{z}$ for the word j and the document i. The model estimates the parameter λ_{ij}, which is a function of a term α_i which is a fixed effect for documents and the term ψ_j which is a fixed effect for the word j - by entering these fixed effects, the fact that some words may appear more times than others is discounted. The parameter of interest is β_j, which captures the importance of each word j to discriminate the positions of i on the latent axis X. Therefore, documents can be grouped based on how similar they are by using certain keywords.

Wordfish has two fundamental assumptions, first, that words always have the same meaning within the text. Second, the texts are ordered by a latent dimension X which is an axis that articulates the ideological differentiation of the documents (Slapin and Proksch, 2008). However, the validity of this assumption is sustained to the extent that the method is robust with other measurements and that the corpus of texts included in the analysis are representative of the dimension.

In political science, some of the works that have used this algorithm with Twitter is that of Andrea Ceron (2017) in which he uses the Wordfish estimates to predict the ideological heterogeneity within Italian political parties, to see which legislator will be selected as minister and the probability that they will leave the party (Ceron, 2017).

We apply the Wordfish algorithm to the DFM class object, specific to Quanteda. We define the direction of parameter θ -the equivalent of β-, in this case that document 3 (FA) is the positive pole and document 1 (CHV) is the negative pole in the estimated ideological dimension. We also use the argument `sparse = T`, which allows working with large amounts of data, saving computational power.

```
wf <- textmodel_wordfish(poltweets_dfm_trimmed,
                        dir = c(3, 1), sparse = T)
```

We plot it in Figure 13.11:

```
df_wf <- tibble(
  # coalition labels:
  coalition = wf[["x"]]@docvars[["coalition"]],
  # then we extract thetas and their SEs from the mode object:
  theta = wf$theta,
  lower = wf$theta - 1.96 * wf$se.theta,
  upper = wf$theta + 1.96 * wf$se.theta
)

df_wf
## # A tibble: 3 x 4
##   coalition  theta  lower  upper
##   <fct>      <dbl>  <dbl>  <dbl>
## 1 ChV         1.07   1.07   1.08
## 2 FA         -0.905 -0.912 -0.897
## 3 LFM        -0.169 -0.180 -0.158

ggplot(data    = df_wf,
       mapping = aes(x = theta, y = fct_reorder(coalition, theta),
                     xmin = lower, xmax = upper)) +
  geom_point() +
```

13.2 Wordfish

```
geom_linerange() +
# add vertical line at x=0:
geom_vline(xintercept = 0, linetype = "dashed") +
scale_x_continuous(limits = c(-1.2, 1.2)) +
labs(y = "")
```

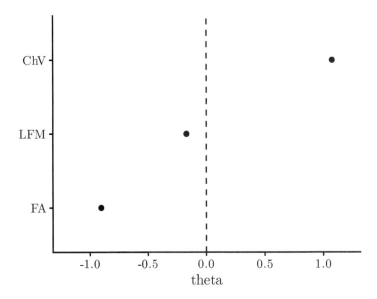

FIGURE 13.11: Classification of coalitions by ideological positioning.

We see that coalitions are grouped along a left-right divide. The interest parameter θ, equivalent to the beta parameter, is the parameter that discriminates the positions of the documents from the word frequencies. We see that this parameter is consistent with how coalitions are grouped politically. The rightmost one, Chile Vamos (ChV), with a θ of 1.07, is located at one end of the X axis, on the contrary, the leftmost one, Frente Amplio (FA), with a θ of -0.91, is located at the opposite end.

Exercise 13A. You can repeat the Wordfish, but now invert the direction of the parameter θ in the `wf` object. How does the distribution of the documents change by inverting the direction of the parameter? Now repeat the exercise by grouping by political parties.

13.2.4 What did we learn from Wordfish?

Using the same tweets as in the previous section, we saw how teets coming from the members of the congress were consistent with an ideological divison, ordered in a left-right axis. In particular, we conclude that the Frente Amplio and Chile Vamos coalitions are at opposite extremes in regards to their tweets during the Chilean protest cycle. Amazingly, the algorithm is capable of locating the Chilean coalitions in a left-right axis taking as input only the tweets of the parliamentarians during the protest cycles, without any manual labelling of the texts. Wordfish is a powerful tool to be used as a spatial positioning method, adding to the repertoire of other political position measurements such as Bayesian ideal point estimations (Barberá, 2015), roll call and cosponsorship (Alemán et al., 2009).

13.3 Structural Topic Modeling

Topic modeling is a computational method for automatically identifying relevant word groupings in large volumes of texts. One of the most popular applications in political science is the Latent Dirichlet Allocation (LDA), developed by Blei et al. (2003) and explained in a didactic way by David Blei at the Machine Learning Summer School 2009 at Cambridge University[7].

Another useful development is the structural topic modeling (STM), a non supervised NLP technique for diving large corpora of texts. The main innovation of the STM is that it incorporates metadata into the topic model, so it allows researchers to discover topics and estimate their relationship to covariates, improving the quality of the inferences and the interpretability of the results. The STM algorithm is available in the `stm` package created by Molly Roberts, Brandon Stewart and Dustin Tingley. For a more detailed review of this method there is a bulk of material in the official site of the package[8].

In this section, we will analize a subset of our tweets to find the most relevant topics and see how they correlate to the gender and coalition variables. Following Julia Silge's lead[9], we will first do all the preprocessing using tidy tools, to then feed a corrected dataset to `stm`.

13.3.1 Pre-processing

We will only employ tweets from May 2018:

```
library(tidyverse)
library(tidytext)
library(stm)
```

---

[7] See https://www.youtube.com/watch?v=DDq30Vp9dNA
[8] See http://www.structuraltopicmodel.com/
[9] See https://juliasilge.com/blog/evaluating-stm/

13.3 Structural Topic Modeling

```
library(quanteda)
library(qdapRegex)


poltweets_onemonth <- poltweets %>%
  filter(created_at >= "2018-05-01" & created_at < "2018-06-01")
```

As mentioned above, we should start by pre-processing the texts. Remember that in the previous subsection we removed strange characters from the text. Next we will create a tokenized version of `poltweets_onemonth`, where every row is a word contained in the original tweet, plus a column with the total number of times that each word is said in the entire dataset (we only keep words that are mentioned ten or more times). Right after doing that, we will we remove stopwords (conjuctions, articles, etc.) using the `stopwords` package. Notice that we will also employ a "custom" dictionary of stopwords, composed by the unique names and surnames of deputies.

```
# obtain unique names and surnames of deputies
names_surnames <- c(poltweets$names, poltweets$lastname) %>%
  na.omit() %>%
  unique() %>%
  str_to_lower() %>%
  f_remove_accent() %>%
  str_split(" ") %>%
  flatten_chr()

poltweets_words <- poltweets_onemonth %>%
  unnest_tokens(word, text, "words") %>%
  # remove stop words:
  filter(!word %in% stopwords::stopwords("es", "stopwords-iso")) %>%
  # remove names/surnames of deputies:
  filter(!word %in% names_surnames) %>%
  # just keep words that are present ten or more times
  add_count(word) %>%
  filter(n > 10)
```

That's it in term of pre-processing! Next we will transform the tokenized dataset into a stm object using the `cast_dfm()` and `convert()` functions.

```
poltweets_stm <- poltweets_words %>%
  cast_dfm(status_id, word, n) %>%
  convert(to = "stm")
```

In order to estimate the relation of the topics and the document covariates, we must add the covariate values into the `poltweets_stm$meta` object. The `metadata` object is a dataframe containing the metadata for every document in the stm object thatn can later be used as the document "prevalence"–or metadata. Notice that for creating the stm_meta object, it is necessary to join by the status_id variable, a column containing a unique identifier for every tweet.

```
metadata <- tibble(status_id = names(poltweets_stm$documents)) %>%
  left_join(distinct(poltweets, status_id, coalition, gender),
            by = "status_id") %>%
  as.data.frame()

poltweets_stm$meta <- metadata
```

Now we have all the necessary ingredients to estimate our structural topic model, stored in the `poltweets_stm` object:

```
summary(poltweets_stm)
##            Length Class      Mode
## documents  5647   -none-     list
## vocab      1378   -none-     character
## meta          3   data.frame list
```

13.3.2 Diagnostics

To estimate a `stm`, one needs to define the number of topics (K) beforehand. However, there is no "right" number of topics, and the appropiate K should be decided looking at the data itself (Roberts et al., 2019). In order to do that, we should train several models and compute diagnostics that will help us decide. What range of K should we consider? In the package manual (Roberts et al., 2019), the authors offer the following advice:

> For short corpora focused on very specific subject matter (such as survey experiments) 3-10 topics is a useful starting range. For small corpora (a few hundred to a few thousand) 5-50 topics is a good place to start. Beyond these rough guidelinesit is application specific. Previous applications in political science with medium sized corpora (10kto 100k documents) have found 60-100 topics to work well. For larger corpora 100 topics is a useful default size. (p. 61)

Our dataset has 5,647 documents, and therefore we will try 5-50 topics. We can use the `searchK()` function from the `stm` package to compute the relevant diagnostics, which we will store in the `stm_search` object. This process is computationally expensive, and might take several minutes on a modern computer. If you do not want to wait,

13.3 Structural Topic Modeling

you can load the object from the book's package (`data("stm_search")`) and keep going.

```r
stm_search <- searchK(documents = poltweets_stm$documents,
                      vocab = poltweets_stm$vocab,
                      data = poltweets_stm$meta,
                      # our covariates, mentioned above:
                      prevalence = ~ coalition + gender,
                      # 5-50 topics range:
                      K = seq(5, 50, by = 5),
                      # use all our available cores (be careful!):
                      cores = parallel::detectCores(),
                      # a seed to reproduce the analysis:
                      heldout.seed = 123)
```

Next we will tidy our newly create object and plot its results using `ggplot2`:

```r
diagnostics <- stm_search$results %>%
  # get a tidy structure to plot:
  mutate_all(flatten_dbl) %>%
  pivot_longer(-K, names_to = "diagnostic", values_to = "value") %>%
  # we will only use some diagnostics:
  filter(diagnostic %in% c("exclus", "heldout", "residual", "semcoh")) %>%
  # give better names to the diagnostics:
  mutate(diagnostic = case_when(
    diagnostic == "exclus" ~ "Exclusivity",
    diagnostic == "heldout" ~ "Held-out likelihood",
    diagnostic == "residual" ~ "Residuals",
    diagnostic == "semcoh" ~ "Semantic coherence"
  ))

ggplot(diagnostics, aes(x = K, y = value)) +
  geom_point() +
  geom_line() +
  facet_wrap(~diagnostic, scales = "free")
```

In Figure 13.12, we present four of the diagnostics obtained with `searchK()`. Both the held-out likelihood and the residuals (characteristics of the models, more information on Roberts et al. (2019)) seem to suggest that increasing the topics is right approach. However, in our experience semantic coherence and exclusivity are the best indicators of K appropriateness. Semantic coherence is perhaps the diagnostic that mostly correlates with human judgement of "topic quality" (Roberts et al., 2019), that is, topics that include terms that make thematic sense. However, Roberts et al. (2014) contend that semantic coherence can be easily achieved with few topics, and therefore should be considered jointly with "exclusivity", how distinctive the terms of the topic are compared to other topics.

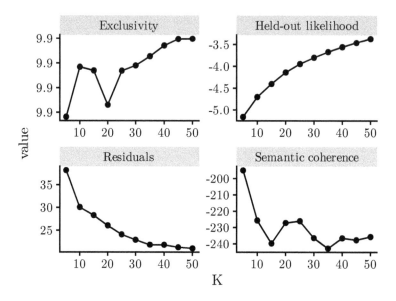

FIGURE 13.12: searchK() diagnosis.

Following these definitions and visually inspecting the plot above, models with K=10 and K=25 seem like good contenders. Let's estimate those separately (if you do not want to wait, you can again load the objects from our `politicalds` package):

```
stm_model_k10 <- stm(documents = poltweets_stm$documents,
                     vocab = poltweets_stm$vocab,
                     data = poltweets_stm$meta,
                     prevalence = ~ coalition + gender,
                     K = 10)

stm_model_k25 <- stm(documents = poltweets_stm$documents,
                     vocab = poltweets_stm$vocab,
                     data = poltweets_stm$meta,
                     prevalence = ~ coalition + gender,
                     K = 25)
```

How should we decide between these two specifications? The exclusivity and semantic coherence presented in the previous plot are summary measures, so perhaps we can get more value by looking at specific values for each topic in the two models. In the following Figure 13.13 we plot semantic coherence against exclusivity, as suggested by Roberts et al. (2019). In the ideal world, we would like for all topics to be as semantically coherent and exclusive as possible. While we are not in that ideal world, it seems like the topics of the K=10 model are better in this sense, as the K=25 model has quite a few topics that position themselves as outliers, which either low semantic coherence or low exclusivity (or both!). Therefore, we will continue our analysis using the K=10 model.

13.3 Structural Topic Modeling

```
# obtain exclusivity and semantic coherence of topics within the models
diagnostics2 <- tibble(
  exclusivity = c(exclusivity(stm_model_k10), exclusivity(stm_model_k25)),
  semantic_coherence = c(
    semanticCoherence(stm_model_k10, documents = poltweets_stm$documents),
    semanticCoherence(stm_model_k25, documents = poltweets_stm$documents)
  ),
  k = c(rep("K=10", 10), rep("K=25", 25))
)

ggplot(data    = diagnostics2,
       mapping = aes(x = semantic_coherence, y = exclusivity, shape = k)) +
  geom_point(size = 2) +
  labs(x = "Semantic coherence", y = "Exclusivity", shape = "")
```

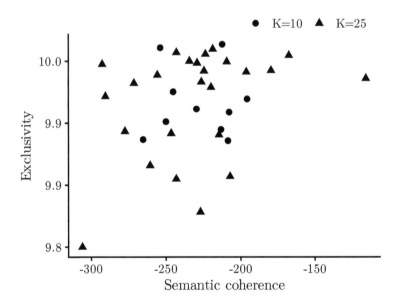

FIGURE 13.13: Semantic coherence and exclusivity of topics within the two models.

13.3.3 Analysis

We we have chosen a model with 10 topics. What does it look like? We can obtain the top terms of each topic (more on what this means in a second) and plot them, as shown in Figure 13.14. The plot will also order the topics based on their expected proportion within the whole set of tweets.

```
model_terms <- tibble(
  topic = as.character(1:10),
```

```
# obtain the top seven terms:
terms = labelTopics(stm_model_k10)$prob %>%
  t() %>%
  as_tibble() %>%
  map_chr(str_c, collapse = ", "),
# expected proportion of each topic in the whole set of tweets:
expected_proportion = colMeans(stm_model_k10$theta)
) %>%
  arrange(-expected_proportion)


ggplot(data    = model_terms,
       mapping = aes(x = expected_proportion,
                     y = fct_reorder(topic, expected_proportion),
                     label = terms)) +
  geom_col() +
  geom_text(size = 3, hjust = "inward") + # to use space better
  labs(x = "Expected proportion", y = "Topic")
```

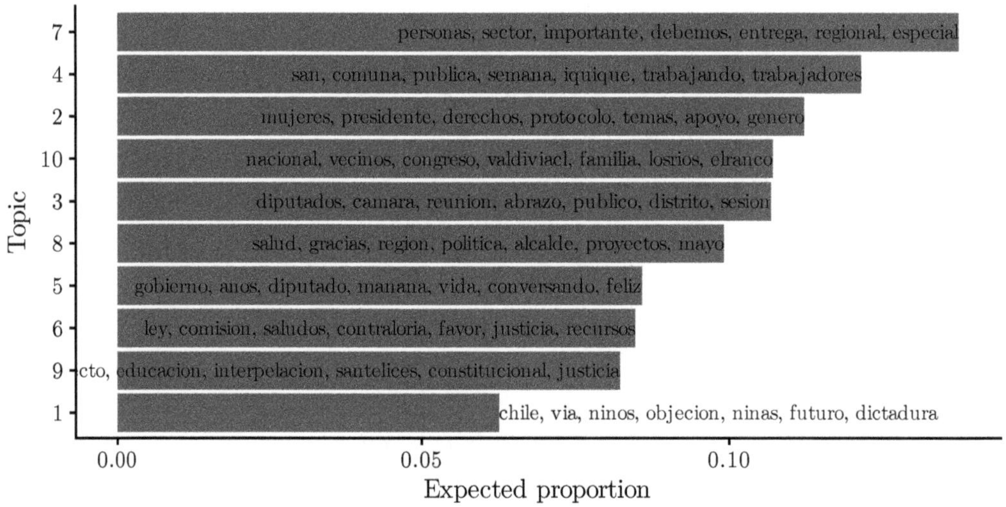

FIGURE 13.14: Topics and their top terms.

Notice that Topic 2 is the one that reffers to gender, including the words "mujeres" (women), "derechos" (rights) and "género" (gender). We can use `labelTopics()` to obtain more terms that characterize this topic, as shown in the output below. "Highest Prob" terms are the same ones we used in the previous plot, the most common ones in the topic. Another interesting measure is "FREX", which picks terms that are distinctive of this topic, when compared to others (in this case, the words are the same). Lastly, "Lift" and "Score" are imported from other packages (Roberts et al., 2019), and can also be useful when describing a topic. In this case, notice how "Lift" top terms include "olafeminista", one of the hashtags we discussed earlier.

13.3 Structural Topic Modeling

```
labelTopics(stm_model_k10,
            topics = 2) # this is the topic we want to analyze

## Topic 2 Top Words:
##
## Highest Prob:
## mujeres, presidente, derechos, protocolo, temas, apoyo, genero
## FREX:
## mujeres, presidente, derechos, protocolo, temas, apoyo, genero
## Lift:
## equidad, olafeminista, utilidades, violencia, apoyo, dejo, discriminacion
## Score:
## mujeres, presidente, derechos, protocolo, genero, apoyo, temas
## NULL
```

With this topic in mind, we can know analyze its relationship with metadata, in our case, the gender and coalition of deputies: are women and left-wing politicians more likely to tweet about the feminist wave? As we said before, the ability to estimate these topic–covariate relationships is a core advantage of structural topic models. The first step is to use `estimateEffect()` to obtain the model coefficients:

```
stm_effects <- estimateEffect(
  # c(1:10) means that we want coefficients for all topics in our model
  formula  = c(1:10) ~ coalition + gender,
  stmobj   = stm_model_k10,
  metadata = poltweets_stm$meta
)
```

In the following we use the tidystm package[10] to extract model effects, via `extract.estimateEffect()`. Let us begin with gender. When the covariate of interest has two levels, the `method = "difference"` argument will obtain the change in topic prevalence shifting from one specific value to another. In our case, the column "estimate" in the `effect_gender` object is the change in topic prevalence when comparing congresswomen to congressmen. There is a relatively large and positive estimate for Topic 2 (as expected), which is significant at the usual confidence levels. It should be noted that for the prevalence of other topics (4, 7, 10), gender seems to have an opposite (although smaller) effect, a finding that might be worth further investigation.

```
library(tidystm)


effect_gender <- extract.estimateEffect(x = stm_effects,
                                        covariate = "gender",
                                        method = "difference",
```

---

[10] tidystm needs to be installed from GitHub: `remotes::install_github("mikajoh/tidystm")`.

```
                                         cov.value1 = "Female",
                                         cov.value2 = "Male",
                                         model = stm_model_k10)
effect_gender %>% arrange(-estimate)
## # A tibble: 10 x 10
##   method topic covariate covariate.value estimate std.error ci.level
##   <chr>  <int> <chr>     <chr>              <dbl>     <dbl>    <dbl>
## 1 diffe~     2 gender    Female-Male       0.0517   0.00949     0.95
## 2 diffe~     8 gender    Female-Male       0.0154   0.00921     0.95
## 3 diffe~     9 gender    Female-Male       0.0145   0.00823     0.95
## # ... with 7 more rows, and 3 more variables
```

It is possible to plot these results using `ggplot2`, a great advantage of the `tidystm` package, shown in Figure 13.15:

```
ggplot(effect_gender,
       aes(x = estimate, xmin = ci.lower, xmax = ci.upper,
           y = fct_reorder(as.character(topic), estimate))) +
  # add line of null effect:
  geom_vline(xintercept = 0, linetype = "dashed") +
  geom_point() +
  geom_linerange() +
  # center the null effect line:
  scale_x_continuous(limits = c(-0.075, 0.075)) +
  labs(x = "Estimate for the Female - Male difference",
       y = "Topic")
```

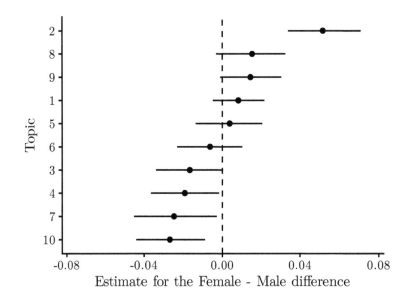

FIGURE 13.15: Effect of gender on topic prevalence.

We can repeat the previous analysis for the other covariate in our model, obtaining the difference in topic prevalence between the most left-wing coalition (FA or Frente Amplio) and the most right-wing coalition (ChV or Chile Vamos). As expected, the model estimates that left-wing politicians were more likely to tweet about the feminist wave (Topic 2).

```
effect_coalition_diff <- extract.estimateEffect(x = stm_effects,
                                                covariate = "coalition",
                                                method = "difference",
                                                cov.value1 = "FA",
                                                cov.value2 = "ChV",
                                                model = stm_model_k10)
effect_coalition_diff %>%
  filter(topic == 2)
## # A tibble: 1 x 10
##   method topic covariate covariate.value estimate std.error ci.level
##   <chr>  <int> <chr>     <chr>              <dbl>     <dbl>    <dbl>
## 1 diffe~     2 coalition FA-ChV            0.0407   0.00988     0.95
## # ... with 3 more variables
```

Exercise 13B. Plot the effects of the coalition covariate, showing the difference between the right-wing coalition (ChV) and the other others (FA and LFM).

Exercise 13C. Add the top 7 terms for each topic to Figure 13.15. Tip: you can use `left_join()` to merge the two datasets of interest.

13.3.4 Concluding remarks

Political comunications in protest cycles have clear implications in the digital domain. In this section we have demonstrated that automated text analysis of political tweets captures variations in tweeting, particularly, how congresswomen appear more correlated to certain topics, in this case the gender related topic 2. STM is among the most recent NLP techniques used in political science for unsupervised text mining. Despite the clear benefit of using text mining, however, these techniques must be used with caution since topics interpretability is sensitive on many desicions like the K (number of topics for a given model) and text cleaning.

14

Networks

Andrés Cruz[1]

Suggested readings

- Newman, M. (2018). *Networks*. Oxford University Press, Oxford, 2nd edition.
- Scott, J. (2013). *Social Network Analysis*. SAGE, Thousand Oaks, CA, 3rd edition.

Packages you need to have installed

- **tidyverse** (Wickham, 2019), **politicalds** (Urdinez and Cruz, 2020), **tidygraph** (Pedersen, 2020), **ggraph** (Pedersen, 2020), **ggcorrplot** (Kassambara, 2019).

14.1 Introduction

It is not an overstatement to say that in politics everything is connected to everything else. For example, let's consider the case of legislators. The connections are obvious: parties, coalitions, commissions, family ties, schools. To better understand these tangled relationships, one of the tools in the social scientists' toolbox is network analysis. Networks can acknowledge that everything is connected to everything else and give clues as to how those connections are built. Networks not only allow us to visualize these connections in a compelling way, but they also allow us to measure how close actors are. In this chapter you will learn the basics of network analysis with R, mainly using the `tidygraph` and `ggraph` packages, which approach network analysis from the precepts of `tidyverse`.

---

[1] Institute of Political Science, Pontificia Universidad Católica de Chile. E-mail: arcruz@uc.cl. Twitter: @arcruz0.

14.2 Basic concepts in a network

14.2.1 Nodes and links

There are two basic concepts that express a certain situation in the shape of a network. The *nodes* (sometimes called *actors*) are the main units of analysis: we want to understand how they relate to each other. The *links* (sometimes called *connections* or *bonds*) show how the nodes are connected to each other. For instance, among legislators (the *nodes*), one possible *link* is "having proposed a law together," also called co-sponsorship.

A network is nothing more than a series of nodes connected by links, as you can see in Figure 14.1. In terms of the previous example, we can imagine that such a graphic network connects legislators A, B, C and D according to their co-sponsorship of projects w, x, y or z. Thus, the network shows that Legislator B has introduced at least one bill with all the other legislators in the network, while Legislator D has only introduced one bill with B. Legislators A and C have two co-sponsorship links each, forming a trio A-B-C.

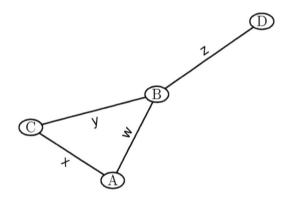

FIGURE 14.1: Co-sponsorship network diagram of four legislators.

14.2.2 Adjacency matrix

Apart from a visual description, like the one in Figure 14.1, it is also possible to represent networks as *adjacent matrices*. Table 14.1 shows the same network as above, yet in matrix format. The "1" indicates that there is a link between two nodes (co-sponsorship, in our example), while the 0 indicates the opposite. Note how the diagonal of the matrix is filled by zeros: this is a useful convention for different mathematical calculations. In addition, this type of matrix for a basic network is symmetric: if we know that node A is linked to node B, we automatically know that node B is linked to node A.

TABLE 14.1: Adjacency matrix, basic network.

	A	B	C	D
A	0	1	1	0
B	1	0	1	1
C	1	1	0	0
D	0	1	0	0

14.2.3 Weights and directions

The nature of our data allows us to create a much more complex network than the example above. Specifically, we can add *weights* and *directions* to the links. To start with the weights, in our current example the legislators are connected if they *ever* have submitted a bill together. However, it is often of interest not only to know the existence of a connection between two actors, but also the strength of that connection: it is not the same that two legislators have reluctantly agreed to introduce a bill together on one occasion, as it is that they have introduced multiple bills together and are political allies. We go back to the adjacency matrix, this time including weights. In this new example, from Table 14.2, legislators A and B have introduced nine bills together, while all other pairs of connected legislators have introduced two bills together. Note how, by convention, the diagonal of the matrix remains full of zeros.

TABLE 14.2: Adjacency matrix, weighted network.

	A	B	C	D
A	0	9	2	0
B	9	0	2	2
C	2	2	0	0
D	0	2	0	0

The second way to add additional information to the links is by adding directions. In some legislatures, bills have a lead author (sponsor), who is joined by other legislators (co-sponsors). In these cases, the co-sponsorship network will naturally have direction: one legislator will "sponsor" another by signing into their bill, without this relationship necessarily being reciprocal. It is possible to include this information in the network, as shown by Fowler (2006) for the U.S. Congress. An adjacency matrix with directions and weights will look like the one in Table 14.3. Note that the matrix is now asymmetrical, as there is more information on the co-sponsorship relationship between the legislators.[2]. While Legislator A sponsored seven bills from Legislator B, Legislator B only reciprocated on two bills. All other previously recorded co-sponsorships imply reciprocity of a bill.

---

[2]This third network, as described in our example, does not record when two legislators jointly support a bill from a different legislator. It only measures *direct* support between pairs of legislators

TABLE 14.3: Adjacency matrix, weighted and directed network.

	A	B	C	D
A	0	7	1	0
B	2	0	1	1
C	1	1	0	0
D	0	1	0	0

14.3 Network datasets

Following the spirit of the previous example, in this chapter we will work with data on co-sponsorship of laws in the Argentine Senate. We will use data from Alemán et al. (2009), specifically for the year 1983, just after the return of democracy. Let's start by loading the datasets:

```
library(tidyverse)
```

```
library(politicalds)
data("cosponsorship_arg")
data("senators_arg")
```

We can check that our datasets were properly loaded with `ls()`:

```
ls()
```

```
## [1] "cosponsorship_arg" "senators_arg"
```

We have two datasets. The first one contains information for the nodes. Each senator is assigned to a unique ID (starting at 1), a necessary condition for the correct functioning of `tidygraph`. Additionally, we have other information about the legislators: name, province, political bloc.

```
senators_arg
## # A tibble: 46 x 4
##    id_sen name_sen          province    political_bloc
##     <int> <chr>             <chr>       <chr>
## 1       1 RAMON A ALMENDRA  SANTA CRUZ  JUSTICIALISTA
## 2       2 JULIO AMOEDO      CATAMARCA   JUSTICIALISTA
## 3       3 RAMON A ARAUJO    TUCUMAN     JUSTICIALISTA
## # ... with 43 more rows
```

14.3 Network datasets

Meanwhile, our second dataset contains information on the links, in this case, the co-sponsorship of laws among senators. In this dataset, the weights and directions are recorded. Thus, it has three variables: two identifying variables (which link the IDs of the senators) and the number of co-sponsored bills in the pair. For example, the first observation indicates that Senator 1 signed a bill from Senator 2.

```
cosponsorship_arg
## # A tibble: 217 x 3
##    id_sen_from id_sen_to n_cosponsorship
##          <int>     <int>           <int>
## 1            1         2               1
## 2            1         9               1
## 3            1        30               1
## # ... with 214 more rows
```

Once we have two the datasets, we can join them in a single `tidygraph` object to start working with them:

```
library(tidygraph)

tg_cosponsorship_arg <- tbl_graph(nodes = senators_arg,
                                  edges = cosponsorship_arg,
                                  directed = T) # our network is directed
```

The `tidygraph` objects join nodes and links which will be useful in following operations:

```
tg_cosponsorship_arg
## # A tbl_graph: 46 nodes and 217 edges
## #
## # A directed simple graph with 1 component
## #
## # Node Data: 46 x 4 (active)
##    id_sen name_sen                   province    political_bloc
##     <int> <chr>                      <chr>       <chr>
## 1       1 RAMON A ALMENDRA           SANTA CRUZ  JUSTICIALISTA
## 2       2 JULIO AMOEDO               CATAMARCA   JUSTICIALISTA
## 3       3 RAMON A ARAUJO             TUCUMAN     JUSTICIALISTA
## 4       4 ALFREDO L BENITEZ          JUJUY       JUSTICIALISTA
## 5       5 ANTONIO TOMAS BERHONGARAY  LA PAMPA    UCR
## 6       6 DEOLINDO FELIPE BITTEL     CHACO       JUSTICIALISTA
## # ... with 40 more rows
## #
## # Edge Data: 217 x 3
##    from    to n_cosponsorship
##   <int> <int>           <int>
## 1     1     2               1
## 2     1     9               1
## 3     1    30               1
## # ... with 214 more rows
```

We can edit the contents of any of the two datasets, activating either of their nodes or links with `activate()`. Let's create two variables that will be of use later on. First, a binary variable for co-sponsorship using `if_else()`, named "d_cosponsorship". Second, an inverted variable for the weights (number of bills), that we will name "n_cosponsorship_inv".

```
tg_cosponsorship_arg <- tg_cosponsorship_arg %>%
  activate("edges") %>% # or "nodes" if you want to edit that dataset
  mutate(d_cosponsorship = if_else(n_cosponsorship >= 1, 1, 0),
         n_cosponsorship_inv = 1 / n_cosponsorship)
```

Now the edges are activated and shown first:

```
tg_cosponsorship_arg
## # A tbl_graph: 46 nodes and 217 edges
## #
## # A directed simple graph with 1 component
## #
## # Edge Data: 217 x 5 (active)
##     from    to n_cosponsorship d_cosponsorship n_cosponsorship_inv
##    <int> <int>           <int>           <dbl>               <dbl>
## 1      1     2               1               1                   1
## 2      1     9               1               1                   1
## 3      1    30               1               1                   1
## 4      1    34               1               1                   1
## 5      1    36               3               1               0.333
## 6      1    37               1               1                   1
## # ... with 211 more rows
## #
## # Node Data: 46 x 4
##    id_sen name_sen         province    political_bloc
##     <int> <chr>            <chr>       <chr>
## 1       1 RAMON A ALMENDRA SANTA CRUZ  JUSTICIALISTA
## 2       2 JULIO AMOEDO     CATAMARCA   JUSTICIALISTA
## 3       3 RAMON A ARAUJO   TUCUMAN     JUSTICIALISTA
## # ... with 43 more rows
```

14.4 Graphic presentation of a network

Having our network object, one of the first actions is to graph it. `ggraph` (the sibling package of `tidygraph`) will allow us to do it using the language of graphs, reusing concepts and functions you learned from `ggplot2` in previous chapters.

```
library(ggraph)
```

14.4 Graphic presentation of a network

The *layout* determines where the nodes will be positioned in the plane. In general, a good visual representation will be able to reveal the structure of the network, optimize the use of space and properly show the relationships between the nodes (Pfeffer, 2017).

Perhaps the most common algorithm for graphical layout is that of Fruchterman and Reingold (1991). It belongs to the family of distance scaling algorithms, which work under the very intuitive premise that the most connected nodes should be closer to each other (Pfeffer, 2017). To use the FR algorithm in our network, we must start by generating it with `create_layout()`. Note that we set a seed with `set.seed()`, as the procedure is iterative and may give different results in different attempts.

```
set.seed(1)
layout_fr <- create_layout(tg_cosponsorship_arg, "fr",
                           weights = n_cosponsorship)
```

Having created the object with the graphic layout, we will proceed to graph it. The `ggraph()` function will be the basis of our command string. Next, we can use `geom_node_point()` to show the points in the plane (Figure 14.2):

```
ggraph(layout_fr) +
  geom_node_point(size = 5) +
  theme_void() # an empty canvas for our network
```

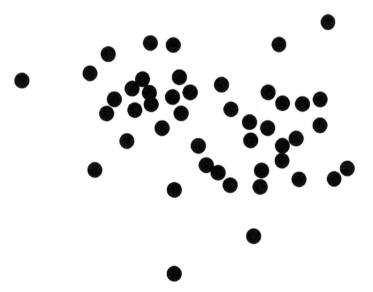

FIGURE 14.2: Network of co-sponsorship in the Argentinian Senate (1983), nodes in the layout.

Figure 14.2 is hardly informative. We can proceed to add the links with `geom_edge_link()`. Considering that our network is directed, we will add arrows with the argument `arrow = arrow(length = unit(3, 'mm'))` for illustrative purposes (although typically they can be a bit confusing when looking for general patterns) (see Figure 14.3).

```
ggraph(layout_fr) +
  geom_edge_link(arrow = arrow(length = unit(3, 'mm')),
                 color = "lightgrey") +
  geom_node_point(size = 5) +
  theme_void()
```

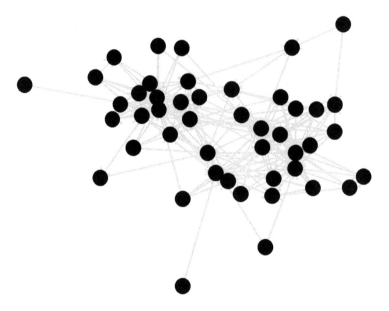

FIGURE 14.3: Network of co-sponsorship in the Argentinian Senate (1983), with nodes and links.

Exercise 14A. Try the `geom_edge_link()` function instead of `geom_edge_arc()`. What is the difference? Which visualization do you think is more clear?

We must remember that our network has weights. This was taken into consideration in the graphical layout when we used the argument `weights = n_cosponsorship` to `create_layout()`. We can incorporate this information explicitly into our graphical links, using the now-familiar `mapping = aes()` argument in our geom. In this case, we'll plot the intensity of the color in the links, using the `alpha =` argument (Figure 14.4).

```
ggraph(layout_fr) +
  geom_edge_link(mapping = aes(alpha = n_cosponsorship_inv)) +
  geom_node_point(size = 5) +
  theme_void()
```

14.4 Graphic presentation of a network

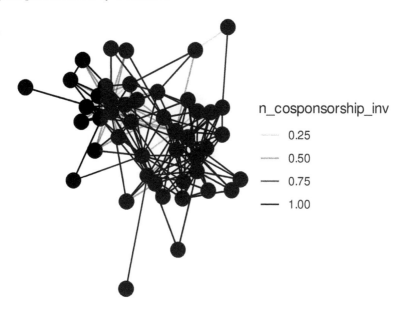

FIGURE 14.4: Network of co-sponsorship in the Argentinian Senate (1983), links with opacity.

The above chart shows a particularly close set of legislators in the upper left corner. It might be interesting to add information to the network nodes. In our dataset, we know the political bloc of each legislator so we will use this information to edit the aesthetic mappings of `geom_node_point()` (Figure 14.5):

```
ggraph(layout_fr) +
  geom_edge_link(mapping = aes(alpha = n_cosponsorship)) +
  geom_node_point(mapping = aes(color = political_bloc,
                                shape = political_bloc),
                  size = 5) +
  scale_shape_manual(values = 15:18) + # edit the shapes
  theme_void()
```

The visual pattern is particularly clear. The network is divided between the two major political blocs: the *Justicialista* and the UCR (*Unión Cívica Radical*). The smaller provincial blocs appear as satellites, with particular co-sponsorship links in some projects. A graphic inspection of the network alone is not enough to draw conclusions.

Exercise 14B. Edit the node dataset in the `tidygraph` object, with a column that distinguishes the metropolitan senators ('CAP FEDERAL' and 'BUENOS AIRES' provinces) from the rest. Then, use this new variable to color the nodes of the visual representation of the network.

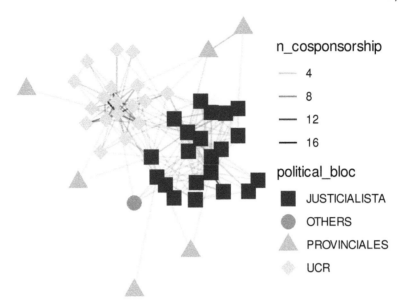

FIGURE 14.5: Network of co-sponsorship in the Argentinian Senate (1983), full version.

14.5 Measures of centrality

When dealing with network data in the social sciences, one of the first questions we ask has to do with the relative influence of each node. It is often relevant for our research questions to discover important figures, particularly well-connected ones, etc. However, developing a centrality estimator is not entirely straightforward. The literature has suggested different alternatives, which highlight varying dimensions of what we can intuitively describe as "centrality". We will now review the most common families of estimators: degree, eigenvector centrality, betweenness and closeness. To conclude, we will you will learn how to implement them in R, a relatively simple task.

14.5.1 Degree

Degree is probably the most intuitive measure of centrality: we will understand a node as central if it is connected to many other nodes. In this way, the degree estimator simply counts the number of connections from each node. Two variations require further elaboration. First, in a network with weights, the degree can also be calculated by adding up all the weights of the existing connections, which is sometimes referred to as the node's strength. Second, in directed networks each node will have two different measurements of degree: the *in-degree* and the *out-degree*. The first will count the times the other nodes connect to the particular node, while the second will record the times the node connects to others.

14.5.2 Eigenvector centrality

One of the most obvious problems of degree is that it is blind to the quality of the connections, perhaps being too unrefined in some contexts. Eigenvector centrality and its derivatives consider the idea that "important people know important people" (Patty and Penn, 2016, p. 155). In other words, it is not only the number of links between the nodes that matters, but also the importance of those who make up those links.

Calculating the eigenvector centrality of a node is a little more complicated than obtaining the degree, and its algebraic explanation is out of the scope of this chapter (see Newman, 2018, pp. 159-163). Intuitively, you should note that it is a somewhat tautological measure, as "[i]nstead of awarding one point for every network neighbor a node has [like degree does], eigenvector centrality awards a number of points *proportional to the centrality scores of the neighbors*" (Newman, 2018, p. 159, original emphasis).

The crude estimator of eigenvector centrality does not work very well with directed networks, along with some other problems (Newman, 2018, pp. 161-163). One of the most popular alternative estimators that refine eigenvector centrality is PageRank, the algorithm used by Google's search engine. In the context of web searches, it "considers (1) the number of in-bound links (i.e., sites that link to your site), (2) the quality of the linkers (i.e., the PageRank of sites that link to your site), and (3) the link propensity of the linkers (i.e., the number of sites the linkers link to)" (Hansen et al., 2019, p. 41). PageRank can be used in weighted and directed networks, such as our example network of co-sponsorship in the Argentine Senate.

14.5.3 Betweenness centrality

In another line, betweenness centrality seeks to measure a different dimension of centrality: how important is the node for the connections between the rest of the nodes. Intuitively, we could consider a node more important if the others in the network "need" it to connect to each other (Patty and Penn, 2016). In general, we interpret that nodes with high values of betweenness centrality work as bridges of connection within the network. To calculate this measure of centrality in its most basic version we will count the pairs of nodes in the network whose shortest path between them (geodetic paths) passes through the node in question:

$$betweenness_i = \sum_{st} \qquad (14.1)$$

where st are the pairs of nodes in the network; and equals 1 if the i node is on the geodetic path between st and 0 if it is not (or there is no such path).

In a directed network, only the links leading to the specific node will be considered when calculating its geodetic paths. If we want to include weights, they will interpreted as distances separating the nodes: higher weights will mean a weaker connection. However, in multiple applications, including our co-sponsorship example, weights are meant to indicate the strength of the relation between two nodes. In those cases, we can invert the weights when calculating the geodetic distances (remember the "n_cosponsorship_inv" variable that we created earlier?), as suggested by Newman (2018).

14.5.4 Closeness

Finally, closeness tries to measure the ease of access of the node in question to the other nodes in the network. Simply put, the idea is that important nodes will have connections good enough to easily access any part of the network. More formally, a node's proximity is the average of the geodetic distances (the lengths of the geodetic paths) it maintains with all the other nodes in the network:

$$closeness_i = \frac{1}{n} \sum d_{ij}$$

where i is the node in question, n is its number of links, j is a node it links to, and d is the geodetic distance.

As it is similarly based on geodetic distances, the same considerations for weighted and directed networks that we reviewed for betweenness apply to closeness.

14.5.5 Application in R

The previous centrality measures try to capture the distance dimensions of the same phenomenon: the importance of each node in the network. The code to calculate them using `tidygraph` is relatively simple. After activating the "nodes" data frame, we will use `mutate()` to add the centrality statistics with the multiple specific functions that start with `centrality_`. Notice that in some cases we will use arguments to specify the mode or identify the variable with weights in the "edges" data frame. Here is the sequence of pipes[3]:

```
tg_cosponsorship_arg_centr <- tg_cosponsorship_arg %>%
  activate("nodes") %>% # the next operations will edit the "nodes" tibble
  mutate(
    # degree
    c_in_degree = centrality_degree(mode = "in"),
    c_out_degree = centrality_degree(mode = "out"),
```

---

[3] R returns a warning for the closeness functions, because the algorithm has problem when dealing with disconnected networks. In order to get the closeness of a particular node, one needs to know its geodetic distance to each other node in the network. What to do when some geodetic paths do not exist is not straightforward, which makes the calculation and interpretation complicated (see Opsahl et al., 2010)

14.5 Measures of centrality

```
  # strength (weighted degree)
  c_in_strength = centrality_degree(mode = "in",
                                    weights = n_cosponsorship),
  c_out_strength = centrality_degree(mode = "out",
                                     weights = n_cosponsorship),
  # weighted pagerank (eigenvector centrality alternative)
  c_pagerank = centrality_pagerank(weights = n_cosponsorship),
  # betweenness, with or without inverted weights
  c_betweenness = centrality_betweenness(),
  c_betweenness_w = centrality_betweenness(weights = n_cosponsorship_inv),
  # closeness, with or without inverted weights
  c_closeness = centrality_closeness(),
  c_closeness_w = centrality_closeness(weights = n_cosponsorship_inv)
)
## Warning in closeness(graph = graph, vids = V(graph), mode = mode, weights
## = weights, : At centrality.c:2784 :closeness centrality is not well-
## defined for disconnected graphs
## Warning in closeness(graph = graph, vids = V(graph), mode = mode, weights
## = weights, : At centrality.c:2617 :closeness centrality is not well-
## defined for disconnected graphs
```

We can see that our nine centrality measures have been added to the nodes data frame:

```
tg_cosponsorship_arg_centr
## # A tbl_graph: 46 nodes and 217 edges
## #
## # A directed simple graph with 1 component
## #
## # Node Data: 46 x 13 (active)
##   id_sen name_sen  province  political_bloc c_in_degree c_out_degree
##    <int> <chr>     <chr>     <chr>                <dbl>        <dbl>
## 1      1 RAMON A~  SANTA C~  JUSTICIALISTA            8            6
## 2      2 JULIO A~  CATAMAR~  JUSTICIALISTA           25            4
## 3      3 RAMON A~  TUCUMAN   JUSTICIALISTA            1            4
## 4      4 ALFREDO~  JUJUY     JUSTICIALISTA            1            4
## 5      5 ANTONIO~  LA PAMPA  UCR                     14            6
## 6      6 DEOLIND~  CHACO     JUSTICIALISTA            9            7
## # ... with 40 more rows, and 7 more variables
## #
## # Edge Data: 217 x 5
##    from    to n_cosponsorship d_cosponsorship n_cosponsorship_inv
##   <int> <int>           <int>           <dbl>               <dbl>
## 1     1     2               1               1                   1
## 2     1     9               1               1                   1
## 3     1    30               1               1                   1
## # ... with 214 more rows
```

Let's begin by building a simple correlation plot with the centrality measures (Figure 14.6):

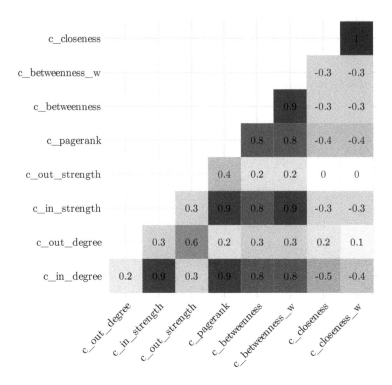

FIGURE 14.6: Correlation plot between centrality scores.

```
library(ggcorrplot)

corr_centrality <- tg_cosponsorship_arg_centr %>%
  select(starts_with("c_")) %>%
  # calculate correlation matrix and round to 1 decimal place:
  cor(use = "pairwise") %>%
  round(1)

ggcorrplot(corr_centrality, type = "lower", lab = T, show.legend = F)
```

Notice that the strongest correlations in the plot exist between the in-degree, in-strength and PageRank measures, which is generally expected. Notably, in-degree does not exhibit that strong of a correlation with out-degree (the same goes for in-strength and out-strength), which suggests that the group of well-sponsorsed legislators is not the same as the group of legislators who generally act as backers.

We can easily get the legislators with the highest in-degree and out-degree from our `tg_cosponsorship_arg_centr` object:

```
tg_cosponsorship_arg_centr %>%
  arrange(-c_in_degree) %>%
  select(id_sen, name_sen, c_in_degree)
## # A tbl_graph: 46 nodes and 217 edges
## #
```

14.5 Measures of centrality

```
## # A directed simple graph with 1 component
## #
## # Node Data: 46 x 3 (active)
##   id_sen name_sen                c_in_degree
##    <int> <chr>                         <dbl>
## 1      2 JULIO AMOEDO                     25
## 2     43 HECTOR J VELAZQUEZ               22
## 3     34 OLIJELA DEL VALLE RIVAS          20
## 4      9 ORALDO N BRITOS                  19
## 5     31 ANTONIO O NAPOLI                 17
## 6      5 ANTONIO TOMAS BERHONGARAY        14
## # ... with 40 more rows
## #
## # Edge Data: 217 x 5
##    from    to n_cosponsorship d_cosponsorship n_cosponsorship_inv
##   <int> <int>           <int>           <dbl>               <dbl>
## 1     9     1               1               1                   1
## 2     9     4               1               1                   1
## 3     9     8               1               1                   1
## # ... with 214 more rows
```

```
tg_cosponsorship_arg_centr %>%
  arrange(-c_out_degree) %>%
  select(id_sen, name_sen, c_out_degree)
## # A tbl_graph: 46 nodes and 217 edges
## #
## # A directed simple graph with 1 component
## #
## # Node Data: 46 x 3 (active)
##   id_sen name_sen                   c_out_degree
##    <int> <chr>                             <dbl>
## 1     12 PEDRO A CONCHEZ                       8
## 2     31 ANTONIO O NAPOLI                      8
## 3      6 DEOLINDO FELIPE BITTEL                7
## 4     10 JORGE A CASTRO                        7
## 5     23 MARGARITA MALHARRO DE TORRES          7
## 6     28 FAUSTINO M MAZZUCCO                   7
## # ... with 40 more rows
## #
## # Edge Data: 217 x 5
##    from    to n_cosponsorship d_cosponsorship n_cosponsorship_inv
##   <int> <int>           <int>           <dbl>               <dbl>
## 1     8    28               1               1                   1
## 2     8    34               1               1                   1
## 3     8    24               1               1                   1
## # ... with 214 more rows
```

The scale of those two measures is noteworthy. The senator with the highest in-degree, Julio Amoedo, received sponsorship from 25 different legislators (more than half of the Senate!). On the other hand, the senator with the highest out-degree, Pedro Conchez, sponsored the bills of only 8 distinct legislators. This might indicate that, while the role of a "main sponsor" is well-defined, the one of a "backer" is not.

Another common way of utilizing centrality scores involves analyzing them according to some of the other relevant variables in our data. This is made easy by the `tidygraph` object structure, which integrates in a data frame all the information for nodes (including descriptive variables and centrality scores). For instance, we can plot the relationship between PageRank/in-degree/in-strength and the two main political blocs with a simple boxplot (Figure 14.7):

```
ggplot(tg_cosponsorship_arg_centr %>%
         as_tibble() %>%
         filter(political_bloc %in% c("UCR", "JUSTICIALISTA")),
       aes(x = political_bloc, y = c_pagerank)) +
  geom_boxplot()
```

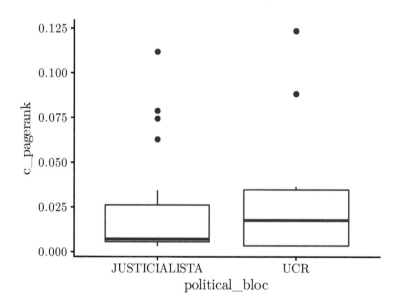

FIGURE 14.7: PageRank by political bloc.

While the range of the PageRank measures is comparable, the shape of the boxplots suggests that the two political blocs behave differently when it comes to co-sponsorship. While the Justicialista bloc has a few notable legislators, who presumably act as well-backed main sponsors, the rest of its senators are not that influential in that regard. On the other hand, while UCR also has a couple of notable legislators, the distribution seems to be more balanced.

14.5 Measures of centrality

Exercise 14C. Create a plot to compare some centrality score by whether the legislator is from the metropolitan area (like in the previous exercise).

Throughout this chapter, we covered the very basic notions of network analysis in social science. We recommend you to read the recommended bibliography at the beginning of the chapter in order to be able to expand your analysis further.

15
Principal Component Analysis

Caterina Labrín and Francisco Urdinez[1]

Suggested readings

- Abeyasekera, S. (2005). Multivariate methods for index construction. In *Household Sample Surveys in Developing and Transition Countries,* pages 367–387. United Nations, New York, NY.

- Bryant, F. B. (2000). Assessing the Validity of Measurement. In Grimm, L. G. and Yarnold, P. R., editors, *Reading and Understanding MORE Multivariate Statistics,* pages 99–146. American Psychological Association, Washington, DC.

- Collier, D., Laporte, J., and Seawright, J. (2008). Typologies: Forming concepts and creating categorical variables. In Box-Steffensmeier, J. M., Brady, H. E., and Collier, D., editors, *The Oxford Handbook of Political Methodology,* pages 152–173. Oxford University Press, Oxford.

- Goertz, G. (2006). *Social Science Concepts: A User's Guide.* Princeton University Press, Princeton, NJ.
    - Chapter 4 - "Increasing Concept-Measure Consistency."

- Jackman, S. (2008). Measurement. In Box-Steffensmeier, J. M., Brady, H. E., and Collier, D., editors, *The Oxford Handbook of Political Methodology,* pages 119–151. Oxford University Press, Oxford.

Packages you need to install

- `tidyverse` (Wickham, 2019), `politicalds` (Urdinez and Cruz, 2020), `FactoMineR` (Husson et al., 2020), `factoextra` (Kassambara and Mundt, 2020), `GGally` (Schloerke et al., 2020), `ggcorrplot` (Kassambara, 2019).

---

[1]Institute of Political Science, Pontificia Universidad Católica de Chile. E-mails: cilabrin@uc.cl and furdinez@uc.cl.

15.1 Introduction

By this point in the book you have learned enough tools to be able to work with social science data and concepts. In this chapter you will learn how to create in a simple way variables that measure complex concepts that many times in social sciences, and especially in political science, we want to use to test hypotheses or simply visualize how they behave.

The relationship between war and democracy or between democracy and development are things that are usually studied in political science. However, many times these concepts are not clearly defined or if they are, we do not know which of the definitions are a better fit for our research. Concepts such as terrorism or populism are used every day by politicians and journalists, many times incorrectly, demonstrating that the problem does not only occur at an academic level but also in all of society, generating quite serious levels of misinformation. This fragility of our concepts becomes even more serious if we have to operate with data, since many times we want to measure complex concepts with indicators that show us the variations or relationships with other concepts in a more quantitative or visual manner.

To illustrate the situation, let's take a hypothetical case: let's pretend that we are working on a project that among other things asks to compare how democratic Latin American countries behave and so must be able to visualize it clearly in a table or graph. Usually, in this case, we would look for some indicator of democracy that V-dem or Freedom House gives us for all Latin American countries. We would assume by practice that these indicators have evident validity and were properly constructed on the basis of clear dimensions. No one would question our choice. Now, let's imagine that we want to investigate a different and more specific concept. For this task we do not have an indicator similar to those of democracy and we must start from scratch: what are we measuring? how do we measure it? What data do we use?

In view of this problem, in this chapter we want to ilustrate the usefulness of the Principal Component Analysis (PCA) since it allow us to generate the different dimensions that make up an abstract concept through the reduction of different previously chosen indicators. This is primarily to analyze how a concept of this style would be composed and to understand what we are exploring. In addition, we will present some ways to operationalize the concepts through indexes that can show us the measurement variability among different analysis units, in order to compare them or include them within a more complex analysis.

15.1.1 Concepts, measurements and validity

When we construct an empirical measurement from an abstract concept (i.e. democracy, populism, development), in addition to accounting for its complexity, we want it to be *valid* (see Bryant, 2000, for an excellent summary on different types of validity). Content validity is what allows us to ensure that the content of the measurement corresponds to what is conceptually defined for that measurement. This must be done every time we want to generate a concept through a theoretical review that places where our concept will go, which will be the parts that will compose it and mainly which variables we will use to measure them. An adequate literature review regarding the particular concept we are working on, which can be placed in a larger theoretical framework, will always be more than desirable when working with abstract and complex concepts.

Construct validity refers to the property of a concept measuring what it claims to measure. This validity must always be questioned, precisely because of what we previously said about the complexity of the concepts we work with. For example, it is very frequent with the measurements that exist of democracy. Even when we have advanced enormously as a discipline, and have comparable measurements between countries and over time, such as Polity and Varieties of Democracy, there are different opinions on how this concept should be operationalized.

There are two ways, however, to gain construct validity for our variable of interest. The first alternative is called discriminative validity, that is, our construct should not be highly correlated to variables that our indicator does not claim to measure. This concept can be well exemplified by the way that V-Dem has chosen to create its democracy index (which they call polyarchy): as the name suggests, they measure varieties of democracy from different dimensions, namely the liberal, participatory, deliberative, electoral and egalitarian dimensions. We would expect that the dimensions are not highly correlated with one another in order to be confident that they truly capture the different faces of this multifaceted concept.

The second alternative is for convergent validity, that is, the correlation of my measure of interest with the measures of others. For example, democracy as measured by V-Dem has a correlation of 0.8 with the way it is measured by Polity. If I create a third measure, whose correlation is 0.70 with Polity and 0.9 with V-Dem, I can be confident that all three variables are similar to the latent variable.

The Principal Component Analysis (PCA) technique helps us define which combination of data best represents the concept we want to measure.

15.2 How PCA works

PCA is a very useful technique when we want to work with different variables that proxy a single concept. By previously choosing the variables we want to work with, the PCA technique gives us different *components*, which are the reduction of the dimensionality of a multivariate dataset condensing the dimensions into one single variable. It literally consists of working with the correlation between variables, and accounting for what they measure in common and what each one measures individually.

PCA brings together the high correlations between variables and the condemnations into a single component. The more similar the variables are (high correlations and in the same direction) the fewer components that will be needed to capture the full complexity of the data.

The PCA will give us a number of "components" equal to the number of variables. However, we will not need all of these to condense de variables and will only keep a few of them. It helps to think of PCA as a smoothie, the variables we run are like the fruit we put into the blender, while the components are the flavors we get from the mixture. For example, if we run the PCA with four variables (X, Y, W, Z) we get four components (always the same number of varaibles and components). We obtain that component 1 is highly and positively correlated to variables X and W, component 2 is highly and negatively correlated to Y and W, component 3 is highly and positively correlated to variable Z and weakly and negatively correlated to X, and component 4 is weakly positively correlated to Y and W. From the analysis we know that if we keep components 1 and 2 our index will be good enough. How do we decide which one we keep?

Components will come with an Eigen Value, which is a score that measures how much variance from the original variables each component explains. The Eigen Value orders the components from highest to lowest explained variance. As a rule of thumb we keep components whose Eigen Values are larger than 1.

The creation of different components due to the PCA will be useful for you to create indexes. We will now provide an example in R.

15.3 Basic notions in R

In Latin America, given the past of dictatorial governments and the recent democratic weakening of many countries, it may be important for possible research to consider public opinion on the institutions of countries. A good question to ask is how can one know the opinion of Latin American citizens regarding the democratic institutions of their countries?

The Latin American Public Opinion Project (LAPOP) - coordinated from Vanderbilt University - specializes in conducting impact assessment studies and producing reports on the attitudes, evaluations and experiences of individuals in Latin American countries. This project provides researchers with different questions that, together, could help us to approximate how much confidence there is in the region regarding democratic institutions in the country corresponding to each individual.

Exercise 15A. Let's assume that you are asked to measure anti-Americanism in Latin America. How would you measure this concept? Choose five or more questions from the LAPOP survey (you can consult the codebook here[2]) that you would use to measure anti-Americanism.

As a first step, it is important to select the variables that will be used to carry out the PCA, leaving aside everything that we do not want to use in the final index. This data contains 12 questions from the LAPOP survey, carried out from a selection of about 7000 people in 10 Latin American countries. The entire questionnaire can be accessed at this link[3]. This first step is essential. This is the step that entails more theoretical work for the researcher, since from here the variables that will be integrated in the analysis (and which will be left out) are defined. Under our analysis, the variables that we will keep are the following:

Name	Description	Source
justify_coup	Dichotomous variable that measures the level of justification of **military coups d'état** in the country of the respondent against a lot of crime.	Based on question "jc10" of LAPOP Survey.
justify_cong_shutdown	Dichotomous variable that measures the level of justification of **congress closure** in difficult situations by the president.	Based on question "jc15a" of LAPOP Survey.
trust_courts	Measures on a scale of 1 to 7 the level of **confidence in the courts** in the country of the respondent.	Based on question "b1" of LAPOP Survey.
trust_institutions	Measures on a scale of 1 to 7 the level of **respect for political institutions** in the respondent's country.	Based on question "b2" of LAPOP Survey.
trust_congress	Measures on a scale of 1 to 7 the level of confidence in the National Congress (legislative branch) of the country of the respondent.	Based on question "b13" of LAPOP Survey.

---

[2]See https://www.vanderbilt.edu/lapop/ab2016/AB2017-v18.0-Spa-170202_W.pdf
[3]See https://www.vanderbilt.edu/lapop/ab2016/AB2017-v18.0-Spa-170202_W.pdf

Name	Description	Source
trust_president	Measures on a scale of 1 to 7 the level of **trust in the president (executive branch)** of the country of the respondent	*Based on question "b21a" of LAPOP Survey.*
trust_parties	Measures on a scale of 1 to 7 the level of **confidence in political parties** in the country of the respondent	*Based on question "b21" of LAPOP Survey.*
trust_media	Measures on a scale of 1 to 7 the level of **confidence in the media** in the country of the respondent	*Based on question "b37" of LAPOP Survey.*
trust_elections	Measures on a scale of 1 to 7 the level of **confidence in the country's elections** of the respondent.	*Based on question "b47a" of LAPOP Survey.*
satisfied_dem	Dicotomic variable that measures the level of **satisfaction with democracy** of the respondents.	*Based on question "pn4" of LAPOP Survey.*
vote_opposers	Measures on a scale from 1 to 7 the satisfaction with the idea that those who are against the government in power can **vote** in the elections of the country of the respondent	*Based on question "d1" of LAPOP Survey.*
demonstrations	Measures on a scale from 1 to 7 the satisfaction with the idea that those who are against the government in power can carry out **peaceful demonstrations** to express their point of view	*Based on question "d2" of LAPOP Survey.*

To be able to select the variables and use the pipes we load 'tidyverse'. Then, we load the dataset from the book's package, `politicalds`.

```
library(tidyverse)
```

```
library(politicalds)
data("lapop")
```

These variables can be explored graphically. For example, we can see the confidence that citizens have in the elections for each country (see Figure 15.1):

15.3 Basic notions in R

```
lapop <- lapop %>%
  group_by(country_name) %>%
  mutate(trust_elections_prom = mean(trust_elections)) %>%
  ungroup()

ggplot(lapop, aes(x = trust_elections)) +
  geom_histogram() +
  scale_x_continuous(breaks = 1:7) +
  labs(title = "Trust in elections",
       x = "The national average is expressed as a dashed line.",
       y = "Frequency")+
  facet_wrap(~country_name) +
  geom_vline(aes(xintercept = trust_elections_prom),
             color = "black", linetype = "dashed", size = 1)
```

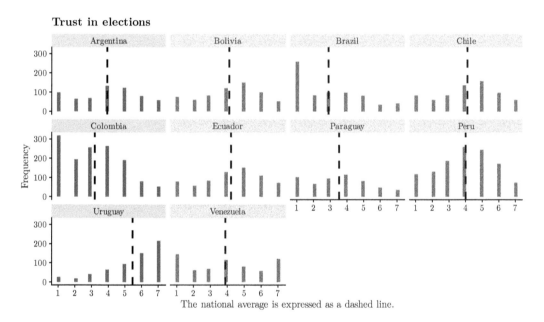

FIGURE 15.1: Histograms of trust in elections per country.

Let's select our variables of interest, this means that we leave out any variable that is not proxy of democracy, for example the variable 'country' that indicates the country of each individual should not be incorporated. For this, we generate a new dataset that contains only the variables that we want to use, in this case, to find out what is the opinion that Latin Americans have of their democratic institutions.

```
lapop_num <- lapop %>%
  select(justify_coup, justify_cong_shutdown, trust_institutions,
         trust_courts, trust_congress, trust_president, trust_parties,
         trust_media, trust_elections, satisfied_dem, vote_opposers,
```

```
                demonstrations) %>%
  mutate_all(as.numeric)

lapop_num
## # A tibble: 7,655 x 12
##     justify_coup justify_cong_sh~ trust_instituti~ trust_courts
##            <dbl>            <dbl>            <dbl>        <dbl>
## 1              0                1                4            6
## 2              1                0                6            4
## 3              0                0                4            4
## # ... with 7,652 more rows, and 8 more variables
```

Having already selected the variables, the first step is to standardize them. The PCA technique accepts variables of different types, but it is important to omit or impute to the means the missing values. In this case, we will omit the possible NA that the dataset has.

```
lapop_num <- lapop_num %>%
  scale() %>%
  na.omit() %>%
  as_tibble()
```

The next step is to observe the correlation between them. This is useful to know how the chosen variables are related, and also to see if there are extremely high correlations between two or more variables. If there is a case where we have two or more variables with high correlation among them, these will have enormous influence on how the components are oriented.

The best alternative to observe correlation between variables is offered by the **ggcorrplot** package (see Figure 15.2). The more intense the color, the stronger the correlation.

```
library(ggcorrplot)

corr_lapop <- lapop_num %>%
  # calculate correlation matrix and round to 1 decimal place:
  cor(use = "pairwise") %>%
  round(1)

ggcorrplot(corr_lapop, type = "lower", lab = T, show.legend = F)
```

What we see is that variables on confidence are positively correlated. That is, those who trust the president tend to trust the media, elections, for example.

Once we look at the correlations, we are ready to do the PCA. For this, we will use two different packages that perform the PCA, giving you options on how to perform this technique in R. The different components of the PCA will allow us to define which

15.4 Dimensionality of the concept

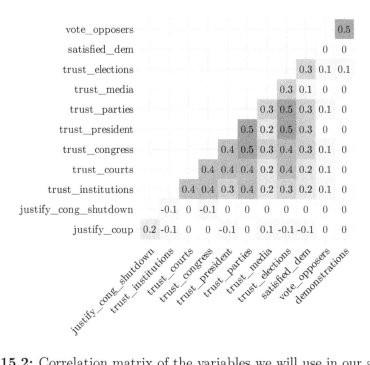

FIGURE 15.2: Correlation matrix of the variables we will use in our anaysis.

dimensions of the concept we'll keep. Then, we'll perfom a simple index that denotes the variation of the concept through different units of analysis.

15.4 Dimensionality of the concept

First, with the subset of data, we generate a PCA with the `stats` package. Then with the `summary()` command we can see what the analysis gives us (here we are only showing the first seven components).

```
library(stats)
pca <- princomp(lapop_num)
```

```
summary(pca, loadings = T, cutoff = 0.3)
```

```
## Importance of components:
##                          Comp.1 Comp.2 Comp.3 Comp.4 Comp.5 Comp.6 Comp.7
## Standard deviation         1.86   1.22   1.11  0.976  0.893  0.882  0.827
## Proportion of Variance     0.29   0.12   0.10  0.079  0.066  0.065  0.057
## Cumulative Proportion      0.29   0.41   0.51  0.593  0.659  0.724  0.781
##
```

```
## Loadings:
##                    Comp.1 Comp.2 Comp.3 Comp.4 Comp.5 Comp.6 Comp.7
## justify_coup                      0.69   0.27   0.24   0.50
## justify_cong_shutdown              0.64  -0.54         -0.39
## trust_institutions  -0.33                       0.53  -0.23  -0.40
## trust_courts        -0.35                       0.34         -0.40
## trust_congress      -0.39                0.26                 0.39
## trust_president     -0.38               -0.27                 0.20
## trust_parties       -0.41                                     0.46
## trust_media         -0.25          0.28  0.46  -0.65         -0.38
## trust_elections     -0.40                      -0.24
## satisfied_dem       -0.27               -0.50          0.67  -0.34
## vote_opposers               -0.69
## demonstrations              -0.70
```

As you can see, there are three components that give us an Eigen Value (denoted by Standard deviation) greater than 1. This cutoff will be our criteria for selecting the components that we will keep. It also shows us how each component is constructed with the variables we selected.

Another way to look at the Eigenvalues of each component is with the `get_eigenvalue()` function from the `factoextra` library. As we got with `summary()`, `get_eigenvalue()` gives us, in addition to the Eigen Value, the combined variance that each component explains.

```
library(factoextra)
eig_val <- get_eigenvalue(pca)
eig_val
## # A tibble: 12 x 3
##    eigenvalue variance.percent cumulative.variance.percent
##         <dbl>            <dbl>                       <dbl>
## 1        3.44             28.7                        28.7
## 2        1.48             12.3                        41.0
## 3        1.23             10.3                        51.3
## # ... with 9 more rows
```

As can be seen, component 1 represents about 28.7% of the total variance of the chosen variables. The variance of the other two important components is 12.3% and 10.3% respectively.

Another way to deliver this information is through graphs. The following command gives us in its simplest form, the percentage of variance explained by each of the components that the PCA gives us (see Figure 15.3).

```
fviz_eig(pca, addlabels = T, ylim = c(0, 50))
```

A small change in the command, also gives us a graph with the Eigen Values of each of the components (see Figure 15.4).

15.4 Dimensionality of the concept

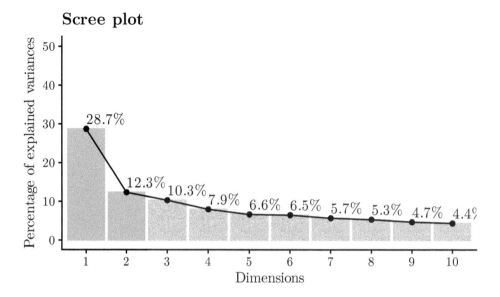

FIGURE 15.3: Percentage of explained variance of each component.

```
fviz_eig(pca, choice = c("eigenvalue"), addlabels = T, ylim = c(0, 3))
```

```
fviz_eig(pca, choice = c("eigenvalue"), addlabels = T, ylim = c(0, 3),
         ggtheme = th, barfill = "darkgray", barcolor  = "darkgray")
```

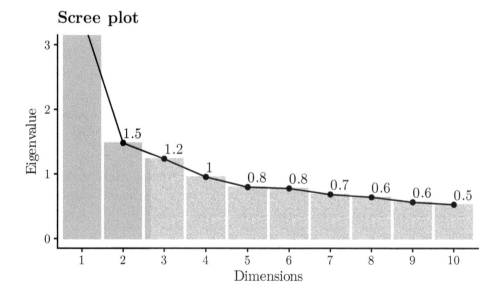

FIGURE 15.4: Eigenvalue of each component.

To know how each of these components is composed, we can generate a Biplot. This type of graph will appear to us as vectors in two dimensions (which will be the first two components of the analysis) (Figure 15.5).

```
fviz_pca_biplot(pca, repel = F, col.var = "black", col.ind = "gray")
```

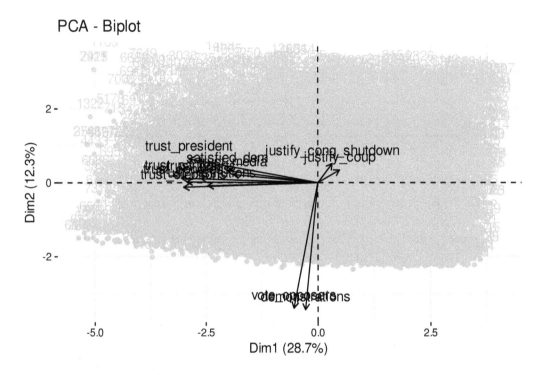

FIGURE 15.5: Biplot with dimensions 1 and 2.

This two-dimensional graph clearly shows that we have three groups of variables. These sets of variables are what are precisely reduced to three components, which can represent three dimensions of opinion on political institutions that we want to measure in Latin America.

We can also explore the different dimensions that will compose the concept we want to measure with the command `fviz_contrib()`, changing the components in each axis (see Figure 15.6):

```
fviz_contrib(pca, choice = "var", axes = 1, top = 10)
fviz_contrib(pca, choice = "var", axes = 2, top = 10)
fviz_contrib(pca, choice = "var", axes = 3, top = 10)
```

For example, the first component is the most diverse but it is largely fed by confidence variables. If you remember the correlation matrix we made with `ggcorrplot()`, all these variables had high correlations with each other. The second component feeds

15.4 Dimensionality of the concept

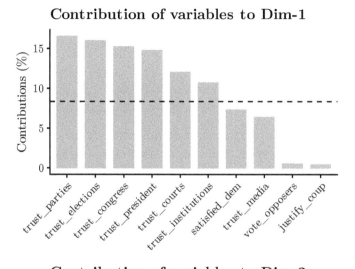

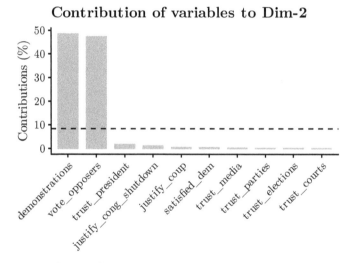

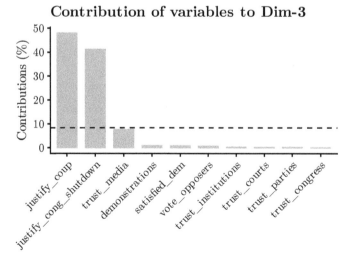

FIGURE 15.6: Variables that contribute to each dimension in the PCA.

on the strong correlation between `demonstrations` and `vote_opposers`. The dotted line expresses the value that would assume a scenario where all variables contribute equally, i.e. 1/12 (8.33%), and serves only as a visual reference.

The composition of each component can help us to name each dimension of the concept we are creating. This will provide more support when it comes to justifying the way in which public opinion oriented towards political institutions is measured.

The first dimension, which has the greatest amount of variance, represents trust in institutions such as state powers, political parties and elections. It is the most diverse but it is largely fed by trust variables. If you remember the correlation matrix we did with `ggcorrplot()`, all these variables had high correlations with each other. This means that when we want to measure the confidence in politics of individuals, what our data will mainly represent is that confidence in the most formal politics, in the most classic concept of politics.

The second of the dimensions is that which denotes the level of political tolerance of dissent, which could orient an opinion towards a way of controlling the rights to vote and peaceful public demonstration. This may denote how prone people are to suspend certain democratic rights of those who think differently.

The third dimension, which represents 10% of the variance, is composed of variables that represent the propensity to justify coups d'état or closures of Congress. Our concept will then have components that measure how individual opinions of democratic breaks are formed, measuring how fragile formal institutions are within public opinion in Latin America.

The qualitative results of each dimension can be represented in a table like this:

Dimension	Label	Variables	Variance explained
1	Trust in institutions	`trust_courts`, `trust_institutions`, `trust_congress`, `trust_president`, `trust_parties`, `trust_elections`	28.7%
2	Democratic rights	`vote_opposers`, `demonstrations`	12.3%
3	Anti-Democratic values	`justify_coup`, `justify_congshutdown`	10.3%

As these new dimensions are created, they can be considered as new variables to be integrated within our original dataset, as shown below:

```
lapop <- bind_cols(lapop, as_tibble(pca$scores))
```

These can be treated as new variables that can be integrated separately into the analysis or, as we will see below, be integrated into an index of confidence in political institutions, which evaluates the three dimensions as a whole.

Exercise 15B. Standardize and omit the NAs from the questions you chose in Part One for the anti-American level. Perform a Principal Component Analysis and answer the following questions: What are the most important components? What variables make them up? What dimensions are involved in the concept of anti-Americanism that you are performing?

15.5 Variation of the concept

Another application of this tool in R will allow to generate an index that characterizes each of the individuals by diferentiating who is more likely to have confidence in the institutions and who is not. From this individual index, we will then be able to make a wide variety of studies, from comparisons between countries, to comparisons using age, gender, income or another variable of interest.

For this part, we will generate a PCA again, but this time we will do it with the PCA command of the `FactoMineR` package. This will give us the same result as the previous package, but we want to show that this tool can be used in several ways in R.

```
library(FactoMineR)

pca_1 <- PCA(lapop_num, graph = F)
```

As we saw before, we must retain the components that contain an Eigen Value greater than 1, which we can see again in the graph (see Figure 15.7):

```
fviz_eig(pca_1, choice = "eigenvalue", addlabels = T, ylim = c(0, 3.5))
```

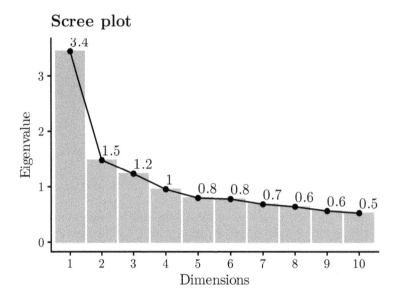

FIGURE 15.7: Eigenvalues of the components. As a rule of thumb we retain components with eigenvalues larger than 1.

As we could see in advance, the components greater than 1 are the first three. We will leave in this case component 4 to add more variables to our index.

In order to condense the chosen components into a single variable, it is necessary to remember how much accumulated variance they represent of the total:

```
eig <- get_eig(pca_1)
```

```
eig
##        eigenvalue variance.percent cumulative.variance.percent
## Dim.1       3.44            28.7                          29
## Dim.2       1.48            12.3                          41
## Dim.3       1.23            10.3                          51
## Dim.4       0.95             7.9                          59
## Dim.5       0.80             6.6                          66
## Dim.6       0.78             6.5                          72
## Dim.7       0.68             5.7                          78
## Dim.8       0.64             5.3                          83
## Dim.9       0.56             4.7                          88
## Dim.10      0.53             4.4                          92
## Dim.11      0.47             3.9                          96
## Dim.12      0.43             3.6                         100
```

We had seen that the four components represented almost 60% of the total variance: the first component 28.7%, the second component 12.3%, the third component 10.3% and the fourth component 7.9%. The next step is to add up these four components,

15.5 Variation of the concept

but weighting each one by the percentage of the variance they represent. We do it in the following way:

```
data_pca <- pca_1$ind$coord%>%
 as_tibble() %>%
 mutate(pca_01 = (Dim.1 * 28.7 + Dim.2 * 12.3 + Dim.3 * 10.3 +
                  Dim.4 * 7.9) / 60)

lapop <- bind_cols(lapop, data_pca %>% select(pca_01))
```

Thus, we have created a single variable, which we call `pca_01`. We are very close to having this variable as our democracy rating indicator! It so happens that the variable `pca_01` is on an unfriendly scale. Ideally we want our indicator to range from 0 to 1, 0 to 10, or 0 to 100 so that it is easier to interpret. We will do this to make it 0 to 100, if you want it to be 0 to 10 or 0 to 1 you have to replace 100 in the formula below with the number you are interested in.

```
lapop <- lapop %>%
  mutate(democracy_index = GGally::rescale01(pca_01) * 100)%>%
  select(democracy_index, everything())
```

Already with our new rescaled index, we can see how its density looks (see Figure 15.8):

```
index_density <- ggplot(data = lapop,
                        mapping = aes(x = democracy_index)) +
  labs(x = "Index of trust in democracy", y = "Density") +
  geom_density()

index_density
```

Now that we have the index ready, we can do all kinds of analysis. For example, we can make comparisons by country. If we had individual variables, we could proceed to regression models with controls for gender, ideology, income, education attainment. To do this, you can use what you learned in Chapters 5 and 8.

Descriptively, we can make a comparison of how our index of confidence in institutions in Latin American countries (see Figure 15.9):

```
lapop <- lapop %>%
  group_by(country_name) %>%
  mutate(democracy_avg = mean(democracy_index)) %>%
  ungroup()

ggplot(lapop, aes(x = democracy_index)) +
```

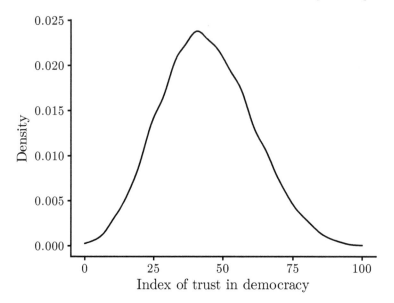

FIGURE 15.8: Density plot of our index of trust in democracy.

```
geom_density() +
labs(title = "Trust in democracy in Latin America (N = 7000)",
     x = "Trust in democracy",
     y = "Density") +
facet_wrap(~country_name) +
geom_vline(aes(xintercept = democracy_avg),
           color = "black", linetype = "dashed", size = 1)
```

Exercise 15C. Using the index of confidence in democracy in Latin America that we have just created, analyze with linear regression models which variables have high explanatory power on this variable are ideology, income or age important variables?

Exercise 15D. With the set of variables you chose, create an index of anti-Americanism along the lines of the chapter.

15.5 Variation of the concept

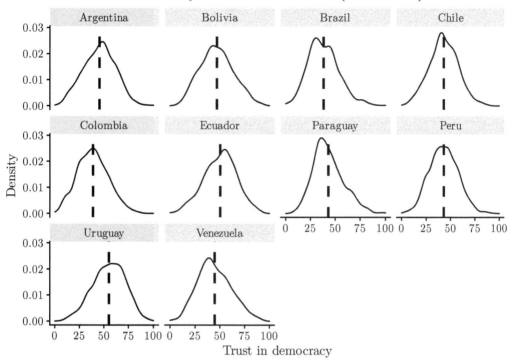

FIGURE 15.9: Density plots of our index of trust in democracy by country. The national average is included as a dashed line.

16
Maps and Spatial Data

Andrea Escobar and *Gabriel Ortiz*[1]

Suggested readings

- Brunsdon, C. and Comber, L. (2015). *An Introduction to R for Spatial Analysis and Mapping.* SAGE, Thousand Oaks, CA.

- Lansley, G. and Cheshire, J. (2016). *An Introduction to Spatial Data Analysis and Visualisation in R.* Consumer Data Research Centre.

- Lovelace, R., Nowosad, J., and Münchow, J. (2019). *Geocomputation with R.* CRC Press, Boca Raton, FL.

- Pebesma, E. (2018). Simple Features for R: Standardized Support for Spatial Vector Data. *The R Journal,* 10(1):439–446.

Packages you need to install

- `tidyverse` (Wickham, 2019), `politicalds` (Urdinez and Cruz, 2020), `sf` (Pebesma, 2020), `ggrepel` (Slowikowski, 2020), `gridExtra` (Auguie, 2017), `rnaturalearthhires` (South, 2020), `spdep` (Bivand, 2019).

16.1 Introduction

In this chapter, we will learn how to work with spatial data using R and the `tidyverse`. We will focus on the richness spatial data has, both as a tool for exploratory data analysis about diverse phenomena, and as a way of visualizing and communicating our findings in an attractive and effective way. As such, this chapter has the following objectives:

a. Describe the format of spatial data.

b. Explain why this data is interesting to be used for analysis in political science.

---

[1]Institute of Political Science, Pontificia Universidad Católica de Chile. E-mails: abescobar@uc.cl and goortiz@uc.cl.

c. Its management and manipulation in R, focusing on its visualization by means of creating static and dynamic maps with `ggplot2`.

d. We will not go into making inferential statistical models from spatial data (*Spatially lagged models*), and neither how to generate geographical data, since we will be working on geographical datasets already created.

The dataset we use in this chapter will be a shapefile (format that we will explain later on) of the states in Brazil, obtained from the *Spatial Data Repository of New York University*[2], which will be merged with data from Freire (2018).

On the one hand, the spatial data repository from NYU provides lots of spatial data for different countries, as well as supra and sub national divisions. In this case, we will use a dataset of the state borders of Brazil in the year 1991, provided by the *Brazilian Institute of Geography and Statistics*. On the other hand, the work of Freire (2018) explains the cause of the decrease in the homicide rate in the state of Sao Paulo as a result of the implementation of state-driven policies for the reduction of crime. This work is useful since it collects state-level data of different socioeconomic indicators such as GDP, Gini Index, average years of schooling, among others, over a period of twenty years (1990-2010). This data will be joined to our shapefile for plotting these variables in maps.

For a secondary example, we will use a shapefile for Chile and its surrounding countries, and data from Freire et al. (2019). We also present a series of exercises that will help you gain experience using a shapefile of South America merged with data of the GDP per capita and the quality of electoral democracy of each country taken from *Varieties of Democracy* dataset.

16.1.1 Spatial Data: What is it?

Spatial data, or rather *geospatial* data, refers to data obtained from geographical locations, that is, areas on the surface of the Earth. The use of this type of data recognizes "the importance of spatial concepts such as distance, locations, proximity, neighborhood and region in human society", allowing us to face phenomena from a multivariate and multidimensional perspective, providing us additional information for our observations (CSISS, 2004).

In political science, the spatial outlook for data analysis has been advanced by works such as Gary King's (1997) on how to make ecological inferences, and Gimpel and Schuknecht's (2003) on accessibility to voting locations. This type of data allows answering questions such as the influence of the proximity of neighboring countries in the diffusion of certain policies, but it also turns out to be a key tool in data exploration for observing the existence of territorial patterns in other studies.

In the last years, spatial data has become widely available, mostly thanks to governments' initiatives that provide data for various social and economic processes of the

---

[2]See https://geo.nyu.edu/catalog/stanford-ys298mq8577

16.1 Introduction

country. New open-access platforms, such as Google Maps or Open Street Map[3], and social networks like Twitter and Facebook also helped in the interest for spatial data. Meanwhile, for investigators and students, datasets like the already-mentioned Spatial Data Repository from NYU provide shapefile files of easy access and manipulation for generating visualizations and joining georeferenced data from other datasets.

16.1.2 Spatial Data Structure

Files that contain spatial data are commonly known as *shapefiles*, which generally are placed in a folder that contains at least three files with the .shp, .dbf and .shx extensions (Figure 16.1). You can download these files, that we will use in this chapter, from this link.[4]

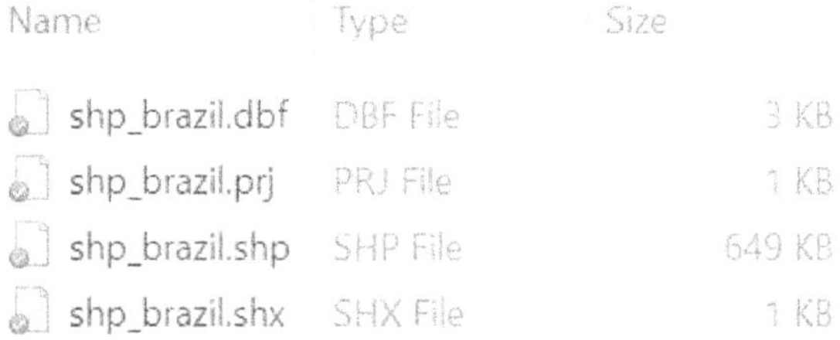

Name	Type	Size
shp_brazil.dbf	DBF File	3 KB
shp_brazil.prj	PRJ File	1 KB
shp_brazil.shp	SHP File	649 KB
shp_brazil.shx	SHX File	1 KB

FIGURE 16.1: Example of a folder containing a shapefile in Windows. Note that the .shp file contains the vast majority of information. Also, this folder contains and additional .prj file.

In R, shapefiles are usually represented by vectors, which consist of the geometry description or *shape* of the objects. They generally contain additional variables, called attributes, over the same data. Using as an example the data that will be analyzed throughout this chapter, our dataset describes the borders of the states of Brazil (geometry), and also contains information of the homicide rate plus other socioeconomic data (attributes). Spatial data are diverse in their characteristics and are usually divided in types of vectors that in every case consist of a set of coordinate pairs (x,y) (see Figure 16.2):

- Points: A single location described by a point as an address of a geocoded house.
- Lines: A segment composed by a series of points connected with each other that do not form a closed object.
- Polygons: A flat surface composed of a series of lines connected to each other that form a closed object.
- Multi-polygons: Multi-surfaces composed of polygons that do not necessarily intersect.

---

[3]See https://www.openstreetmap.org/
[4]See https://arcruz0.github.io/politicalds/spatial_data_politicalds.zip

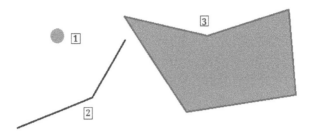

FIGURE 16.2: Types of forms in spatial data. (1) Point, (2) Line, (3) Polygon.

Tip. As we have stated, the Spatial Data Repository from NYU[5] can be a good alternative if you want to start working with spatial data from your country, state or municipality. You can easily search for shapefiles entering the webpage of the Repository and selecting "Polygon" in the "Data Type" category. After that, you have to introduce the name of the area of interest in the search engine, select a result and press "Download Shapefile" in the top right corner of the page. Another advantage of this data is that files are often lightweight but of sufficient quality, which is convenient in the case of hardware with limitations for processing large data files.

16.2 Spatial Data in R

In the last years, R has made huge progress in its tools for manipulating and analyzing geographical data in a way that it is compatible with the syntax and functionalities of other packages in the program. In the past, these tasks were a little too challenging. In this chapter, we will focus on the sf package, which was created in 2016 over the base functionalities of three previous packages: sp, rgeo and rgdal, and which implements the standard model for open code *simple features*[6]. It allows for the representation of real world objects in the computer, which possess both spatial and non spatial attributes, based on 2D geometry with linear interpolation between vertices (Pebesma, 2018).

16.2.1 Simple Features in R

The main advantage that the sf package provides is allowing us to work with spatial data inside the tidyverse, that is, we can manage spatial data just as if they were

---

[5]See https://geo.nyu.edu/

[6]Simple features refers to a format standard (ISO 19125-1:2004) that describes how objects in the real world can be represented in computers. See r-spatial.github.io/sf/articles/sf1.html.

any other type of dataset. We can do this by functions from R contained in the metapackage `tidyverse`, such as `ggplot2` and `dplyr`, in alignment with what we have learned throughout the book. Some alternative options for constructing maps in R are the `tmap` and `cartogram` packages, which are explained in depth in Chapter 8 of "Geocomputation with R" by (Lovelace et al., 2019).

It is important to have the latest `tidyverse` version to run the geoms of `ggplot2` that will be used in this chapter. You can update it to the last version with the `install.packages("tidyverse")` command:

```
library(tidyverse)
```

Now we need to install the `sf` package. In order for this to work on Mac and Linux you need the latest versions of FDAL, GEOS, Proj.4 and UNDUNITS already installed[7]. Once this is done we can install the `sf` package:

```
remotes::install_github("r-spatial/sf")
library(sf)
```

16.2.2 Structure

To see the structure of the `sf` type objects we load our dataset in the shapefile format, use the `read_sf()` function (note that we are loading the `shp_brazil.shp` *file* contained in the `shp_brazil` *folder*):

```
shp_brazil <- read_sf("shp_brazil/shp_brazil.shp")
```

In the following image, we detail the parts of our shapefile with its names:

As you can observe in the previous figure, the simple features are saved by `sf` in a data.frame format and with an additional `sf` class format. Then, just as stated before, we can address this dataset just as any other inside the `tidyverse`, using its functions, pipes (%>%), etc. Here, every row consists of a simple feature, and every column consists of an attribute. The difference with a normal dataset is that here we find a new column of simple feature geoms (*simple feature geometry column* - `sfc`), resulting in a data.frame with an extra column for spatial information.

We can verify this by checking the type of file we have loaded with the `class()` function. Note that `shp.brazil` is, at the same time, a `sf` type file and a `data.frame`:

```
class(shp_brazil)
## [1] "sf"          "tbl_df"      "tbl"         "data.frame"
```

---

[7]For more information, check https://github.com/r-spatial/sf#installling

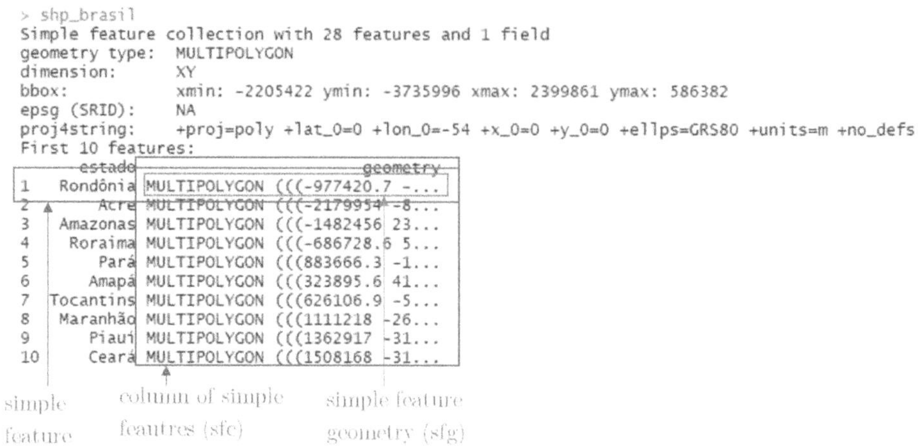

```
> shp_brasil
Simple feature collection with 28 features and 1 field
geometry type:  MULTIPOLYGON
dimension:      XY
bbox:           xmin: -2205422 ymin: -3735996 xmax: 2399861 ymax: 586382
epsg (SRID):    NA
proj4string:    +proj=poly +lat_0=0 +lon_0=-54 +x_0=0 +y_0=0 +ellps=GRS80 +units=m +no_defs
First 10 features:
      estado                        geometry
1   Rondônia  MULTIPOLYGON (((-977420.7 -...
2       Acre  MULTIPOLYGON (((-2179954 -8...
3   Amazonas  MULTIPOLYGON (((-1482456 23...
4    Roraima  MULTIPOLYGON (((-686728.6 5...
5       Pará  MULTIPOLYGON (((883666.3 -1...
6      Amapá  MULTIPOLYGON (((323895.6 41...
7  Tocantins  MULTIPOLYGON (((626106.9 -5...
8   Maranhão  MULTIPOLYGON (((1111218 -26...
9      Piauí  MULTIPOLYGON (((1362917 -31...
10     Ceará  MULTIPOLYGON (((1508168 -31...
```

simple   column of simple   simple feature
feature   feautres (sfc)   geometry (sfg)

FIGURE 16.3: Structure of a sf object.

Now, if we explore our data in greater detail we will find that this standard can be implemented into different types of geometry in our dataset:

- A numerical vector for just one point (POINT)
- A numerical matrix (every row is a point) for a series of points (MULTIPOINT o LINESTRING)
- A list of matrices for a set of point sets (MULTIINESTRING, POLYGON)
- A list of matrices (MULTIPOLYGON) which becomes the most used when geographical data is represented as form and location of countries or other administrative units of these.
- A list of any of the elements previously mentioned (GEOMETRYCOLLECTION)

Using the st_geometry() function we can see the geometry included in our shapefile:

```
st_geometry(shp_brazil)
## Geometry set for 28 features
## geometry type:  MULTIPOLYGON
## dimension:      XY
## bbox:           xmin: -2200000 ymin: -3700000 xmax: 2400000 ymax: 590000
## projected CRS:  Polyconic
## First 5 geometries:
```

Once our shapefile is loaded, we can start generating maps. Note that the way to do this is based on `ggplot2`; we just need to select our dataset and add the `geom_sf()` geom[8] (see Figure 16.4).

---

[8]You might see the coordinates in the axis. This can be useful for exploratory data analysis, but you should remove them if you want to distribute your map. This can be done with the addition of ... + theme_void().

16.3 Spatial Data Management

```
ggplot(data = shp_brazil) +
  geom_sf()
```

FIGURE 16.4: Map of the states of Brazil.

Exercise 16A.

1. Download the shapefile of South America from ArcGIS[9] and load it into R using `read_sf()`. Select just the "CNTRY_NAME","ISO_3DIGIT" and "geometry" variables.
2. Filter the dataset by the 'CNTRY_NAME' variable to eliminate the "South Georgia & the South Sandwich Is." and "Falkland Is." observations.
3. Plot the shapefile using `ggplot()` and `geom_sf()`.

16.3 Spatial Data Management

We have already learned how to load shapefiles with `read_sf()`. Now that we have loaded the file, we will proceed to learn what type of modifications we can make for

---

[9]See https://www.arcgis.com/home/item.html?id=d3d2bae5413845b193d038e4912d3da9

generating new data from the information we already have. We will also show you how to join data from other datasets into our shapefile.

16.3.1 Modifications

There are two ways of making modifications into our georeferenced dataset. The first consists in applying the techniques we have already learned using the `tidyverse`, while the second consists in using functions incorporated into the `sf` package. In this package, all functions start with `st_`, so they can be easily grasped in the autocomplete tool of RStudio. These functions are primarily used for transforming and making geographical operations. In these sections we will combine both techniques for generating new variables and data associated to our dataset.

16.3.1.1 Filter and select by geographical units

One of the first functions we can use are the `dplyr` functions for selecting data inside our dataset. For example, we can occupy `filter()` for selecting some Brazilian states, such as São Paulo:

```
shp_brazil %>%
  filter(state == "São Paulo")
## # A tibble: 1 x 2
##   state                                                      geometry
##   <chr>                                            <MULTIPOLYGON [m]>
## 1 São Paulo (((719692 -2716734, 719500 -2717678, 719210 -2716889, 719362~
```

A common challenge we often run into when working with shapefiles from entire countries is the existence of insular regions or zones that are part of the administrative territory but that are geographically isolated, for example, Easter Island in Chile or Galapagos Islands in Ecuador. For various reasons, data for these regions is limited, and are often left out of analysis. Thus, we are interested in getting these regions out of our shapefile. In Brazil, the State District of Fernando de Noronha, an archipelago of 21 isles located in the Atlantic Ocean, is one of these type of cases (you can possibly identify it as a small dot in the upper right part of the previous map). We can easily eliminate this type of data from our shapefile, again, with the `filter()` function.

```
shp_brazil <- shp_brazil %>%
  filter(state != " State District of Fernando de Noronha (PE)")
```

This will be reflected in the following maps:

```
ggplot(data = shp_brazil)+
  geom_sf()
```

16.3 Spatial Data Management

FIGURE 16.5: Map of Brazilian states except the State District of Fernando de Noronha.

16.3.1.2 Generate new units with `st_union()`

Another interesting option is to generate new variables that group multiple geographical units. By doing this, we will be effectively generating new geographical units beyond the initial information contained in our shapefile. For example, in 1969 the *Brazilian Institute of Geography and Statistics* divided the country into five regions that group the 27 states of the country. Since this is a division with academic purposes, and not recognized in political-administrative terms, it is not possible to find shapefile type files that show these regions. Nonetheless, using `mutate()` and `case_when()`, we can easily generate this category ourselves:

```
shp_brazil <- shp_brazil %>%
  mutate(region = case_when(
    state %in% c("Goiás", "Mato Grosso", "Mato Grosso do Sul",
                "Distrito Federal") ~ "Center-West",
    state %in% c("Acre", "Amapá", "Amazonas", "Pará", "Rondônia", "Roraima",
                "Tocantins") ~ "North",
    state %in% c("Alagoas", "Bahia", "Ceará", "Maranhão", "Paraíba",
                "Distrito estadual de Fernando de Noronha (PE)",
                "Pernambuco", "Piauí", "Rio Grande do Norte",
                "Sergipe") ~ "Northeast",
    state %in% c("Espírito Santo", "Minas Gerais", "Rio de Janeiro",
                "São Paulo") ~ "Southeast",
    state %in% c("Paraná", "Rio Grande do Sul",
                "Santa Catarina") ~ "South")
)
```

Once generated, we can incorporate this variable into our map (see Figure 16.6)::

```
ggplot(data = shp_brazil)+
  geom_sf(aes(fill = region)) +
  labs(fill = "")
```

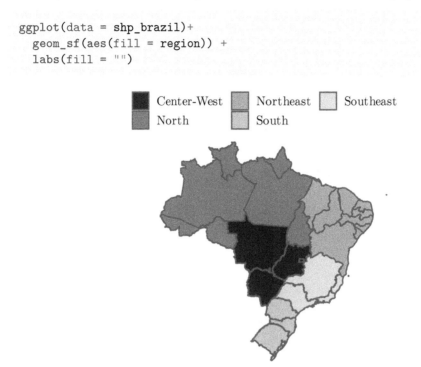

FIGURE 16.6: Map of Brazilian states grouped by region.

Better yet, we can use this category to generate a new geometry with `group_by()`, `summarize()` and `st_union()`:

```
shp_brazil_regions <-  shp_brazil %>%
  group_by(region) %>%
  summarize(geometry = st_union(geometry)) %>%
  ungroup()
```

We can graph this result with `ggplot()` (see Figure 16.7):

```
ggplot(shp_brazil_regions) +
  geom_sf(aes(fill = region)) +
  labs(fill = "")
```

Note that this does not generate completely "flat" objects, that is, we still can observe some lines inside of them, probably because the polygons of our shapefile do not overlap perfectly. This is a common hardship when doing these types of operations, and it can even occur when working with sophisticated shapefiles. Despite the pitfalls, these operations are beneficial when, for example, elaborating electoral districts consisting of various provinces or municipalities.

16.3 Spatial Data Management

FIGURE 16.7: Map of Brazilian regions.

16.3.1.3 Create new shapefiles with `st_write()`

We can save this new shapefile with the `st_write()` functions, in which we only have to select the object we want to save and the path where we want it to be saved. In this case, we will save the `shp_brazil_regions.shp` file in a folder with the same name. This will automatically generate not only the .shp, but also all the other files that constitute the shapefile:

```
dir.create("shp_brazil_regions/")
st_write(shp_brazil_regions, "shp_brazil_regions/shp_brazil_regions.shp")
```

> Warning: The `st_write()` command cannot overwrite existing files and trying to do so will automatically report an error. If you want to modify an already generated shapefile, you have to manually erase it from its folder before generating the new files.

16.3.2 Add data from other datasets with `left_join()`

First, we will learn how to add data from other datasets into our shapefile, since we will usually want to map *attributes* from our geographical locations to generate exploratory or statistical inference analysis. As an example of it, we will load Freire's dataset (we previously eliminated some data to simplify the dataset):

```
library(politicalds)
data("brazil_states")
```

```
head(brazil_states)
## # A tibble: 6 x 6
##   state_code state year homicide_rate  gini population_extreme_poverty
##        <dbl> <chr> <dbl>         <dbl> <dbl>                      <dbl>
## 1         12 Acre   1990          15.8 0.544                      48965
## 2         12 Acre   1991          25.1    NA                         NA
## 3         12 Acre   1992          24.7 0.560                      60910
## # ... with 3 more rows
```

Note that our dataset contains the `state` variable. This same variable is found in our shapefile:

```
head(brazil_states$state)
## [1] "Acre" "Acre" "Acre" "Acre" "Acre" "Acre"
head(shp_brazil$state)
## [1] "Rondônia" "Acre"     "Amazonas" "Roraima"  "Pará"     "Amapá"
```

We have previously made sure that the observations are coded in the same way both in the dataset and in the shapefile. By doing this, we can join both datasets with the `left_join()` command:

```
shp_brazil_data <- shp_brazil %>%
  left_join(brazil_states)
```

Using `head()` we can see the result of this operation:

```
head(shp_brazil_data)
## # A tibble: 6 x 8
##   state               geometry region state_code  year homicide_rate
##   <chr>     <MULTIPOLYGON [m]> <chr>       <dbl> <dbl>         <dbl>
## 1 Rond~ (((-977421 -892385, -975~ North      11  1990          51.3
## 2 Rond~ (((-977421 -892385, -975~ North      11  1991          43.3
## 3 Rond~ (((-977421 -892385, -975~ North      11  1992          34.6
## # ... with 3 more rows, and 2 more variables
```

Once Freire's information is incorporated into our shapefile, this data will be useful for mapping variables.

Exercise 16B.

1. Select the countries of the Southern Cone (Chile, Argentina and Uruguay) and graph them.

2. Generate a new shapefile with subregions of South America. We suggest the following:
   - Caribbean Coast, corresponding to Colombia, Venezuela, Suriname, Guiana and French Guiana.
   - Andean Region, corresponding to Bolivia, Ecuador, Perú and Colombia.
   - Oriental Region, corresponding to Brazil and Paraguay.
   - Southern Cone, corresponding to Chile, Argentina and Uruguay.
3. Download the extended dataset ('Country-Year: V-Dem Full+Others') from V-Dem[10] and select just the following variables: 'country_name', 'country_text_id', 'year','v2x_polyarchy','e_migdppc'. Filter them to consider just the period between 1996 and 2016 (the last 20 years for which data is available for all the variables).
4. Using `left_join()`, add to the original shapefile the loaded dataset from the previous exercise. Tip: use the `by.x="ISO_3DIGIT"` and `by.y="country_text_id""`) arguments. Check the dataset. You will notice that one country is missing. Which one is it?

16.4 Mapping in R

Maps have been historically the main technique for storing and communicating spatial data, and the objects and its attributes can be easily shaped in a way that the human eye can rapidly recognize patterns and anomalies in a well-designed map. (see Spatial Analysis Online)[11]. In this section, you will learn how to make different types of maps (both static and animated) from geographical data and its attributes, using the `ggplot2` format as a base.

16.4.1 Generating centroids

The first option is to generate new variables associated with our units using the `mutate()` command. A common action is to generate what are called "centroids", that is, points located in the center of our units. For generating centroids, we need to create the following variables associated to our geoms: `centroid`, `coords`, `coords_xy` and `coords_y`. We can do this with the `map()` and `map_dbl()` commands, from the `purrr` package, and the `st_centroid()` and `st_coordinates()` commands, from the `sf` package.

---
[10]See https://www.v-dem.net/en/data/data-version-9/
[11]See http://www.spatialanalysisonline.com/HTML/index.html

```
shp_brazil <- shp_brazil %>% mutate(centroid = map(geometry, st_centroid),
    coords = map(centroid, st_coordinates), coords_x = map_dbl(coords, 1),
    coords_y = map_dbl(coords, 2))

head(shp_brazil)
## # A tibble: 6 x 7
##   state                      geometry region centroid coords coords_x
##   <chr>             <MULTIPOLYGON [m]>  <chr>   <list>  <list>    <dbl>
## 1 Rond~ (((-977421 -892385, -975~  North  <XY [2]> <dbl[~  -9.68e5
## 2 Acre  (((-2179954 -836549, -20~  North  <XY [2]> <dbl[~  -1.81e6
## 3 Amaz~ (((-1482456 230568, -147~  North  <XY [2]> <dbl[~  -1.19e6
## # ... with 3 more rows, and 1 more variable
```

Once generated, we can graph these variables with `ggplot()` and the `ggrepel` package for generating text (see Figure 16.8):

```
library(ggrepel)


ggplot(data = shp_brazil) +
  geom_sf()+
  geom_text_repel(mapping = aes(coords_x, coords_y, label = state),
                  size = 4, min.segment.length = 0)+
  labs(x = "", y = "")
```

FIGURE 16.8: Map of Brazil with states' names.

16.4.2 Mapping variables

Since we have already added the data from Freire (2018) into our shapefile, we can use these variables to exemplify the differences among our geographical units. For example, we can show the Gini index for every state with the `fill =` argument. The data corresponds to the average of the years observed. The lighter the color, the more unequal a state is (see Figure 16.9).

```
ggplot(shp_brazil_data) +
  geom_sf(aes(fill = gini))+
  scale_fill_gradient(low = "white", high = "black") +
  theme(legend.key.width = unit(2, "cm")) + # make legend readable
  labs(fill = "Gini index")
```

FIGURE 16.9: Map of the states of Brazil according to the Gini index.

We can also combine this with `filter()` to select only some years from our dataset. For example, we can show the amount of people in extreme poverty in the year 2009 in the different states (see Figure 16.10):

```
ggplot(shp_brazil_data %>% filter(year == 2009)) +
  geom_sf(aes(fill = gini))+
  scale_fill_gradient(low = "white", high = "black") +
  theme(legend.key.width = unit(2, "cm")) + # make legend readable
  labs(fill = "Gini index")
```

FIGURE 16.10: Map of the states of Brazil according to the population in extreme poverty.

16.4.3 Mapping points

If, apart from the polygons, we have data that denote the locations of events that have occurred inside the space that covers our entire shapefile, it is also possible to graph all that information into a group. In this case, we will learn two ways to perform this task; one using `geom_sf()`, and the other, using a geom we have already seen previously in the book: `geom_point`.

For this example, we will use an adapted example from Freire et al. (2019) that corresponds to a study that geo-references the fatal victims of Augusto Pinochet's coup between 1973 and 1990. This package, under the name of `pinochet`, is uploaded to the CRAN depository. To start, let's install this package in conjunction with the `rnaturalearthhires` package, which provides us with the shape of all countries around the world from the Natural Earth Data's[12] web page:

```
install.packages("pinochet")
install.packages("rnaturalearthhires",
                 repos="http://packages.ropensci.org", type = "source")
```

> Tip: The `rnaturalearthhires` package can be of great use if you possess a dataset with variables that can be compared between countries, using the techniques you have already learned.

---

[12] See http://www.naturalearthdata.com/

16.4 Mapping in R

Next, we will filter the dataset to select Chile and its neighboring countries:

```
chile <- rnaturalearthhires::countries10 %>%
  st_as_sf() %>%
  filter(SOVEREIGNT %in% c("Chile", "Argentina", "Peru", "Bolivia"))
```

Following this, we load the `pinochet` dataset and select the variables to be used, which correspond to the type of location where the victim was attacked, as well as the latitude and longitude of the location. It also contains a description of the locations, which will be used later in an exercise.

```
pinochet <- pinochet::pinochet %>%
  select(last_name, first_name, place_1, longitude_1, latitude_1,
         location_1)
```

```
head(pinochet)
## # A tibble: 6 x 6
##    last_name    first_name     place_1 longitude_1 latitude_1 location_1
##    <chr>        <chr>          <chr>         <dbl>      <dbl> <chr>
## 1 Corredera R~ Mercedes del~  In Pub~       -70.7      -33.5 Calle Gran A~
## 2 Torres Torr~ Benito Herib~  Home          -70.7      -33.4 Santiago
## 3 Lira Morales Juan Manuel    In Pub~       -70.6      -33.5 La Legua sha~
## # ... with 3 more rows
```

Next, we use the `st_as_sf()` function to transform the longitude and latitude variables into geometries of the `sf` type. The crs argument represents the type of Coordinate Reference System (CRS) according to its code called EPSG. In this case, we will use the code "4326", which corresponds to the standard WGS84, the most common. On the other hand, we use the `remove = F` argument for the function not to eliminate the original latitude and longitude variables of the dataset, in such a way that they can be used to generate points with `geom_points`. We will also previously remove the NA values.

Tip: When replicating this exercise, take note that in the argument `coords` you always have to put first the longitude, and the latitude. Otherwise, your variables can have their coordinates reversed.

```
pinochet_sf <- pinochet %>%
  filter(place_1 != "NA" & !is.na(longitude_1) & !is.na(latitude_1)) %>%
  st_as_sf(coords = c("longitude_1", "latitude_1"), crs = 4326, remove = F)
```

```
head(pinochet_sf)
## # A tibble: 6 x 7
##   last_name first_name place_1 longitude_1 latitude_1 location_1
##   <chr>     <chr>      <chr>         <dbl>      <dbl> <chr>
## 1 Correder~ Mercedes ~ In Pub~       -70.7      -33.5 Calle Gra~
## 2 Torres T~ Benito He~ Home          -70.7      -33.4 Santiago
## 3 Lira Mor~ Juan Manu~ In Pub~       -70.6      -33.5 La Legua ~
## # ... with 3 more rows, and 1 more variable
class(pinochet_sf)
## [1] "sf"         "data.frame"
```

Note that with this action we have transformed the dataset into a shapefile. Next, we generate the map with `geom_sf()`, using it in two instances: first, for mapping the polygons of the countries, and second, for graphing the spatial points. We use the `size` attribute for the size of the point. Consider that we used the argument `coords_sf` to delimit the area that will be shown in the map. By doing this, we can also identify the victims attacked outside the national territory of Chile, without losing the focus on this country (Figure 16.11).

```
ggplot() +
  geom_sf(data = chile) +
  geom_sf(data = pinochet_sf, size = 1) +
  coord_sf(xlim = c(-75.6, -67), ylim = c(-55, -19)) +
  scale_x_continuous(breaks = c(-76, -67)) # also clean the x scale
```

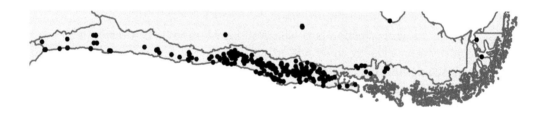

FIGURE 16.11: Map of the locations of Pinochet's dictatorship victims in Chile and its neighboring countries.

We can obtain a similar result with `geom_point()`. In this case, we added the `color =` argument to show the type of place where the victim was attacked (Figure 16.12).

```
ggplot() +
  geom_sf(data = chile) +
  geom_sf(data    = pinochet_sf,
          mapping = aes(shape = place_1),
          size    = 1) +
```

16.5 Inference from Spatial Data

```
coord_sf(xlim = c(-75.6, -67), ylim = c(-55, -19)) +
scale_x_continuous(breaks = c(-76, -67)) +
scale_shape_manual(values = 0:6) +
labs(shape = "")
```

FIGURE 16.12: Map of the locations of Pinochet's dictatorship victims in Chile and its neighbouring countries, shape-coded by type of place.

Exercise 16C.

1. Generate centroids for the countries (Tip: use `CNTRY_NAME`).
2. Generate a map using the `fill` argument of `ggplot()` to use a different color for each country.
3. Graph a map combining the attributes of the two previous exercises.
4. Create a map with the GDP per capita (`e_migdppc`) of every country in the year 2016. Which are the countries that don't have data for 2016?
5. Create a grid with the value of Electoral Democracy (`v2x_polyarchy`) of every country in the years 2013, 2014, 2015 and 2016.

16.5 Inference from Spatial Data

Beyond the exploratory and descriptive analysis that can be made with spatial data, these are also of great use for making inference over the relation of various phenomena. Inference based on spatial data starts by recognizing that spatial observations cannot be assumed as mutually independent, since observations that are close to one another tend to be similar. Thus, we need to pay close attention to the different association patterns that exist in the phenomena we are studying.

These spatial patterns (*spatial autocorrelation*), measure the influence of distance over a particular variable, and can be used as relevant information of the types of influence that have not yet been observed or considered (Bivand et al., 2013, p. 11).

16.5.1 Local indicator of spatial association (LISA)

In this section you will find the basic mechanisms for getting into spatial correlation, based on the Local Indicator of Spatial Association (LISA), introduced by Luc Anselin (1995). These allow indicating the existing similitude (correlation) between observations that are close to each other, that is, if they are grouped (*clustered*) spatially. For this, Anselin indicates that:

- The LISA for each observation gives an indicator of the degree of significant spatial clustering of similar values around that observation.
- The sum of the observations' LISAs is proportional to a global spatial correlation indicator.

Through the testing of statistical significance for these indicators we can identify locations where there is a significant degree of clustering (or spatial repulsion) (Brunsdon and Comber, 2015, p. 249). It is said that variables have a positive spatial correlation when similar values tend to be closer to each other than the different values (Lansley and Cheshire, 2016, p. 77).

16.5.2 Spatial Weights Matrix

The first step for doing this type of analysis is to determine the set of neighborhoods for each observation, that is, to identify the polygons that share boundaries between them. Then, we need to assign a weight for each neighboring relationship, which allows defining the force of this relationship based on proximity. In the weight matrices, neighbors are defined in a binary way [0, 1] in every row, indicating if a relationship exists or not.

For doing this type of analysis, first we need to load the `spdep` package, and save the coordinates of the units in our map:

```
library(spdep)


coords <- coordinates(as((shp_brazil), 'Spatial'))
```

16.5 Inference from Spatial Data

In this section, we will also work with the shapefile to which we have added data from Freire's (2018) dataset, `shp_brazil_data`, but we will only use the data from the last available year, 2009:

```
shp_brazil_data <- shp_brazil_data %>% filter(year == 2009)
```

There are three different criteria for calculating neighborhoods:

16.5.2.1 Rook criterion

The Rook criterion considers as neighbors any pair of cells that share an edge.

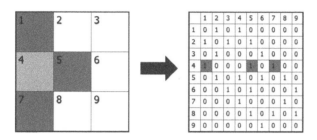

FIGURE 16.13: Rook neighborhood criterion.

To generate this criterion, we use the `poly2nb()` function from the `spdep` package. We need to ensure that the `queen =` argument is set to F.

```
rook_brazil <- poly2nb(as(shp_brazil_data, 'Spatial'), queen = FALSE)
```

We can graph this over our map with `ggplot()`. First, we need to pass our coordinates into a dataframe with this function from Maxwell B. Joseph[13].

```
nb_to_df <- function(nb, coords){
  x <- coords[, 1]
  y <- coords[, 2]
  n <- length(nb)

  cardnb <- card(nb)
  i <- rep(1:n, cardnb)
  j <- unlist(nb)
  return(data.frame(x = x[i], xend = x[j],
                    y = y[i], yend = y[j]))
}
```

---

[13] See https://mbjoseph.github.io/2015/06/15/nb_ggmap.html

We generate the dataframe:

```
rook_brazil_df <- nb_to_df(rook_brazil, coords)
```

Now, we can generate the graph with the geom_point() and geom_segment() geoms (Figure 16.14):

```
ggplot(shp_brazil_data) +
  geom_sf()+
  geom_point(data = rook_brazil_df,
             mapping = aes(x = x, y = y))+
  geom_segment(data = rook_brazil_df,
               mapping = aes(x = x, xend = xend, y = y, yend = yend))+
  labs(x = "",y = "")
```

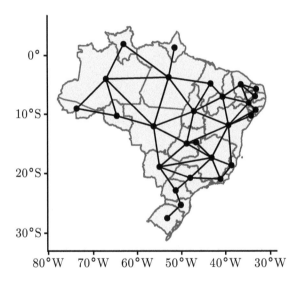

FIGURE 16.14: Map of neighbors, Rook criterion.

16.5.2.2 Queen criterion

The Queen criterion considers as neighbors any pair of cells that share an edge or a point

To generate this criterion, we also use the poly2nb() function, and then the dataframe:

```
queen_brazil <- poly2nb(as(shp_brazil_data, 'Spatial'), queen = T)

queen_brazil_df <- nb_to_df(queen_brazil, coords)
```

16.5 Inference from Spatial Data

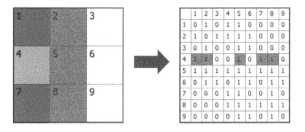

FIGURE 16.15: Queen neighborhood criterion.

Now we can generate the graph, which we do directly over our map of Brazil (Figure 16.16):

```
ggplot(shp_brazil_data) +
  geom_sf()+
  geom_point(data = queen_brazil_df,
             mapping = aes(x = x, y = y))+
  geom_segment(data = queen_brazil_df,
               mapping = aes(x = x, xend = xend, y = y, yend = yend))+
  labs(x = "", y = "")
```

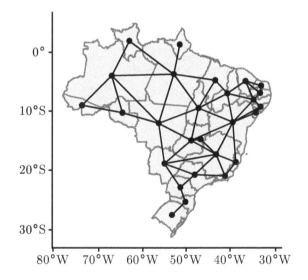

FIGURE 16.16: Map of neighbors, Queen criterion.

16.5.2.3 K-nearest criterion

In the K-nearest criterion neighborhoods are generated based on the distance between neighbors, where 'k' refers to the number of neighbors of a particular location, calculated as the distance between the center points of the polygons. It is usually applied when the areas have different sizes, so that every location contains the same number of neighbors, independently of how big the neighboring areas are (Source: https://geodacenter.github.io/glossary.html).

In this case, we will use the six closest neighbors. We can do this with the `knn2nb()` command. We also immediately generate the dataframe:

```
kn_brazil <- knn2nb(knearneigh(coords, k = 6))
kn_brazil_df <- nb_to_df(kn_brazil,coords)
```

Then, generate the map with `ggplot()`(Figure 16.17):

```
ggplot(shp_brazil_data) +
  geom_sf()+
  geom_point(data = kn_brazil_df, mapping = aes(x = x, y = y))+
  geom_segment(data = kn_brazil_df,
               mapping = aes(x = x, xend = xend, y = y, yend = yend))+
  labs(x = "", y = "")
```

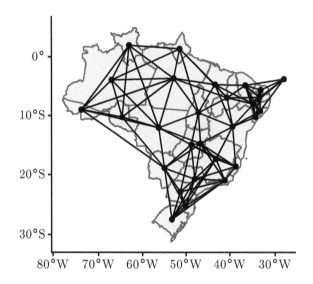

FIGURE 16.17: Map of neighbors, K-nearest criterion.

Note that, while Rook and Queen generate, for our case, similar results in terms of neighborhood in our map, the K-nearest model adds many more relationships.

16.5.3 Moran's I

The Moran's I is the most used statistic for identifying spatial correlation:

$$I = \frac{n}{\sum_{i=1}^{n}(yi - \bar{y})^2} \frac{\sum_{i=1}^{n}\sum_{j=1}^{n} wij(yi - \bar{y})(yj - \bar{y})}{\sum_{i=1}^{n}\sum_{j=1}^{n} wij}$$

16.5 Inference from Spatial Data

This formula, although it looks complex, is not more than an expanded version of the formula for computing the correlation coefficient, to which a spatial weight matrix is added[14]. Through it, we can test and visualize the presence of spatial autocorrelation. First, we will run a global test that creates a single measure of spatial correlation. This Moran's test creates a correlation between -1 and 1, in which:

- 1 determines a perfect positive spatial correlation (it indicates that our data is grouped in clusters).

- 0 means that our data is randomly distributed.

- -1 represents negative spatial correlation (dissimilar values that are close to each other).

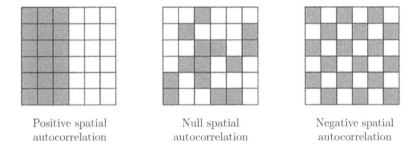

FIGURE 16.18: Illustration of spatial autocorrelation.

Using a Monte Carlo simulation, where the values are randomly assigned to the polygons for computing the Moran's I, the simulation is repeated many times, establishing a distribution of the expected values. After this, the observed Moran's I value is compared to the simulated distribution to see how likely it is that observed values can be considered random, allowing to determine if there is significant spatial autocorrelation (Rspatial).

To run the Moran's I test we use the `moran.test()` command. The variable for which we will diagnose spatial correlation is the Gini index, and we will use a Queen criterion. Before, we need to generate a `listw` (weigh matrix) type object, based on the Queen criterion.

```
queen_brazil_lw <- nb2listw(queen_brazil)
```

With this object we can run the Moran's I test, where we select the weight matrix we just created:

```
moran.test(shp_brazil_data$gini, listw = queen_brazil_lw)
##
##  Moran I test under randomisation
```

---

[14]See RSpatial[15]

```
## 
## data:  shp_brazil_data$gini
## weights: queen_brazil_lw
## 
## Moran I statistic standard deviate = 2, p-value = 0.01
## alternative hypothesis: greater
## sample estimates:
## Moran I statistic      Expectation         Variance
##             0.284           -0.038            0.020
```

The result of Moran's I test shows that there is a slight positive correlation relationship, versus an expectation of a slight negative relationship. The tests is also statistically significant since it has a p-value over the significance threshold of 0.05. Therefore, we can indicate that the Gini score presents certain levels of spatial autocorrelation when analyzing the data at the state level in Brazil. If your test results aren't statistically significant, you can try using a smaller unit of analysis like microregions within states or provinces and see if they show spatial correlation for your variable of interest.

We also create a scatter plot for visualizing the sign and strenght of the spatial correlation. For generating a scatter plot we use the `moran.plot()` command (Figure 16.19):

```
moran.plot(shp_brazil_data$gini, listw = queen_brazil_lw,
           xlab = "Gini",
           ylab = "Gini (spatially lagged)")
```

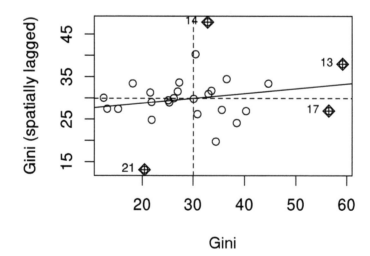

FIGURE 16.19: Spatial correlation of Gini index in Brazil at state level.

16.5 Inference from Spatial Data

In this map, the solid line of the graph indicates the Moran's I value, that is, the global measure of spatial autocorrelation in our data. Just as we saw in the previous test, it is slightly positive. The horizontal axis of the graph shows the Gini score data at state level in Brazil, and the vertical axis shows the same data, but spatially lagged. The four quadrants of the graph describe the value of the observation in relationship with its neighbors: high-high, low-low (positive spatial autocorrelation), low-high or high-low (negative spatial autocorrelation), and the graph also points out the values considered outliers in this relationship, like the Federal District or Acre[16].

While the Moran's I test allows us to identify the existence of clustering at a global lever, it does not allow identifying if local significant clusters exist in the variable we are analyzing (Lansley and Cheshire, 2016, p. 78). For this reason, we want to run a Moran's I test at a local level, where local indicators of spatial association are calculated for each unit of our data, and the relationship is tested to prove its statistical significance. This gives us data on the geographic pattern of the relationship of spatial data, and whether there are local deviations from the global patterns of spatial autocorrelation.

After creating a new `listw` object of neightbors, this type with the `style = "B"` argument to indicate a binary classification, we use `localmoran()` to execute the Moran's I test. We add its results (estimate and p-value) to our existing `sf` object with the familiar `mutate()`:

```
queen_brazil_b_lw <- nb2listw(queen_brazil, style = "B")

shp_brazil_data <- shp_brazil_data %>%
  mutate(lmoran = localmoran(x = gini, listw = queen_brazil_b_lw)[, 1],
         lmoran_pval = localmoran(x = gini, listw = queen_brazil_b_lw)[, 5]
  )
```

Next, we map the results with `ggplot2` (Figure 16.20):

```
ggplot(shp_brazil_data) +
  geom_sf(aes(fill = lmoran))+
  labs(fill = "Local Moran Statistic") +
  scale_fill_gradient(low = "white", high = "black")
```

This map enables us to observe the variation of autocorrelation throughout the space, but it does not allow to identify if the patterns of autocorrelation are clusters with high or low values. This would allow analyzing the type of spatial autocorrelation that exists and its significance. For this reason we need to create a map of LISA clusters, which will create a label based on the types of relation that is shared with its neighbors (high-high, low-high, insignificant, etc) in concordance with the values of the variable we are analyzing (Gini).

---

[16]Source: https://geodacenter.github.io/glossary.html

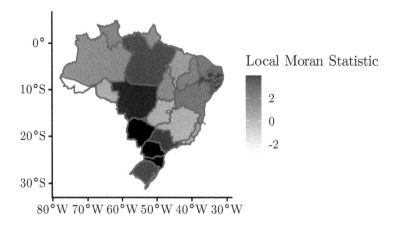

FIGURE 16.20: Moran's I at local level for different Gini values.

In order to do this, we need to make a series of transformations we detail below:

```
shp_brazil_data <- shp_brazil_data %>%
  mutate(
    # Standardize the Gini and Local Moran's to their means:
    st_gini = gini - mean(gini),
    st_lmoran = lmoran - mean(lmoran),
    # Create the new categorical variable:
    quadrant = case_when(
      lmoran_pval > 0.05 ~ "Insignificant",
      st_gini > 0 & st_lmoran > 0 ~ "High-High",
      st_gini < 0 & st_lmoran < 0 ~ "Low-Low",
      st_gini < 0 & st_lmoran > 0 ~ "Low-High",
      st_gini > 0 & st_lmoran < 0 ~ "High-Low"
    )
  )
```

Now we can generate the graph with ggplot(). Note that we use scale_fill_manual() for changing the colors for them to reflect the intensity of the autocorrelation (Figure 16.21):

```
ggplot(shp_brazil_data, aes(fill = quadrant)) +
  geom_sf()+
  labs(fill = "Quadrant")+
  scale_fill_manual(values = c("white", "lightgray", "black"))
```

This map provides us with greater information about the geographic patterns of spatial autocorrelation. This map shows us the existence of clusters; thus, it shows clustered regions rather than individual locations. It is important to note that these maps are no significant, but they allow us finding locations or relations that can be interesting for further analysis.

16.5 Inference from Spatial Data

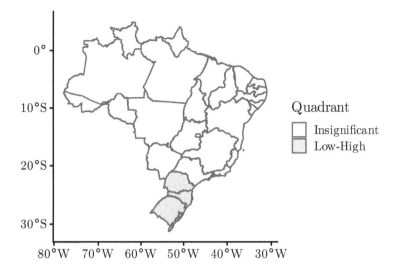

FIGURE 16.21: Geographic patterns of clustering for different values of Gini.

In this case, we can interpret that in most parts of Brazil, there is no spatial correlation for the Gini values at a state level, that is, its value is not influenced by the proximity of other similar values. Nonetheless, in some states located in the northeast of Brazil, we can observe the existence of a cluster where high Gini values are concentrated, which have neighbors who also have high values in the same variable (*hot spots*). This indicates that these regions contribute significantly to a global positive spatial autocorrelation. On the other hand, in the states that are colored in light gray we find clusters that concentrate low values of Gini where its neighbors have high values of the same variables and, therefore, contribute significantly to a negative global spatial autocorrelation (since this happens when dissimilar values are next to each other).

Exercise 16D.

1. Generate and object with the coordinates of the shapefile using the `coordinates()` function.

2. Filter the dataset for only using data from 2018.

3. Generate a weight matrix following the Queen criterion using `poly2nb`. Generate the dataframe using `nb_to_df()` and graph it using `geom_point()` and `geom_segment()`.

4. Perform the Moran's I test with the `moran.test()` command using the dataset and the variable of Electoral Democracy. Graph it using `moran.plot()`.

5. Perform the local Moran's I test with the `localmoran()` command (use the parameters of the previous exercise), bind it to the dataset with `cbind()` and graph the result with `ggplot2`.

Final comments: References for further analysis of spatial data

As we stated at the beginning, the objective of this chapter was to make an introduction for spatial data analysis in R, focusing on the creation of maps as a valuable tool for descriptive analysis of political data. However, spatial data analysis does not begin nor end with what was explained in this chapter. If you are interested in learning more about certain topics like statistical learning and geographic dataset modeling, the interaction with other softwares of geographical data analysis like QGIS, and the application of these in areas such as transportation or ecology, we recommend you take a look at Geocomputation in R (Lovelace et al., 2019), a book where you will find a guide for introduction to these topics. Also, we recommend you look into[17] the `highcharter` R wrapper for the *Highcharts* JavaScript library if you want to find new ways of visualizing spatial data and creating more complex maps.

---

[17] See http://jkunst.com/highcharter/index.html

Part IV

Bibliography and Index

Bibliography

Abeyasekera, S. (2005). Multivariate methods for index construction. In *Household Sample Surveys in Developing and Transition Countries*, pages 367–387. United Nations, New York, NY.

Adler, D., Gläser, C., Nenadic, O., Oehlschlägel, J., and Zucchini, W. (2020). *ff: Memory-Efficient Storage of Large Data on Disk and Fast Access Functions*. R package version 2.2-14.2.

Agresti, A. (2007). *An Introduction to Categorical Data Analysis*. Wiley-Interscience, Hoboken, NJ, 2nd edition.

Alemán, E., Calvo, E., Jones, M. P., and Kaplan, N. (2009). Comparing Cosponsorship and Roll-Call Ideal Points. *Legislative Studies Quarterly*, 34(1):87–116.

Allison, P. D. (2014). *Event History and Survival Analysis*. SAGE, Thousand Oaks, CA, 2nd edition.

Altman, D. (2019). *Citizenship and Contemporary Direct Democracy*. Cambridge University Press, Cambridge.

Amorim Neto, O. (2012). *De Dutra a Lula: A Condução e Os Determinantes Da Política Externa Brasileira*. Elsevier / Konrad Adenauer Stiftung / Campus, Rio de Janeiro.

Angrist, J. D. and Pischke, J.-S. (2008). *Mostly Harmless Econometrics: An Empiricist's Companion*. Princeton University Press, Princeton, NJ.

Angrist, J. D. and Pischke, J.-S. (2015). *Mastering 'Metrics: The Path from Cause to Effect*. Princeton University Press, Princeton, NJ.

Anselin, L. (1995). Local Indicators of Spatial Association-LISA. *Geographical Analysis*, 27(2):93–115.

Arel-Bundock, V. (2020). *countrycode: Convert Country Names and Country Codes*. R package version 1.2.0.

Auguie, B. (2017). *gridExtra: Miscellaneous Functions for "Grid" Graphics*. R package version 2.3.

Bailey, D. and Katz, J. N. (2011). Implementing Panel-Corrected Standard Errors in R: The pcse Package. *Journal of Statistical Software*, 42(1):1–11.

Barberá, P. (2015). Birds of the Same Feather Tweet Together: Bayesian Ideal Point Estimation Using Twitter Data. *Political Analysis*, 23(1):76–91.

Barrett, M. (2020). *ggdag: Analyze and Create Elegant Directed Acyclic Graphs*. R package version 0.2.2.

Beck, N. (2001). Time-Series-Cross-Section Data : What Have We Learned in the Past Few Years? *Annual Review of Political Science*, 4(1):271–293.

Beck, N. (2008). Time-series Cross-section Methods. In Box-Steffensmeier, J. M., Brady, H. E., and Collier, D., editors, *The Oxford Handbook of Political Methodology*, pages 475–493. Oxford University Press, Oxford.

Beck, N. and Katz, J. N. (1995). What To Do (and Not to Do) with Time-Series Cross-Section Data. *American Political Science Review*, 89(3):634–647.

Beck, N. and Katz, J. N. (2011). Modeling Dynamics in Time-Series–Cross-Section Political Economy Data. *Annual Review of Political Science*, 14(1):331–352.

Benoit, K., Muhr, D., and Watanabe, K. (2020a). *stopwords: Multilingual Stopword Lists*. R package version 2.0.

Benoit, K., Watanabe, K., Wang, H., Müller, S., Perry, P. O., Lauderdale, B., and Lowe, W. (2020b). *quanteda.textmodels: Scaling Models and Classifiers for Textual Data*. R package version 0.9.1.

Benoit, K., Watanabe, K., Wang, H., Nulty, P., Obeng, A., Müller, S., Matsuo, A., Lua, J. W., Kuha, J., and Lowe, W. (2020c). *quanteda: Quantitative Analysis of Textual Data*. R package version 2.0.1.

Berry, W. D., DeMeritt, J. H. R., and Esarey, J. (2010). Testing for Interaction in Binary Logit and Probit Models: Is a Product Term Essential? *American Journal of Political Science*, 54(1):248–266.

Bivand, R. (2019). spdep: Spatial Dependence: Weighting Schemes, *Statistics*. R package version 1.1-3.

Bivand, R. S., Pebesma, E., and Gómez-Rubio, V. (2013). *Applied Spatial Data Analysis with R*. Springer, New York, NY, 2nd edition.

Blei, D. M., Ng, A. Y., and Jordan, M. I. (2003). Latent Dirichlet Allocation. *Journal of Machine Learning Research*, 3(Jan):993–1022.

Box-Steffensmeier, J. M. and Jones, B. S. (2004). *Event History Modeling: A Guide for Social Scientists*. Cambridge University Press, Cambridge.

Brambor, T., Clark, W. R., and Golder, M. (2006). Understanding Interaction Models: Improving Empirical Analyses. *Political Analysis*, 14(1):63–82.

Broström, G. (2012). *Event History Analysis with R*. CRC Press, Boca Raton, FL.

Brunsdon, C. and Comber, L. (2015). *An Introduction to R for Spatial Analysis and Mapping*. SAGE, Thousand Oaks, CA.

Bryant, F. B. (2000). Assessing the Validity of Measurement. In Grimm, L. G. and Yarnold, P. R., editors, *Reading and Understanding MORE Multivariate Statistics*, pages 99–146. American Psychological Association, Washington, DC.

Calvo, E. (2015). *Anatomía Política de Twitter En Argentina: Tuiteando #Nisman*. Capital Intelectual, Buenos Aires, Argentina.

Ceron, A. (2017). Intra-party politics in 140 characters. *Party Politics*, 23(1):7–17.

Chan, S. (2005). Is There a Power Transition between the U.S. and China? The Different Faces of National Power. *Asian Survey*, 45(5):687–701.

Clark, T. S. and Linzer, D. A. (2015). Should I Use Fixed or Random Effects? *Political Science Research and Methods*, 3(2):399–408.

Collier, D., Laporte, J., and Seawright, J. (2008). Typologies: Forming concepts and creating categorical variables. In Box-Steffensmeier, J. M., Brady, H. E., and Collier, D., editors, *The Oxford Handbook of Political Methodology*, pages 152–173. Oxford University Press, Oxford.

Croissant, Y. and Millo, G. (2018). *Panel Data Econometrics with R*. John Wiley & Sons, Hoboken, NJ.

Croissant, Y., Millo, G., and Tappe, K. (2020). *plm: Linear Models for Panel Data*. R package version 2.2-3.

Cruz, A. (2020). *inexact: fuzzy join supervisor*. R package version 0.0.3.

CSISS (2004). *Spatial Social Science–for Research, Teaching, Application, and Policy*.

Cunningham, S. (2018). Causal Inference: The Mixtape. v. 1.7.

Dowle, M. and Srinivasan, A. (2019). *data.table: Extension of 'data.frame'*. R package version 1.12.8.

Dunning, T. (2012). *Natural Experiments in the Social Sciences: A Design-Based Approach*. Cambridge University Press, Cambridge.

Elwert, F. (2013). Graphical causal models. In Morgan, S. L., editor, *Handbook of Causal Analysis for Social Research*, chapter 13, pages 245–273. Springer, New York.

Fowler, J. H. (2006). Connecting the Congress: A Study of Cosponsorship Networks. *Political Analysis*, 14(4):456–487.

Fox, J., Weisberg, S., and Price, B. (2020). *car: Companion to Applied Regression*. R package version 3.0-8.

Freire, D. (2018). Evaluating the Effect of Homicide Prevention Strategies in São Paulo, Brazil: A Synthetic Control Approach. *Latin American Research Review*, 53(2):231.

Freire, D., Skarbek, D., Meadowcroft, J., and Guerrero, E. (2019). Deaths and Disappearances in the Pinochet Regime: A New Dataset. Preprint, SocArXiv.

Fruchterman, T. M. J. and Reingold, E. M. (1991). Graph drawing by force-directed placement. *Software: Practice and Experience*, 21(11):1129–1164.

Gerber, A. S. and Green, D. P. (2012). *Field Experiments: Design, Analysis, and Interpretation*. W. W. Norton, New York.

Gerring, J. (2006). *Case Study Research: Principles and Practices*. Cambridge University Press, Cambridge.

Gimpel, J. and Schuknecht, J. (2003). Political participation and the accessibility of the ballot box. *Political Geography*, 22(5):471–488.

Glasgow, G. and Alvarez, R. M. (2008). Discrete Choice Methods. In Box-Steffensmeier, J. M., Brady, H. E., and Collier, D., editors, *The Oxford Handbook of Political Methodology*, pages 513–529. Oxford University Press, Oxford.

Goertz, G. (2006). *Social Science Concepts: A User's Guide*. Princeton University Press, Princeton, NJ.

Golub, J. (2008). Discrete Choice Methods. In Box-Steffensmeier, J. M., Brady, H. E., and Collier, D., editors, *The Oxford Handbook of Political Methodology*, pages 530–546. Oxford University Press, Oxford.

Graham, B. A. T. and Tucker, J. R. (2019). The international political economy data resource. *The Review of International Organizations*, 14(1):149–161.

Greenhill, B., Ward, M. D., and Sacks, A. (2011). The Separation Plot: A New Visual Method for Evaluating the Fit of Binary Models. *American Journal of Political Science*, 55(4):991–1002.

Greenhill, B. D., Ward, M. D., and Sacks, A. (2020). *separationplot: Separation Plots*. R package version 1.3.

Hansen, D., Shneiderman, B., Smith, M. A., and Himelboim, I. (2019). *Analyzing Social Media Networks with NodeXL: Insights From a Connected World*. Morgan Kaufmann Publishers, Burlington, MA, 2nd edition.

Henningsen, A. and Henningsen, G. (2019). Analysis of Panel Data Using R. In Tsionas, M., editor, *Panel Data Econometrics: Theory*, pages 345–396. Academic Press, London.

Henshaw, A. L. and Meinke, S. R. (2018). Data Analysis and Data Visualization as Active Learning in Political Science. *Journal of Political Science Education*, 14(4):423–439.

Hernán, M. A. and Robbins, J. M. (2020). *Causal Inference: What If*. CRC Press, Boca Raton, Florida.

Hester, J. (2020). *glue: Interpreted String Literals*. R package version 1.4.1.

Hester, J., Csárdi, G., Wickham, H., Chang, W., Morgan, M., and Tenenbaum, D. (2020). *remotes: R Package Installation from Remote Repositories, Including 'GitHub'*. R package version 2.1.1.

Ho, D., Imai, K., King, G., and Stuart, E. (2018). *MatchIt: Nonparametric Preprocessing for Parametric Causal Inference*. R package version 3.0.2.

Honaker, J. and King, G. (2010). What to Do about Missing Values in Time-Series Cross-Section Data. *American Journal of Political Science*, 54(2):561–581.

Hothorn, T., Zeileis, A., Farebrother, R. W., and Cummins, C. (2019). *lmtest: Testing Linear Regression Models*. R package version 0.9-37.

Huber, E., Nielsen, F., Pribble, J., and Stephens, J. D. (2006). Politics and Inequality in Latin America and the Caribbean. *American Sociological Review*, 71(6):943–963.

Husson, F., Josse, J., Le, S., and Mazet, J. (2020). *FactoMineR: Multivariate Exploratory Data Analysis and Data Mining*. R package version 2.3.

Im, K. S., Pesaran, M., and Shin, Y. (2003). Testing for unit roots in heterogeneous panels. *Journal of Econometrics*, 115(1):53–74.

Ince, J., Rojas, F., and Davis, C. A. (2017). The social media response to Black Lives Matter: How Twitter users interact with Black Lives Matter through hashtag use. *Ethnic and Racial Studies*, 40(11):1814–1830.

Jackman, S. (2008). Measurement. In Box-Steffensmeier, J. M., Brady, H. E., and Collier, D., editors, *The Oxford Handbook of Political Methodology*, pages 119–151. Oxford University Press, Oxford.

Jackman, S., with contributions from Alex Tahk, Zeileis, A., Maimone, C., Fearon, J., and Meers, Z. (2020). *pscl: Political Science Computational Laboratory*. R package version 1.5.5.

Johannesson, M. (2020). *tidystm: Extract Effect from estimateEffect in the stm Package*. R package version 0.0.0.9000.

Kassambara, A. (2019). *ggcorrplot: Visualization of a Correlation Matrix using 'ggplot2'*. R package version 0.1.3.

Kassambara, A., Kosinski, M., and Biecek, P. (2020). *survminer: Drawing Survival Curves using 'ggplot2'*. R package version 0.4.7.

Kassambara, A. and Mundt, F. (2020). *factoextra: Extract and Visualize the Results of Multivariate Data Analyses*. R package version 1.0.7.

Kastellec, J. P. and Leoni, E. L. (2007). Using Graphs Instead of Tables in Political Science. *Perspectives on Politics*, 5(4).

Kearney, M. W. (2020). *rtweet: Collecting Twitter Data*. R package version 0.7.0.

King, G. (1997). *A Solution to the Ecological Inference Problem: Reconstructing Individual Behavior from Aggregate Data*. Princeton University Press, Princeton, NJ.

King, G. and Nielsen, R. (2019). Why propensity scores should not be used for matching. *Political Analysis*, 27(4):435–454.

Lall, R. (2016). How Multiple Imputation Makes a Difference. *Political Analysis*, 24(4):414–433.

Lansley, G. and Cheshire, J. (2016). *An Introduction to Spatial Data Analysis and Visualisation in R*. Consumer Data Research Centre.

Le Pennec, E. and Slowikowski, K. (2019). *ggwordcloud: A Word Cloud Geom for 'ggplot2'*. R package version 0.5.0.

Leeper, T. J. (2018). *margins: Marginal Effects for Model Objects*. R package version 0.3.23.

Leeper, T. J. (2019). *prediction: Tidy, Type-Safe 'prediction()' Methods*. R package version 0.3.14.

Leifeld, P. (2020). *texreg: Conversion of R Regression Output to LaTeX or HTML Tables*. R package version 1.37.1.

Levin, A., Lin, C.-F., and James Chu, C.-S. (2002). Unit root tests in panel data: Asymptotic and finite-sample properties. *Journal of Econometrics*, 108(1):1–24.

Lewis-Beck, C. and Lewis-Beck, M. (2016). *Applied Regression: An Introduction*. SAGE, Thousand Oaks, CA.

Lieberman, E. S. (2005). Nested Analysis as a Mixed-Method Strategy for Comparative Research. *American Political Science Review*, 99(3):435–452.

Long, J. A. (2020). *jtools: Analysis and Presentation of Social Scientific Data*. R package version 2.0.5.

Long, J. S. (1997). *Regression Models for Categorical and Limited Dependent Variables*. SAGE, Thousand Oaks, CA.

Lovelace, R., Nowosad, J., and Münchow, J. (2019). *Geocomputation with R*. CRC Press, Boca Raton, FL.

Mainwaring, S. and Pérez-Liñán, A. (2013). *Democracies and Dictatorships in Latin America: Emergence, Survival, and Fall*. Cambridge University Press, Cambridge.

Miller, J. E. (2013). *The Chicago Guide to Writing about Multivariate Analysis*. University of Chicago Press, Chicago, IL, 2nd edition.

Monogan, J. E. (2015). *Political Analysis Using R*. Springer, New York, NY.

Morgan, S. L. and Winship, C. (2007). *Counterfactuals and Causal Inference: Methods and Principles for Social Research*. Cambridge University Press, Cambridge.

Neto, O. A. and Malamud, A. (2015). What Determines Foreign Policy in Latin America? Systemic versus Domestic Factors in Argentina, Brazil, and Mexico, 1946–2008. *Latin American Politics and Society*, 57(4):1–27.

Newman, M. (2018). Networks. Oxford University Press, Oxford, 2nd edition.

Newman, M. E. J. (2001). Scientific collaboration networks. II. Shortest paths, weighted networks, and centrality. *Physical Review E*, 64(1):016132.

Opsahl, T., Agneessens, F., and Skvoretz, J. (2010). Node centrality in weighted networks: Generalizing degree and shortest paths. *Social Networks*, 32(3):245–251.

Patty, J. W. and Penn, E. M. (2016). Network Theory and Political Science. In Victor, J. N., Montgomery, A. H., and Lubell, M., editors, *The Oxford Handbook of Political Networks*. Oxford University Press, Oxford.

Pearl, J., Glymour, M., and Jewell, N. P. (2016). *Causal Inference in Statistics: A Primer*. Wiley, Hoboken, New Jersey.

Pearl, J. and Mackenzie, D. (2018). *The Book of Why: The New Science of Cause and Effect*. Basic Books, New York.

Pebesma, E. (2018). Simple Features for R: Standardized Support for Spatial Vector Data. *The R Journal*, 10(1):439–446.

Pebesma, E. (2020). *sf: Simple Features for R*. https://r-spatial.github.io/sf/, https://github.com/r-spatial/sf/.

Pedersen, T. L. (2020a). *ggraph: An Implementation of Grammar of Graphics for Graphs and Networks*. R package version 2.0.3.

Pedersen, T. L. (2020b). tidygraph: A Tidy API for Graph Manipulation. R package version 1.2.0.

Pfeffer, J. (2017). Visualization of Political Networks. In Victor, J. N., Montgomery, A. H., and Lubell, M., editors, *The Oxford Handbook of Political Networks*. Oxford University Press, Oxford.

Proksch, S.-O. and Slapin, J. B. (2010). Position Taking in European Parliament Speeches. *British Journal of Political Science*, 40(3):587–611.

Rainey, C. (2016). Compression and Conditional Effects: A Product Term Is Essential When Using Logistic Regression to Test for Interaction. *Political Science Research and Methods*, 4(3):621–639.

Reyes-Housholder, C. (2019). A Theory of Gender's Role on Presidential Approval Ratings in Corrupt Times. *Political Research Quarterly*, OnlineFirst.

Rinker, T. (2017). *qdapRegex: Regular Expression Removal, Extraction, and Replacement* Tools. R package version 0.7.2.

Roberts, M., Stewart, B., and Tingley, D. (2019a). *stm: Estimation of the Structural Topic Model*. R package version 1.3.5.

Roberts, M. E., Stewart, B. M., and Tingley, D. (2019b). Stm: An R Package for Structural Topic Models. *Journal of Statistical Software*, 91(1):1–40.

Roberts, M. E., Stewart, B. M., Tingley, D., Lucas, C., Leder-Luis, J., Gadarian, S. K., Albertson, B., and Rand, D. G. (2014). Structural Topic Models for Open-Ended Survey Responses. *American Journal of Political Science*, 58(4):1064–1082.

Robinson, D. (2017). *unvotes: United Nations General Assembly Voting Data*. R package version 0.2.0.

Robinson, D. and Hayes, A. (2020). *broom: Convert Statistical Analysis Objects into Tidy Tibbles*. R package version 0.5.6.

Robinson, D. and Silge, J. (2020). *tidytext: Text Mining using 'dplyr', 'ggplot2', and Other Tidy Tools*. R package version 0.2.4.900.

Robitzsch, A., Grund, S., and Henke, T. (2020). *miceadds: Some Additional Multiple Imputation Functions, Especially for 'mice'*. R package version 3.9-14.

Rodrigues, P., Urdinez, F., and de Oliveira, A. (2019). Measuring International Engagement: Systemic and Domestic Factors in Brazilian Foreign Policy from 1998 to 2014. *Foreign Policy Analysis*, 15(3):370–391.

Rohrer, J. M. (2018). Thinking clearly about correlations and causation: Graphical causal models for observational data. *Advances in Methods and Practices in Psychological Science*, 1(1):27–42.

Rudis, B., Bolker, B., and Schulz, J. (2017). *ggalt: Extra Coordinate Systems, 'Geoms', Statistical Transformations, Scales and Fonts for 'ggplot2'*. R package version 0.4.0.

Sachs, M. C. (2018). *plotROC: Generate Useful ROC Curve Charts for Print and Interactive Use*. R package version 2.2.1.

Salganik, M. J. (2017). *Bit by Bit: Social Research in the Digital Age*. Princeton University Press, Princeton, NJ.

Schloerke, B., Cook, D., Larmarange, J., Briatte, F., Marbach, M., Thoen, E., Elberg, A., and Crowley, J. (2020). *GGally: Extension to 'ggplot2'*. R package version 2.0.0.

Scott, J. (2013). *Social Network Analysis*. SAGE, Thousand Oaks, CA, 3rd edition.

Seawright, J. (2016). *Multi-Method Social Science: Combining Qualitative and Quantitative Tools*. Cambridge University Press, Cambridge.

Shalizi, C. R. (2019). *Advanced Data Analysis from an Elementary Point of View*.

Shpitser, I. and Pearl, J. (2008). Complete identification methods for the causal hierarchy. *Journal of Machine Learning Research*, 9:1941–1979.

Shugart, M. S. and Carey, J. M. (1992). *Presidents and Assemblies: Constitutional Design and Electoral Dynamics*. Cambridge University Press, Cambridge.

Signorell, A. (2020). *DescTools: Tools for Descriptive Statistics*. R package version 0.99.36.

Silge, J. and Robinson, D. (2017). *Text Mining with R: A Tidy Approach*. O'Reilly, Sebastopol, CA.

Singer, J. D., Bremer, S., and Stuckey, J. (1972). Capability Distribution, Uncertainty, and Major Power War, 1820-1965. In Russett, B., editor, *Peace, War, and Numbers*, pages 19–48. SAGE, Thousand Oaks, CA.

Skocpol, T. (1979). *States and Social Revolutions: A Comparative Analysis of France, Russia, and China*. Cambridge University Press, Cambridge.

Slapin, J. B. and Proksch, S.-O. (2008). A Scaling Model for Estimating Time-Series Party Positions from Texts. *American Journal of Political Science*, 52(3):705–722.

Slowikowski, K. (2020). *ggrepel: Automatically Position Non-Overlapping Text Labels with 'ggplot2'*. R package version 0.8.2.

South, A. (2020). *rnaturalearthhires: High Resolution World Vector Map Data from Natural Earth used in rnaturalearth*. https://docs.ropensci.org/rnaturalearthhires, https://github.com/ropensci/rnaturalearthhires.

Spinu, V., Grolemund, G., and Wickham, H. (2020). *lubridate: Make Dealing with Dates a Little Easier*. R package version 1.7.8.

Steinert-Threlkeld, Z. C. (2018). *Twitter as Data*. Cambridge University Press, Cambridge.

Stock, J. H. and Watson, M. W. (2019). *Introduction to Econometrics*. Pearson, New York, NY, 4th edition.

Textor, J. and van der Zander, B. (2016). *dagitty: Graphical Analysis of Structural Causal Models*. R package version 0.2-2.

Therneau, T. M. (2020). *survival: Survival Analysis*. R package version 3.1-12.

Tierney, N., Cook, D., McBain, M., and Fay, C. (2020). *naniar: Data Structures, Summaries, and Visualisations for Missing Data*. R package version 0.5.1.

Trott, V. (2018). Connected feminists: Foregrounding the interpersonal in connective action. *Australian Journal of Political Science*, 53(1):116–129.

Tufte, E. R. (2006). *Beautiful Evidence*. Graphics Press, Cheshire, Conn.

Urdinez, F. and Cruz, A. (2020). *politicalds: Package of the book 'Political Data Science: A Practical Guide'*. R package version 1.0.

van Buuren, S. and Groothuis-Oudshoorn, K. (2020). *mice: Multivariate Imputation by Chained Equations*. R package version 3.9.0.

van der Loo, M. (2019). *stringdist: Approximate String Matching and String Distance Functions*. R package version 0.9.5.5.

Waring, E., Quinn, M., McNamara, A., Arino de la Rubia, E., Zhu, H., and Ellis, S. (2020). *skimr: Compact and Flexible Summaries of Data*. R package version 2.1.1.

Welbers, K., Van Atteveldt, W., and Benoit, K. (2017). Text Analysis in R. *Communication Methods and Measures*, 11(4):245–265.

Wickham, H. (2014). Tidy Data. *Journal of Statistical Software*, 59(1):1–23.

Wickham, H. (2019a). *rvest: Easily Harvest (Scrape) Web Pages*. R package version 0.3.5.

Wickham, H. (2019b). *tidyverse: Easily Install and Load the 'Tidyverse'*. R package version 1.3.0.

Wickham, H. and Bryan, J. (2019). *readxl: Read Excel Files*. R package version 1.3.1. Wickham, H. and Grolemund, G. (2016). *R for Data Science: Import, Tidy, Transform, Visualize, and Model Data*. O'Reilly, Sebastopol, CA.

Wickham, H. and Miller, E. (2020). *haven: Import and Export 'SPSS', 'Stata' and 'SAS' Files*. R package version 2.3.1.

Wooldridge, J. M. (2016). *Introductory Econometrics: A Modern Approach*. Cengage Learning, Boston, MA, 6th edition.

Xuetong, Y. (2006). The Rise of China and its Power Status. *The Chinese Journal of International Politics*, 1(1):5–33.

Zeileis, A. and Lumley, T. (2019). *sandwich: Robust Covariance Matrix Estimators*. R package version 2.5-1.

Zerán, F. (2018). *Mayo Feminista: La Rebelión Contra El Patriarcado*. LOM Ediciones, Santiago.

Index

basic R
    functions, 7
    objects, 6

causal inference
    average treatment effect (ATE), 240
    causaleffect, 247
    causation(), 237
    directed acyclic graphs (DAGs), 238
        backdoors, 243
        ggdag, 247

data exploration
    countrycode, 220
    dplyr
        arrange(), 24, 179
        count(), 18
        desc(), 24
        filter(), 22
        glimpse(), 17, 41
        group_by(), 27
        if_else(), 31
        mutate(), 25
        pivot_longer(), 33
        pivot_wider(), 34
        rename(), 21
        select(), 19
        summarize(), 27
    mutate(), 167

data loading
    data.table, 85
    ff, 86
    haven
        read_spss(), 78
        read_stata(), 78
    janitor
        clean_names(), 84
    readr, 73

    class(), 75
    load(), 76
    read_csv, 74
    read_csv2(), 75
    read_rds(), 76
    readxl
        read_excel(), 81
    shapefiles
        read_sf(), 399
        st_write(), 405
    utils, 73
data visualization
    data visualization
        sf, 67
    ggalt
        geom_dumbbell(), 220
    ggcorrplot, 93, 117, 164, 185, 369, 382
    ggparliament, 14, 40, 67
    ggplot, 95, 127
        annotate(), 66
        time trends, 151
    ggplot2
        coordinates, 47
        facets, 44
        geom, 43
        group, 50
        mapping=aes(), 42
        scales, 45
        themes, 48
    ggpubr, 151
        ggdensity, 129
    ggraph, 68
    ggrepel, 58
    lubridates, 149
    patchwork, 68
    plot(), 120
    spatial data

centroids, 407
ggplot2, 400
ggrepel, 408
LISA, 414
spdep, 414
data wrangling
    merging
        countrycode, 283
        inexact, 286
        left_join(), 281
        stringdist, 285
    missing values, 289, 291
    standardizing
        countrycode, 283

exercise
    1A. basic R, 3
    1B. basic R, 10
    1C. basic R, 14
    2A-B. data management, 23
    2C-D. data management, 26
    2E-G. data management, 29
    2H-I. data management, 32
    2J. data management, 34
    3A. data visualization, 60
    3B. data visualization, 62
    3C-E. data visualization, 69
    4A-B. data loading, 86
    5A. linear models, 96
    5B. linear models, 101
    5C. linear models, 104
    5D. linear models, 106
    5E. linear models, 110
    5F. linear models, 130
    6A. case selection, 140
    6B. case selection, 143
    6C. case selection, 145
    7A. panel data, 157
    7B. panel data, 163
    7C. panel data, 169
    7D. panel data, 171
    8A. logistic models, 182
    8B. logistic models, 189
    8C. logistic models, 196
    8D. logistic models, 206
    8E. logistic models, 208
    9A. survival models, 219
    9B. survival models, 225
    9C. survival models, 226
    9D. survival models, 233
    10A. causal inference, 255
    10B. causal inference, 271
    11A. advanced data management, 282
    11B. advanced data management, 288
    11C. advanced data management, 305
    12A-B. web scraping, 314
    12C. web scraping, 321
    12D. web scraping, 321
    13A. quantitative text analysis, 345
    13B-C. quantitative text analysis, 355
    14A. networks, 364
    14B. networks, 365
    14C. networks, 373
    15A. principal component analyisis, 378
    15B. principal component analyisis, 389
    15C-D. principal component analysis, 392
    16A. maps and spatial data, 401
    16B. maps and spatial data, 406
    16C. maps and spatial data, 413
    16D. maps and spatial data, 423

networks
    adjacency matrix, 358
    centrality, 366
    directions, 359
    links, 358
    nodes, 358
    visualization, 362
    weights, 359

politicalds
    data(), 16
    installation, 12

principal component analysis
    factoextra, 384
    FactoMineR, 389
    ggplot, 391
    stats, 384

quantitative text analysis
    descriptive analysis, 328
    structural topic modeling, 346
    Wordfish, 343

regression analysis
    ggplot2
        fitted.values, 112
        residuals, 112
    linear
        case selection, 142
        causal inference, 262
        lm(), 98
        panel data, 153, 158, 169
        prediction, 107
        qqPlot, 129

logit
    ggplot, 190
    glm, 183
    jtools, 194
    margins, 190, 199
    plot, 192
    prediction, 190
matching
    MatchIt, 263
propensity scores, 144
survival models, 217
    cox.zph, 231
    ggforest, 228
    stcox, 227
    survfit, 225

tidyverse, 16
    %>%, 29

web scraping
    glue, 310
    gtrendsR, 307
    rtweet, 318
    rvest, 310

For Product Safety Concerns and Information please contact our EU representative GPSR@taylorandfrancis.com
Taylor & Francis Verlag GmbH, Kaufingerstraße 24, 80331 München, Germany

www.ingramcontent.com/pod-product-compliance
Ingram Content Group UK Ltd.
Pitfield, Milton Keynes, MK11 3LW, UK
UKHW031042080625
459435UK00013B/558